APPLIED
ECOTOXICOLOGY

APPLIED ECOTOXICOLOGY

Jörg Römbke
Johann F. Moltmann

Deutsche Gesellschaft für
Technische Zusammenarbeit (GTZ) GmbH

LEWIS PUBLISHERS

Boca Raton New York London Tokyo

Library of Congress Cataloging-in-Publication Data

Römbke, Jörg.
Applied ecotoxicology / Jörg Römbke, Johann F. Moltmann.
p. cm.
Includes bibliographical references and index.
ISBN 0-56670-070-1 (alk. paper)
1. Pollution--Environmental aspects. I. Moltmann, Johann Felix, 1954- . II. Title.
QH545.A1R65 1995
363.7--dc20 95-41713
CIP

Direct all inquiries to CRC Press, Inc., 2000 Corporate Blvd., N.W., Boca Raton, Florida 33431.

Lewis Publishers is an imprint of CRC Press

International Standard Book Number 0-56670-070-1
Library of Congress Card Number 95-41713
Printed in the United States of America 1 2 3 4 5 6 7 8 9 0
Printed on acid-free paper

ACKNOWLEDGMENTS

We wish to thank the many friends and colleagues who have encouraged and assisted us in putting this book together. In particular, we thank the editor, J. Lewis, for his patience and suggestions; D. L. Jones for doing the difficult job of translating the German version into English; H.-J. Schallnaß who has drawn many of the figures on PC, as well as T. Knacker, C. Bauer and L. Beck, who were always prepared to discuss the matter with us. For the earlier GTZ version of the text we have received many helpful comments, in particular from F. Heimbach, A. Marschner, K. Mathes and W. Mücke. On behalf of the Deutsche Gesellschaft für Technische Zusammenarbeit (GTZ), which has provided financial support for this book, we thank the project managers involved, S. Krall and G. Vaagt, for their inputs. And last but not least, our long-suffering wives, Gerti Pfaff-Römbke and Beate Brehmer, as well as our children, deserve most of our thanks for putting up with our preoccupation on evenings and weekends while we were working on this project.

The Deutsche Gesellschaft für Technische Zusammenarbeit (GTZ) GmbH is a federal corporation with a development-policy mission: to improve the fundamentals of human existence in the South and East and to stabilize the natural conditions for life. The GTZ operates within the scope of the development policy formulated by its owner, the German Federal Government. It was set up in 1975 as an instrument to implement development cooperation and as such concluded a general agreement with the German Federal Ministry for Economic Cooperation and Development (BMZ) under which it was assigned responsibility for implementing technical cooperation activities. It also supports development and reform processes on behalf of other German government departments as well as partner-country governments and international organizations. The GTZ performs its tasks on a public-benefit basis. Any surpluses obtained are used exclusively for development-related purposes.

Promoting development and reform processes

Technical cooperation is designed to raise the performance capability of manpower and organizations in partner countries by developing and mobilizing skills and expertise or improving the conditions for their use. It aims to strengthen individual initiative in these countries, enabling people to improve their living conditions through their own efforts. In many cases this means first taking steops to prevent conditions from deteriorating further. Where acute need jeopardizes human survival, the GTZ responds with programs to provide emergency relief and assist refugees. Even short-term relief measures, however, are designed to enhance the potential for self-help.

The GTZ operates in more than 120 countries, with experts and management personnel working in projects, project administration services and project service offices. It has threee decades of experience in tailoring problem solutions to the specific conditions in partner countries.

GTZ services include:

- providing advice to organizations in partner countries on the planning implementation and assessment of their projects and programs;
- selecting, preparing and assigning expert personnel and attending to their professional and personal welfare during their period of assignment;
- planning and implementing project-oriented training and upgrading measures;
- specification planning and procurement of materials and equipment for the projects;
- granting and dispersing non-repayable financial contributions from Technical Cooperation funds.

Wordwide cooperation

The GTZ joins forces with partner countries to design project concepts and measures that draw upon both local knowledge and Western expertise tailored to the needs of the situation. The emphasis is on combining regional, sector-specific, commercial and logistical know-how. The GTZ works closely with other national and international organizations engaged in development cooperation activities. It also avails itself of the entire spectrum of know-how available within the private sector and professional public institutions, both in the Federal Republic of Germany and abroad, wherever and to the extent that this appears expedient and cost-effective. The GTZ thus brings together contributions by various project participants so as to achieve optimal results.

LIST OF TABLES

LIST OF FIGURES

ABBREVIATIONS

AHA	Aquatic Hazard Assessment Database
ANOVA	Analysis of Variance
ASTM	American Society for Testing and Materials
BBA	Biologische Bundesanstalt für Land- und Forstwirtschaft (German Federal Research Centre for Agriculture and Forestry)
BgVV	Bundesinstitut für gesundheitlichen Verbraucherschutz und Veterinärmedizin (Federal Institute for Health Protection of Consumers and Veterinary Medicine)
CPP	Crop Protection Product
DDT	1,1,1-Trichloro-2,2-bis[4′-chlorophenyl]ethane (IUPAC)
DIN	Deutsche Industrienorm (German Industrial Standard)
DM	Deutsche Mark, approx. 0.65 US $
DT_{50} or DT_{90}	Amount of time it takes for 50% or 90% of the applied concentration (day 0 = 100%) of a substance to be broken down (Degradation Time)
EC	European Community
EC_{50}	Effect Concentration for 50% of the test animals
ECETOC	European Centre for Ecotoxicology and Toxicology of Chemicals
ECPA	European Crop Protection Association
EPA	Environmental Protection Agency (USA)
EPPO	European Plant Protection Organization of the FAO
EU	European Union (since 1994)
FAO	Food and Agriculture Organization of the United Nations
FDA	Food and Drug Administration (USA)
FIFRA	Federal Insecticide, Fungicide and Rodenticide Act (USA)
GEMs	Genetically Engineered Organisms
GIFAP	Groupement International des Associations Nationales de Fabricants de Produits Agrochimiques (International Group of National Associations of Manufacturers of Agrochemical Products)
GLP	Good Laboratory Practice

GTZ	Deutsche Gesellschaft für Technische Zusammenarbeit mbH (German Agency for Technical Cooperation)
HC_5	Five percentile Hazard Concentration
IOBC	International Organization for Biological and Integrated Control of Noxious Animals and Plants
ISO	International Organization for Standardization
IVA	Industrieverband Agrar e.V. (German Agrochemical Association)
LC_{50}	Lethal Concentration for 50% of the test animals
LD_{50}	Lethal Dose for 50% of the test animals
LOEC	Lowest Observed Effect Concentration
MAB	Man and Biosphere research program of the UNESCO
MERL	Marine Ecosystem Research Laboratory (University of Rhode Island, USA)
NOEC	No Observed Effect Concentration
OECD	Organization for Economic Cooperation and Development
PCP	Pentachlorophenol
PEC	Predicted Environmental Concentration
PIC	Prior Informed Consent process within the FAO Code of Conduct on the Distribution and Use of Pesticides
PNEC	Predicted No Effect Concentration
QSAR	Quantitative Structure-Activity Relationships
TME	Terrestrial Model Ecosystem
UBA	Umweltbundesamt (German Federal Environment Agency)
USDA	United States Department for Agriculture
WHO	World Health Organization of the United Nations

CONTENTS

1 INTRODUCTION

Ecotoxicology became necessary when human beings began introducing chemicals into the environment. No one was concerned about the fate of these chemicals until they turned up in unexpected places or had adverse effects on organisms in the environment. Some substances have an immediate, lethal effect, whereas others may have impacts which do not become manifest until later, e.g. when a population declines as a result of reduced fertility.

Ecotoxicology became necessary when many new substances which did not originally exist in nature were synthesized in the industrialized countries. Many of these artificial substances have noticeable effects, even when highly diluted. However, ecotoxicology also deals with substances which have always existed in nature, but are released into the environment for specific purposes or are translocated and concentrated in large quantities through human activity.

Ecotoxicology became necessary because of the fact that industrialized society, and the developing world as well in the meantime, not only derives a great deal of benefit from the use of chemicals of all kinds, but has also become aware of the potential inherent risks of these substances. Despite the use of efficient and environmentally sound technologies, global population growth has made it necessary to employ chemicals on an ever broader scale. Yet, little if anything is known about the amount of stress which ecosystems can take. In short, ecotoxicology is an outgrowth of the necessity for human beings to protect themselves and to safeguard the basis for their survival — the balance of nature — against the effects of their own products. This is true with regard to crop protection products in particular.

2 TASKS, OBJECTIVES, AND DEFINITIONS

2.1 ECOTOXICOLOGY

The task of ecotoxicology is to assess, monitor and predict the fate and effects of foreign substances in the environment (MORIARTY, 1988). The aim of ecotoxicology is to provide a scientific basis which will allow such substances to be assessed with a reasonable amount of effort and expense (Figure 2.1). This, in turn, provides the basis for evaluations to decide which substances can be released into the environment, or are environmentally tolerable, and in what quantities.

The aim of ecotoxicology is motivated by the desire to maintain the natural structure and function of ecosystems (ecosystem integrity; HARRIS et al., 1990). In other words, ecotoxicological research is also ecosystem research (MATHES et al., 1991). However, one should never lose sight of the fact that ecosystems are dynamic. It would be unnatural to preserve a status quo.

The following three definitions describe the currently accepted scope of this science:

- Ecotoxicology is concerned with the toxic effects of chemical and physical agents on living organisms, especially on populations and communities within defined ecosystems: it includes the transfer pathways of those agents and their interactions with the environment (TRUHAUT, 1977; BUTLER, 1978).
- Ecotoxicology investigates the effects of substances on organisms. The hazard to animal and plant populations can be determined by using survey data (retrospective) or by performing specific tests (prospective) (original definition in German, RUDOLPH & BOJE, 1986).
- The science which seeks to predict the impacts of chemicals on ecosystems (LEVIN et al., 1989b).

When the effects of a substance are investigated, quantitative evidence of its occurrence must also be provided. This is the task of analytical chemistry (= residue analysis). The detection of a substance, however, is not an indication

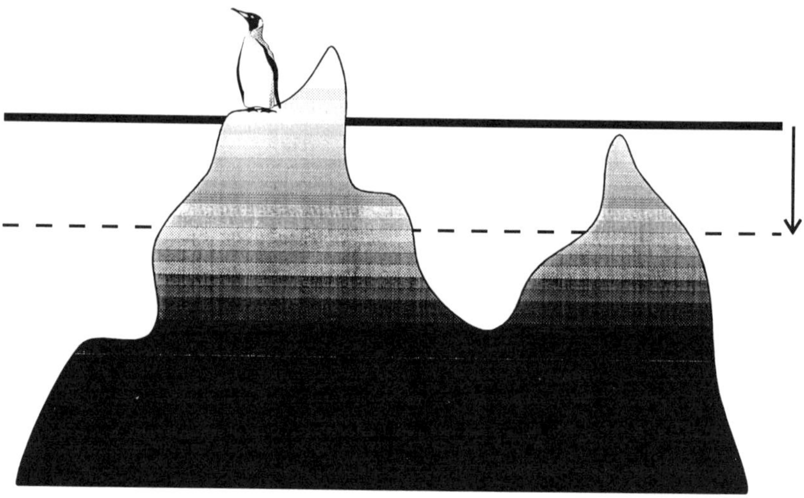

Figure 2.1 "The tip of the iceberg" syndrome (with permission from K. van de Guchte, Lelystad, The Netherlands). A limited set of parameters can be used to assess the environmental hazard potential as a whole. Measurement of additional parameters indicates whether this assessment is valid or not. If not, additional parameters should be incorporated in the decision-making framework to enable adequate remediation.

of its potential effects. Advances in technology, especially in measurement equipment, have now made it possible to detect infinitesimal quantities of substances in the residue analysis. In many cases, the quantities detected are lower than the known biological effect thresholds. Consequently, the ecotoxicological relevance of residue data is still a subject of controversy (KOCH, 1989).

In recent years, an attempt has been made — especially in Germany, Austria and Switzerland — to distinguish between the fate of chemicals including their abiotic impact (environmental or ecological chemistry), on the one hand, and their biological effects (ecotoxicology in the technical sense), on the other (e.g. KORTE, 1980, 1987). As such, environmental chemistry focuses on the chemicals themselves, their entry and occurrence, their transport, fate and transformation processes (e.g. in BOURDEAU et al., 1989, PAASIVIRTA, 1991). Chemodynamics focuses even more directly on the fate of the chemicals from the time of their application until they reach a receptor (STEINEMANN et al., 1990).

A similar distinction is made by the Council of Experts for Environmental Issues in Germany (Sachverständigenrat für Umweltfragen, SRU, 1987), which refers to the study of a chemical's fate as "ecotoxicokinetics", and the study of its effects as "ecotoxicodynamics". However, this combination of terms has not gained currency.

Although the Environmental Impact Assessments (Umweltverträglichkeitsprüfungen) make use of ecotoxicological methods to some extent, they are essentially a tool of the regional planning process, which takes land development claims and socio-economic trends into account, as well as the ecological impact (HEINRICH & HERGT, 1990). Therefore, these assessments will not be discussed any further.

2.2 THE ORIGINS OF ECOTOXICOLOGY

The science of ecotoxicology is an outgrowth of the link between toxicology, ecology and chemistry. The term "ecotoxicology" was coined by TRUHAUT in 1969 (BUTLER, 1984), when the scope of investigation into the harmful effects of substances on a single organism (the human being) was expanded to include entire ecosystems. The fundamental goal of toxicology, which is to protect human health against damaging influences from the environment, was thus extended to include the preservation of plant and animal life. These natural resources are not assessed merely in terms of their economic, aesthetic, educational, scientific or recreational value for human beings. Individual organisms, as well as ecosystems as a whole, are assigned a value of their own.

In toxicology, dosage is the criterion which determines the toxic effect of a substance on an individual. In ecotoxicology, we are dealing with different populations (animals, plants or microorganisms) which not only exhibit varying degrees of sensitivity to a substance but are also exposed to it in very different degrees. But no single measurement of the concentration of a substance in the environment is sufficient in itself to evaluate the stress on the ecosystem. The way in which a substance is released into the environment and the rate at which it is degraded determine whether it ever even reaches an organism or affects it in any way — ecotoxicological effects are a function of bioavailability. The potential effects and exposure must both be considered in order to assess the ecotoxicological relevance of a foreign substance.

From the standpoint of toxicology, the human being itself is a complicated system of organs, functions and developmental processes. When, in addition, it is a matter of studying the effects of a foreign substance on a larger group of organisms (quantitative aspect), and on other compartments of the environment (qualitative aspect), including long-term processes in ecosystems, the consequences become less transparent and the research becomes even more complex.

Public concern over the large number of synthetic chemicals in the environment was provoked above all by the side effects of pesticides as well as the effects of accumulated industrial pollutants. At the same time, it was less surprising to find that such substances indeed had side effects than the forms in which they were manifested: the pathway and persistence in the environment as well as the sensitivity of non-targets — which, in many cases, were actually beneficial organisms — had not been foreseen (MORIARTY, 1988; HOFFMAN et al., 1990). Classical examples of such effects are (SCHÜEPP et al., 1990):

- near extinction of predatory birds due to accumulation of chlorinated hydrocarbons in the food web (e.g. DDT);
- large-scale dispersion of the herbicide Atrazine through evaporation or leaching into the groundwater;
- strong adsorption of the pesticide Paraquat to soil particles, resulting in a high level of persistence, but not necessarily bioavailability;
- accumulation of copper in the topsoil of vineyards;
- sharp increase in the numbers of previously negligible pests after insecticide application.

In England and the United States, the problem of surface-water pollution from household and industrial effluents (especially from mines) led to the first studies referred to as "ecotoxicological" today (e.g. CARPENTER, 1927; HART et al., 1945). It was not until some time later that a discussion got underway in Germany about the possible side effects of pesticides (e.g. RICHTER, 1953; BAUER, 1964). Comparable studies on environmental chemicals were not conducted in larger numbers until the 1970s (e.g. KORTE, 1980). Since these experiments focused on acute effects, the impact of chemicals on the ecosystem were often underestimated. Yet, there has never been very much contact between "applied" biologists (= ecotoxicologists) and "classical" ecologists. This is at least one of the reasons for the current deficit of ecosystemic approaches for the assessment of foreign substances.

The science of ecology provides the bases for examining the effect of a foreign substance on the diversity of organisms as well as the possible relationships between them and with the environment. Nature is extremely complex but not chaotic. A hierarchical order prevails in nature, and it is the task of ecology to decipher it (ODUM, 1971). Ecotoxicology, which focuses specifically on foreign substances, draws on nearly all of the subdisciplines of ecology, from genetics and ecophysiology to the mathematical simulation of whole ecosystems.

The methods employed by ecotoxicology were shaped by governmental requirements in the industrialized countries on a scale that is perhaps unprecedented in the history of science (CONNELL, 1986; CAIRNS, 1989). Whereas legislative sanctions and regulations have always lagged behind the technical advances in other fields of scientific endeavor (which is particularly evident in the case of bioengineering, for example), ecologists, toxicologists and chemists have been faced with the task of providing legislators with legally verifiable criteria and parameters which can be used to assess and evaluate the risks and benefits of chemicals before the fact.

Threshold toxicity values are an example of such criteria. However, values defined for individuals are not necessarily valid for an entire population.

Ecotoxicology must allow for the inherent inadequacies and uncertainties of ecology as a stochastic and highly complex science. Consequently, test guidelines are often a reflection of the attempt to simulate processes in real but not directly accessible ecosystems through the use of models or model organisms, invariably with a view to minimizing the inputs in terms of time, effort and expense.

2.3 "POLLUTANT" AND "FOREIGN SUBSTANCE"

In the context of this discussion, pollutants or stress-producing substances are understood to mean chemical substances which were released into the environment primarily as a result of human activity and have damaging or adverse effects on living organisms (MORIARTY, 1988).

"Damage" in this context is regarded as the interaction between a substance and a biological system. The substance's potential to cause damage is weighed against the protective potential inherent in the biological system (e.g. excretion or metabolic reactions, adaptation or regeneration). In that sense, "ecotoxicity" is not a characteristic which is specific to a substance intrinsically. It also depends on the biological system in which the substance occurs as well as environmental influences on the system (i.e. bioavailability, NUSCH, 1991).

MacKAY (1988) proposed that chemicals be classified in terms of three different modes of action on biological systems: "disruptives" have an unspecific effect which depends primarily on the quantity or concentration. "distributives" are substances which have an extremely selective effect; in many cases, they may accumulate in organisms or in food webs, with potentially long-term damage, even in low concentrations. The third group, referred to as "directives", comprises substances such as radioisotopes, which are harmful in any concentration.

Ecotoxicology is primarily devoted to studying substances which have negative effects in the environment. The effects of a pesticide — an agent which, by definition, is introduced into the environment to control a "harmful" organism — may be considered "negative" or not, depending on whose point of view is involved. As a matter of principle, it has to be assumed that there are no "neutral" substances and that every substance has an effect of one kind or another. However, since it is impossible to assess or evaluate all of the conceivable potential effects, it makes more sense to speak in terms of "foreign substances" (xenobiotics) rather than "pollutants" (noxes). It is initially irrelevant for the evaluation whether the observed effect represents a positive or negative deviation from a prescribed control level. In any case, the temporal dynamics of the effect should also be taken into consideration when assessing whether any "damage" has occurred (MATHES et al., 1991). Evidence of a substance does not, by that fact alone, automatically mean that it is a pollutant. This distinction is often blurred in non-scientific contexts.

It is important to bear in mind that ecotoxicology is a specialized science which tends to assign too much importance to its object of study — foreign substances — in the overall dynamics of ecosystems. The fact is, in relation to other natural and anthropogenic factors (e.g. climate and soil quality or, as the case may be, land development and tillage), stress from foreign substances usually plays only a minor role (HALTRICH, 1991). In many cases, the influence of stress from xenobiotics does not become visible until other influences have been eliminated in a standardized laboratory test. Nonetheless, human beings must still endeavor to minimize the entries and effects of foreign substances and limit their use to the degree necessary. For, when it comes to foreign substances as opposed to other ecological factors, human beings really do have some control over what they choose to do or not to do. This is probably also why ecotoxicological issues have become the focus of such widespread public attention.

The term "stress" in the context of stress on individuals or ecosystems is also difficult to define in a neutral sense. Stress is understood to mean that a metabolism (also on ecosystem scale) must be concerned with eliminating a substance from the body which does not belong there. This requires the expenditure of at least a certain amount of energy. But if this substance only occurs in traces and there is no recognizable effect on the organism's sense of well-being, this could hardly be referred to as stress in any negative sense. The problem is how to establish a link between a microscopic metabolic process and a macroscopic state of the organism as a whole. Such multidimensional questions are typical of ecology and ecotoxicology. In this book, "stress" is used for the most part in the neutral sense of "occurrence", as opposed to "contamination", which has inherently negative connotations.

2.4 ECOLOGY

One could say that ecotoxicology is applied ecology for a specific purpose. Thus, the scope of research is bound by the need to generate unequivocal results. Another constraint faced by ecotoxicology is that many ecological methods do not permit any clear-cut legally valid conclusions to be drawn about the effects of foreign substances in populations and ecosystems. For both theoretical and practical reasons, they entail a number of statistical uncertainties. In order to overcome this dilemma and limit the amount of investigative input to a reasonable degree, ecotoxicology must take a pragmatic approach. It must make use of the existing ecological knowledge to create foundations that will enable the benefits and risks of a substance for human beings and the environment to be assessed (KETTRUP et al., 1991). Thus, a basic understanding of ecology itself is absolutely essential in order to select the right method or measurement parameter for the assessment of a foreign substance.

The task of ecology is to analyze environmental effects on individuals and, conversely, their countereffects on the environment, and to assess the patterns of these effects. KREBS (1985) defines ecology as follows:

> Ecology is the scientific study of interactions that determine the distribution and abundance of organisms.

This comprehensive definition of ecology covers nearly everything from the field of biology. Nonetheless, it does provide a better point of departure than HAECKEL's (1866) general "theory of the balance of nature" by pointing to the questions of *where* organisms are found, *how many* of them occur there and *why* they exist or are able to survive at a specific site. Within biology, ecology is closely associated with physiology, behavioral research as well as genetics and evolutionary research. Chemistry, hydrology, and soil science are the three main outside disciplines related to ecology.

The classification and description of communities, or biocoenoses, and their respective biotopes marked the historical origin of ecology. It was not until some time later that an attempt was made to explain the composition of characteristic biocoenoses in terms of the effect of environmental factors and relationships between the species in the community. This has added a physiological, functional component to the descriptive, systematic-morphological approach of ecology. The morphological structures of organisms, their form and organization is primarily determined by genes, whereas the functional characteristics of organisms are influenced by their interaction with environmental factors. In addition to these two aspects, namely, the external environmental factors and the genetic foundation, the factor of time is also fundamental to life processes.

The terms "structure" and "function", which go hand in hand, form the essential means of access for evaluating the state of ecological units. From the standpoint of ecotoxicology, it is ultimately a matter of preserving or restoring certain functions in an ecosystem (e.g. decomposition, productivity) which have been altered by a foreign substance in a way that is unfavorable from the human point of view. In many cases, structures (e.g. behavioral pattern, species composition) have a certain function which, in turn, must be protected or restored, and can thus serve as indicators of these functions.

3 BASIC ECOLOGICAL PROBLEMS OF ECOTOXICOLOGY

3.1 THE LEVELS OF ORGANIZATION OF BIOLOGICAL SYSTEMS

The objects of ecological study are organisms, each of which represents a complex system of cells and organs. These organisms, in turn, are organized into larger units (populations, communities, ecosystems). Nature's division into different levels of organization (Figure 3.1) is not arbitrary and has implications for the acquisition of knowledge in all areas of biology (O'NEILL et al., 1986). With each higher level of organization (cell → organism → population → ecosystem), it becomes increasingly difficult to establish clear-cut relationships based on cause and effect, or to maintain an overview of the growing abundance of such relationships.

Although the biochemical reactions in cells are manifold and complex, they can be analyzed at this level and are similar in many cells. Interactions between organs can also be explained at the level of individual organisms. But any prognosis of individual development and behavior is necessarily imprecise. For all practical purposes, populations can only be analyzed statistically. On the population level, integrative parameters (e.g. fertility) are being observed to a greater degree in experimental settings, although the changes in such parameters cannot be explained in detail; nor is it necessary to do so. This approach cannot be applied ad infinitum at the ecosystem level because the number of ecosystems which could reasonably be compared on the basis of statistical data is too low. The individuality of ecosystems, even if classified by type, only allows for rough comparisons.

As the scale of observation rises, experimentation yields an increasing number of statistical rather than causal explanations, with a corresponding decline in the precision and reproducibility of the experiments.

This is also apparent from the fact that the methodology derived from biochemistry and toxicology, as the one origin of ecotoxicology, is much more

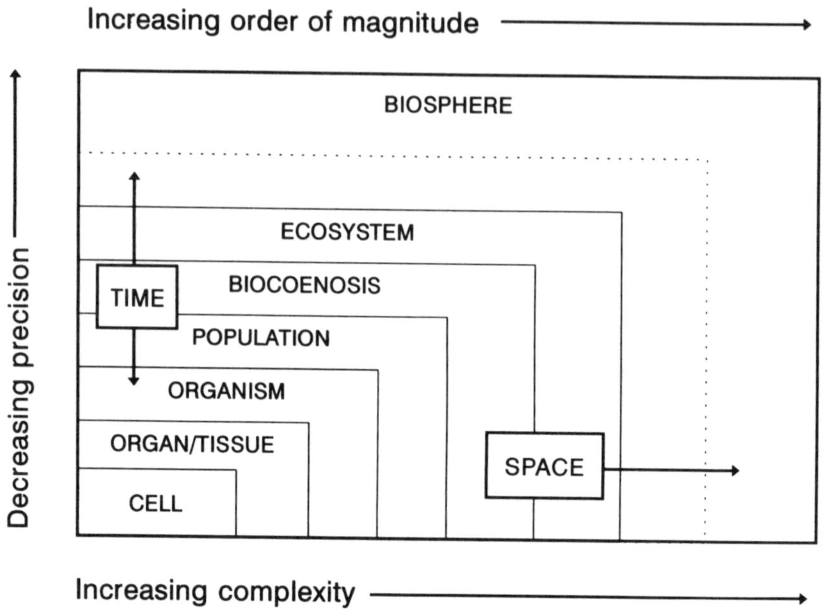

Figure 3.1 Biological levels of organization. The dimensions of time and space are less important for the investigation up to the levels of populations and biocoenoses.

advanced than that of its other source, ecology in general, and ecosystems research in particular.

The goal of ecotoxicology is to assess the impact of foreign substances on systemic levels higher than that of the individual. Resistance, for example, is an individual, genetically determined characteristic. However, this characteristic does not become relevant until the relative number of resistant individuals in a population increases after a foreign substance (e.g. a pesticide) has taken effect. At the level of the population, the effect of a foreign substance is verifiable by a shift in the allele frequencies, i.e. an intervention at the level of organisms is measured in terms of a population parameter although it is actually manifested at the molecular level. In any case, it does not seem possible to make predictions for more than two levels at a time (cf. Chapter 3.3).

3.2 INDIVIDUALS, POPULATIONS, BIOCOENOSES, ECOSYSTEMS

Individuals

Organisms interact with their environment. In other words, abiotic factors in their environment (light, temperature, moisture, foreign substances, etc.) and biotic factors (members of the species, prey, enemies, competitors) determine whether an organism can survive in a specific site. Conversely, an organism influences its environment, for example, by physically changing the soil structure or by absorbing and giving off substances. In the long-term perspective, geomorphological influences (e.g. isolation through the formation of islands) and climatic influences (e.g. glaciation) determine the distribution of organisms and species, aside from the added factors of extinction or displacement as a result of human intervention.

The individuals of a species make certain demands of the factors in their environment. In the event of a change in these factors, they also have a certain capacity for reaction which is genetically determined and may vary from one member of the species to another. A subdiscipline of ecology, namely, physiological ecology, examines the regulatory achievements of individual organisms in relation to environmental factors.

Populations

Individuals of the same species form populations. The interaction between internal, genetically prescribed conditions and external influences, regardless of whether they are induced by the locality as such (abiotic) or by other organisms (biotic), determines the abundance of individuals in a population within the limits of its distribution. The sum of all factors which make it possible for a species (or more precisely, a population) to exist in a certain habitat is called an "ecological niche". It is practically impossible to define the ecological niche of a species precisely. For one thing, the optimal state of any given factor depends on the influence of other factors and, for another, the genetic constitution of a population changes under the influence of environmental factors over longer periods of time (selection pressure). Ultimately, however, it is not the individual who is important but the species or the population. The attempt to provide a mathematical and empirical description of the fluctuations in a population, as well as the population genetics, is the subject of population ecology.

Although the time factor plays a role on all levels of organization, it does not become a major aspect of consideration until attention is focused on the population dynamics. All biological processes are coordinated with one another in the sense of timing, a fact which is particularly clear in individual development. Processes at higher levels of organization also unfold within a continuous flow of time (evolution).

Biocoenoses

The organisms of one species coexist with members of other species and, under the pressure of internal and external factors, form more or less characteristic communities (biocoenoses). The corresponding, special composition of abiotic factors makes up the space in which the organisms live (the *biotope* of a community, or the *habitat* of a particular population).

The dual dimensions of structure and function are also analyzed at the level of biocoenoses (communities), albeit in a more abstract sense. At this level, the relationships between individual organisms are no longer the important factor but rather those which exist between the populations of various species. Biocoenoses are characterized by the specific composition of species comprising them as well as the relative sizes of the populations. Whereas plant communities (phytocoenoses) can often be correlated with the regional climate and soils (KÜCHLER & ZONNEVELD, 1988), the limits to the distribution of animals often extend beyond the particular phytocoenoses. For example, communities of microorganisms and animals in the soil must be considered separately from, but not independently of, the animal and plant communities above ground (GRAEFE, 1993).

Ecosystems

Together, biocoenoses and biotopes make up an ecosystem. Space and climate are the two key structural elements which come into view at the point of transition from population ecology to ecosystem research. Processes in inanimate nature are studied by the geosciences and possess their own dynamics.

For the interpretation of ecosystems, individual populations are assigned functional roles and survival strategies in the overall dynamics. Structures such as heterogeneity and stability are useful for a comparative classification of ecosystems (cf. Chapter 4.2). Thus, the concept "ecosystem" is not only used to refer to a specific section of nature but also, in the figurative sense, for models that are derived from it (STÖCKER, 1979).

Biosphere

The global level of organization, which encompasses all vital processes and integrates the laws and systemic properties of all levels below it, is called the biosphere. Due to the complexity of the individual chains of effect, the laws of which are not yet known, this level cannot be studied directly. Even estimates of global processes, such as the effects of CO_2 emissions in the atmosphere, pose serious problems. For this reason, the biosphere is not considered as a level for experimental investigation.

Linking the Levels of Organization

Ecotoxicology draws on various subdisciplines of biology, depending on which level of organization is involved (Figure 3.2). These subdisciplines, along with the respective methods and issues, roughly correspond to the levels of organization of biological systems, but with an approach which focuses in varying degrees on descriptive (structural), comparative (functional) or evolutionary

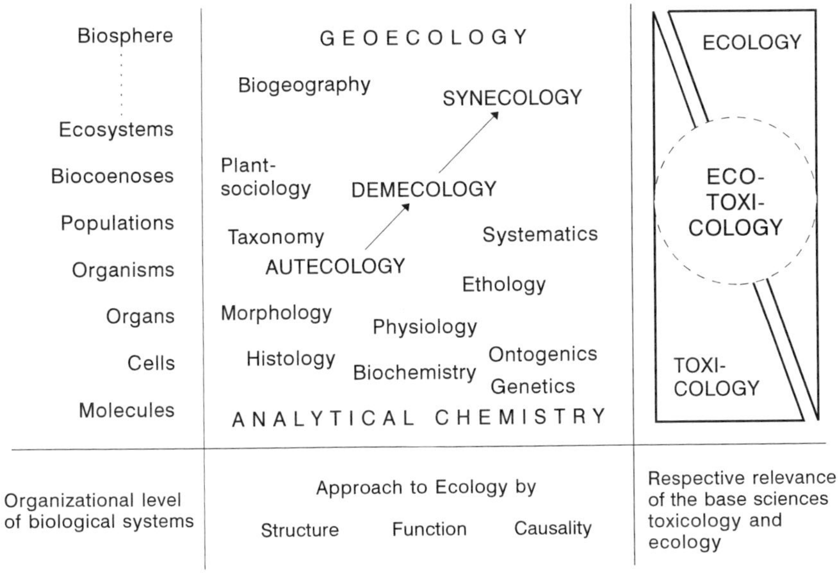

Figure 3.2 Classification of the biological and ecological subdisciplines in a system of coordinates based on the levels of organization of biological systems and the three approaches to ecology. Toxicology focuses primarily on phenomena at the lower levels of organization, whereas classical ecology deals with those on a larger scale. Ecotoxicology comes between the two, integrating aspects of both disciplines.

(causal) aspects. The two overlapping sciences, ecology and toxicology, are relevant to ecotoxicology in inverse proportion. A toxicological approach is applied primarily on the smaller-scale levels of organization, but on any larger scale, the problems take on an increasingly ecological dimension. Ecotoxicology is not merely a compound of two subjects; it projects toxicological questions onto the ecosystem level. When parameters are selected to assess the effect of a foreign substance, the levels of organization should not be confused in order to ensure that the data obtained at each level can be properly assessed and put in relation to other parameters.

3.3 EXTRAPOLATION FROM ONE LEVEL TO ANOTHER

In many cases, it is necessary to extrapolate from the knowledge acquired at the biochemical or individual level and apply it to higher levels of organization. Although a certain amount of generalization is unavoidable, it is not possible to draw any deterministic conclusions about higher-level processes from what is known about the processes on lower levels. There are practical reasons for this — the number of additional relationships becomes so great as to be no longer manageable. However, there are also theoretical reasons, because events with a statistical probability (stochastic events) influence the course of a process to an increasing degree. In other words, the relevance or accuracy of the data yielded at a certain level is characterized by a growing factor of uncertainty as one proceeds to higher levels. This is not only a result of inadequacies in the methods employed but is also immanent to the system of every succeeding level of organization.

For this reason, there is a contradiction between the desire to assess the effects of foreign substances on the ecosystem level and the methods available for such assessments, given the fact that such methods describe the effects on individuals or, at most, on populations. This contradiction is also called the "Dilemma of Ecotoxicology" (cf. KIMBALL & LEVIN, 1985). Despite initial attempts to devise appropriate field tests (ecosystem level), laboratory testing (individual level) is still the basis for assessing the effects of substances. By the same token, these tests, which are relatively simple and easy to reproduce, have "no relevance whatsoever" for conditions in the field; the results of lab tests are always open to question or in need of further verification (CAIRNS, 1989). Therefore, the most serious problem is posed by the need to develop methods which will make it possible to recognize ecosystem effects. In particular, the reciprocal effects between the various species, the continuity of the functions and the history of the respective ecosystem must also be taken into consideration. This includes the question of how much can be learned from model simulations or sections of the ecosystem (KÖNIG, 1985; HEIDLER, 1987).

The degree to which test results obtained in the more or less comprehensive, easily manageable experimental context of the laboratory are relevant for the real conditions of an ecosystem is the central problem of ecotoxicology. This issue becomes clear in Chapters 6 to 12, where it is always a matter of judging the role of certain methods or parameters in terms of what they can tell us about the distribution and effect of chemical substances in the environment.

3.4 STRUCTURE AND FUNCTION

The effects of foreign substances are assessed by analyzing the structure and function on different ecological levels (individuals, populations, ecosystems) over the course of time. Any categorical distinction between structural and functional elements is academic, for there is no such thing as structure without function (by virtue of selection) and functions always take place on structures. Both structure and function are characteristic of every system at a given level of organization. If anything, a distinction can be made between elements which lend themselves to static analysis (formal system characteristics) and those which describe processes or their results (functional system characteristics) (cf. SCHUBERT, 1991b). A few examples are listed in Table 3.1.

Table 3.1 Examples of Structural and Functional Characteristics at the Various Levels of Biological Systems

Level	Structure	Function
	Description of forms in certain time segments	Assessment over a longer time period
Cell/Organ	Molecules, membranes, tissues, metabolic pathways, types of cells, types of organs	Growth, metabolism, energetic conversion, detoxification
Organism	Species, habit, age, sex	Metabolism, growth and development, reproduction, health, death, concentration in the tissue, behavior
Population	Occurrence, number of individuals, density, dispersion, age pyramid, sex ratio	Fertility, reproduction, yield/ production, metabolism
Biocoenosis	Number and composition of species, dominance ratios, diversity	Regulation, guild function
Ecosystem	Climate, geographic location, type, succession, balance	Resilience, stability, degradation, filtration, primary production, accumulation, bioretention

From all of the many elements which constitute organisms and ecosystems, those elements are selected to serve as measurement parameters (Chapter 6) which will help to achieve the aims of ecotoxicology. Certain elements are chosen over others because they are easy to monitor for the widest variety of reasons, or because they are particularly relevant for assessing the fate and effects of foreign substances to preserve the ecosystem's integrity. All of the rules which have been established on the basis of empirical research and observation are taken into consideration in the selection of parameters. A few examples of such rules are:

- Fat-soluble substances tend to accumulate in the animal body; for this reason, the partition coefficient log P_{ow} is an important parameter.
- An additional mortality factor may be crucial for the affected population; for this reason, toxicity is an important parameter.
- Species diversity is put in relation to ecosystem stability; this is why changes in diversity are an important, but not easily manageable, parameter.

- The adsorption to soil components is one of the decisive contributing factors in bioavailability and degradation of chemicals; this is why soils that approximate reality are important when conducting experiments.

As a result, ecotoxicological research automatically generates a great deal of basic knowledge about the structure and function of organisms and ecosystems. The more one knows about the ecosystems, the better one can assess the tolerance to stress. Therefore, ecotoxicological research is always ecosystem research as well (MATHES et al., 1991).

Ecotoxicology is devoted to the study of structural and functional system characteristics at all levels of organization, but these characteristics are extremely disparate. Which of the many characteristics are ultimately selected to assess the impact of a foreign substance strongly depends on the state of scientific knowledge and the methods available to conduct the assessment. Although ecotoxicology in the true sense of the word is aimed at more highly integrated systems, it also seeks to identify and protect individual targets or habitats (e.g. endangered species).

At the level of cells and organs, these units themselves represent the structures on which the corresponding functions take place, many of which have already been thoroughly examined. At this level, the function is usually not regarded as being dependent on external factors. Instead, it is more a matter of defining standard functions and reproducing them for the purposes of experimentation.

An organism is a basic unit which depends on the harmonious functioning of all internal organs. The malfunction of an organ (i.e. disease) has an effect on the organism as a whole. An undisturbed process of development in the early life stages is vital to ensure that an organism will be able to reproduce, which, biologically speaking, is the purpose of its ontogeny. Different substructures are important or especially sensitive to foreign substances at various points in the course of an individual's development.

Population dynamics tries to explain fluctuations in the numbers of individuals in a population, first in terms of statistics, and then in terms of cause and effect. The descriptive component (What and Where?) and the functional component (How?) of ecology are supplemented by the genetic-evolutionary (causal) components (When, or Why at that time?); this is when the temporal sequence of events is taken into account. The essential characteristic of a population is its gene pool — a structural term which in itself is highly immaterial and abstract in character. As a result, the development of a population over time and the changes in its gene pool have, as a whole, gained in importance relative to the individual fate.

MORIARTY (1988) suggests putting the focus in ecotoxicology on the population level because it becomes increasingly difficult to overcome the element of uncertainty in the predictability of reactions, an element which is immanent to the ecosystem level. Above and beyond MORIARTY's argumentation, however, the assessment of the impact of foreign substances on ecosystems is important because the effects on lower levels of organization are integrated at this level; consequently, qualitatively new kinds of reactions can occur (HOPE, 1993). Even so, one cannot expect to find any universal indicators which will be valid in every case, i.e. regardless of the type and state of the ecosystem involved, the respective chemical, its concentration or exposure period. The fact is — as MATHES et al. (1991) have emphasized — ecosystems show non-specific reactions.

Depending on which approach — descriptive-structural or functional — is taken towards biocoenoses and ecosystems, the parameters may be selected to serve more or less as structural parameters (e.g. species composition, age structure) or as functional parameters (CAIRNS & PRATT, 1986). Changes in metabolism and energy flows, for instance, would be examples of functional parameters.

The question of whether structural or functional parameters are preferable to analyze the effects of foreign substances revolves around two basic problems: the question of priorities — What do we consider important? — and the question of practicability — What is the best way to identify the impact? MORIARTY (1988) — makes a convincing case for the fact that it makes more sense to study structure rather than function at both levels (the population as well as the ecosystem). Therefore, in his opinion, it is more fruitful to monitor selected, well-known species (cf. Chapter 10.3). However, since it can be very difficult to attribute changes in the structure of a biocoenosis to a systemic value per se (How does the elimination of a collembola species affect the biocoenosis of the soil in a forest?), it does not seem justified to limit the approach to only one of the two aspects. In other words, structural as well as functional parameters should be used for the assessment of ecotoxicological hazard potential. The use of both would also seem warranted on the basis of theoretical considerations, which indicate that there is no direct, generally valid link between the structures and functions of an ecosystem (FORBES & FORBES, 1993).

Certain structural parameters devised for ecosystem research are based on the implicit assumption that ecosystems are a kind of supra-organism (cf. Chapter 12.1, the concept of "ecological health", KARR, 1993; SUTER, 1993) with a defined sequence of development (succession) starting from a pioneer stage with an unbalanced, rapidly changing community structure. This is followed by a spatial and structural differentiation, which ultimately makes the transition to a stable climax stage (e.g. in Central Europe: from clear-felled areas to beechwood forests) with a highly diversified biocoenosis.

Foreign substances represent an intervention which disturbs this directed process by eliminating individual components of the ecosystem by unnatural means. The process of succession is reversed (= retrogression). The change in structural parameters (in the direction of development, the state of equilibrium and stability) is used as a parameter to gauge the tolerance level of an ecosystem. The ecosystem's resilience — the capacity and speed with which it returns to its original state after being disturbed — is also considered an important tolerance criterion.

More recent findings, however, have made it necessary to depart from these generalizing hypotheses and take local parameters as well as cyclical processes into greater account. Thus, succession no longer corresponds to the developmental process of a supra-organism. Rather, it is based on two factors: the probability that pioneer populations of various species will be present and their varying degree of capacity to survive each stage of succession (MORIARTY, 1988).

4 GENERAL ASPECTS OF FATE AND EFFECTS

4.1 MEDIA AND COMPARTMENTS

The ecotoxicological investigation of foreign substances focuses on the fate of a substance as well as its effects. It is important to determine the quantity of a substance, or the level of exposure to which individuals may be affected by it, at the potential locus of effect. This is the only way in which observed effects can be correlated with their trigger. There are a vast number of loci at which an effect may occur. They can be classified by substrate, by medium or compartment as well as in terms of their biotic or abiotic composition in ecosystems.

The roughest abstract classification is by reduction to the three media water, land and air. The sum of all organisms is sometimes also given as a fourth medium ("biota"). This division is especially appropriate for the application of simple exposure models if, for example, the purpose is to examine the relative distribution of a substance in the environment in order to obtain reference data for more in-depth tests (cf. Chapter 8.1). The fate of a chemical is subject to wide variations in different media, but even segments of the same medium (= compartments) may be so disparate that it will not be possible to predict the fate of a substance in a specific field situation from the general type of medium alone.

The terrestrial medium comprises the soil (from the C-horizon to the litter layer), with the vegetation (from herbaceous to canopy level) and ambient air. It is distinguished from the aquatic medium by different methodological requirements. Although most freshwater bodies are very small (in toto they make up only 0.6% of the earth's total volume of water), they still shape the physical environment and, above all, are indispensable for vital processes in the terrestrial biocoenoses. The marine milieu is almost as different from fresh water as it is from the terrestrial medium, due to the osmotic effect of the salt.

The media water and air are often considered to be much more homogeneous than the terrestrial medium. This assumption, which is justified at first glance, is not tenable on closer examination. If one takes the compartment of

Table 4.1 Classification of the Environment by Medium and Compartment

Medium	Compartments
Air	Atmosphere
	Airborne particles
	Clouds
	Biota
Water	
(aquatic medium)	Surface waters
	Sediments
	(Organic) Suspended solids
	Groundwater
	Biota
Land	
(terrestrial medium)	Soil
	Soil pore water
	Soil atmosphere
	Vegetation layer and other biota

Note: In the aquatic medium, a distinction can be made between limnic and marine (possibly also brackish) compartments.

sediments and suspended matter or airborne particles into account, aspects of fate as well as effect become much more complex and no longer essentially differ from the processes in the terrestrial medium. It is still much easier to apply biological and residue-analysis methods in water than it is in more complex media (FRANSSON, 1985; CAIRNS, 1989). Due to this predominance of aquatic methods, the body of test results that have been compiled for this medium (e.g. PILLI et al., 1988) is much more voluminous than that available for terrestrial systems (e.g. DONKER et al., 1994).

Every medium is divided into different compartments (Table 4.1). The term compartment refers to a context in which a given substance exhibits a uniform environmental fate (STREIT, 1991) — a definition which only partially covers the functional characteristics of a compartment.

Certain compartments have historically received more attention than others in ecotoxicology, not only because they are easier to test or because tests were readily available for them, but also and especially because they were directly relevant for human beings. Most data have come from limnic waters, followed by marine waters and soils (including the animals and (small) plants living on the soil surface). Sediments are only gradually receiving the attention commensurate with their importance, whereas the other compartments have hardly been studied at all.

Compared to the most important compartments of the air and water media (the atmosphere and surface waters, respectively), the most important terrestrial compartment (soil) is characterized by a greater degree of spatial diversification and fundamentally different conditions of production. Moreover, foreign substances have different effects in aquatic and terrestrial food webs. On land, for example, persistent organochlorine insecticides are primarily taken up through the food, whereas aquatic organisms take up such chemicals directly from the surrounding body of water.

4.2 ECOSYSTEM TYPES

The environment can also be classified by ecosystems which, in turn, can be grouped under specific ecosystem types. In many cases, these ecosystem types encompass more than one medium or compartment. An ecosystem approach is especially important for ecotoxicological investigations for two reasons:

- ecosystems, not media or compartments, are the real target level of ecotoxicological issues;
- since it is not possible to examine every site or every real ecosystem, a classification by ecosystem type provides a basis on which to assess the relevance of test results for different ecosystems.

Ecosystems can be classified either in terms of vegetation geography or in terms of their functional characteristics. With the former (older) approach, a distinction is made between terrestrial and aquatic systems (geo- and hydrobiosphere).

4.2.1 Terrestrial Ecosystem Types

According to WALTER & BRECKLE (1983 ff.), the geobiosphere is divided into large ecological units which form overviewable units of landscape based on climatic (e.g. central European hardwoods), orographic (e.g. Kilimanjaro massif) or pedographic (e.g. Nullarbor area in Australia) characteristics.

Vegetation is most useful for the definition of smaller ecological units (cf. KÜCHLER & ZONNEVELD, 1988). There are only minor historical differences within the plant communities (phytocoenoses); the macroclimate is uniform; exposure, gradient, groundwater, etc. are determining features; and the vegetation reacts sensitively to these factors. The terrestrial ecosystems that have been studied more closely to date correspond to the ecological units of the plant communities, which also include the respective animal world (zoocoenosis) and the biotope. However, it can be shown in gradients of environmental factors that there are continuous transitions from one community to another in the case of plants as well. This diminishes the role of indicator species, whose occurrence is influenced to a relatively strong degree by competitive relationships. Approaches which also make use of animal communities in the soil to classify sites are relatively rare (VOLZ, 1962; GRAEFE, 1993).

There is still a very big gap between ecosystems and larger ecological units of landscape which have actually been examined. This gap is difficult to

Table 4.2 Classification of Terrestrial Ecosystem Types (From SCHUBERT, 1991b)*

A. Forest ecosystems
- I. Evergreen tropical rainforests
- II. Tropical deciduous forests
- III. Tropical thorny succulent ecosystems
- IV. Savannas
- V. Laurel woods
- VI. Etesian sclerophyllous woods
- VII. Temperate deciduous forests
- VIII. Boreal conifer forests

B. Shrub and dwarf shrub ecosystems
- I. Subpolar shrub and dwarf shrub heaths
- II. Alpine shrub and dwarf shrub heaths
- III. Subtropical dry shrub ecosystems
- IV. Etesian shrub and dwarf shrub (Macchien-Garriguen) ecosystems
- V. Temperate shrub and dwarf shrub heaths

C. Grassland and herb layer ecosystems
- I. Tropical mountainous grassland ecosystems
- II. Alpine grasslands
- III. Tundras
- IV. Steppes
- V. Temperate meadow ecosystems

D. Desert ecosystems
- I. Deserts
- II. Cold deserts

E. Pioneer community ecosystems
- I. Dunes
- II. Erosion plant communities
- III. Rock plant communities

F. Anthropogenous ecosystems
- I. Field agroecosystems
- II. Artificial high-productivity ecosystems
- III. Urban-industrial ecosystems

* Main types are classified according to climatic zones. They are classified by soil characteristics, altitude and anthropogenous influence.

bridge from the standpoint of methodology. Future research will have to identify medium-scale ecological units — which form a certain ecological sequence in space or in time (succession series).

WALTER & BRECKLE (1983 ff.) define ecological units strictly in terms of climatic zones and geographic (orographic, pedographic) site factors, or vegetation communities. Initial inventories and descriptions are available for a number of terrestrial ecosystems (Table 4.2).

Table 4.3 Division of the Aquatic Medium into Zones

Zone	Description
Marine zones	
Intertidal	Beach zone with frequent wide variations in salt content
Neritic	Shallow water in the area of the continental shelf
Pelagic	Open sea
Bathyal	Abyssal realm, void of light
Abyssal	Level of the ocean floor
Limnic zones	
Littoral	Trophogenic coastal zone
Profundal	Sediment in deeper zones, usually without plants
Pelagic	Area of open water

Note: The sediment zone of a body of water is also referred to as benthic.

4.2.2 Aquatic Ecosystem Types

About two-thirds of the earth's surface is covered with water. This makes up the hydrobiosphere. Due to the mobility of this medium, there are no clear-cut boundaries between individual ecosystems (aside from small lakes). For this reason, the aquatic medium cannot be divided up into ecosystem types like the terrestrial medium has been on the basis of plant distribution.

Nevertheless, the aquatic ecosystems can be classified in the following two main categories:

Limnic:	Streams and rivers; lakes and reservoirs
Marine:	Intertidal systems; coral reefs; wadden sea; continental shelf; abyssal realm

Aside from that, all systems influenced by anthropogenic activity may be summarized in a group of their own (cf. RUDOLPH & BOJE, 1986).

Aside from the rough division into marine and limnic bodies of water based on salt content, the hydrobiosphere is commonly split up into biotopes which are defined primarily in terms of lakes and oceans, regardless of the scale observed and the population diversity (Table 4.3). In the limnic zone, this includes flowing water such as streams or rivers, which can in turn be subdivided into a large number of different zones (e.g. as a function of the flow rate). In the latter case, a classification based on the organisms living there, fish in particular, is possible (e.g. the trout region = cold, fast-flowing upstream segments).

4.2.3 Functional Classification of Ecosystems

Parallel to the geographical division by vegetation, ecosystems can be regarded as functional units of organisms and their environment (ODUM, 1971; BICK et al., 1984). The following parameters can be described and quantified over time (e.g. Figure 4.1 and Chapter 6.6):

- energy flows
- bio/geochemical cycles of elements, e.g. nutrients
- species spectrum
- food webs
- development and evolution
- regulation (cybernetics)

The functions of the individual units are most important, whereas the respective species are assigned a "designated ecological position". Most of the substance and energy conversion takes place directly between the primary producers (the plants) and the destruents (mostly in the soil or sediment). The primary consumers (herbivores) and other links in the food web have an essential regulatory function. The methodological foundations for functional ecosystem research exist (e.g. ELLENBERG et al., 1986; the Solling Project), but the interpretation of the results related to the stability and self-regulation of ecosystems has not progressed very far as yet. For the time being, the study of ecological effects of foreign substances at ecosystem level is only considered worthwhile for sites that are well defined (e.g. certain soils or small lakes). Certain important parameters, such as substance and energy flows, can be quantified at such sites, even without determining all species occurring there. These parameters can then be used to identify changes caused by foreign substances.

4.3 ENTRY PATHWAYS

Chemicals can enter ecosystems in very different ways. The most important entry pathways are:

- undesirable but tolerated emissions (e.g. from landfills or dry-cleaning), or unintentional emissions during the production, utilization or waste treatment of environmental chemicals. The emissions may be diffuse (many sources with low quantities, e.g. dioxins in the tire abrasion near roads) or punctual (defined spots with often high concentrations; e.g. polyaromatic hydrocarbons in the soil of production sites);
- accidents involving environmental chemicals or pesticides are a special case of local, unintentional emission;
- targeted emission (= application) of pesticides in agriculture or to combat disease vectors. Producers and regulatory agencies recommend the appropriate quantities (indicated in g/ha) and a time pattern for the application.

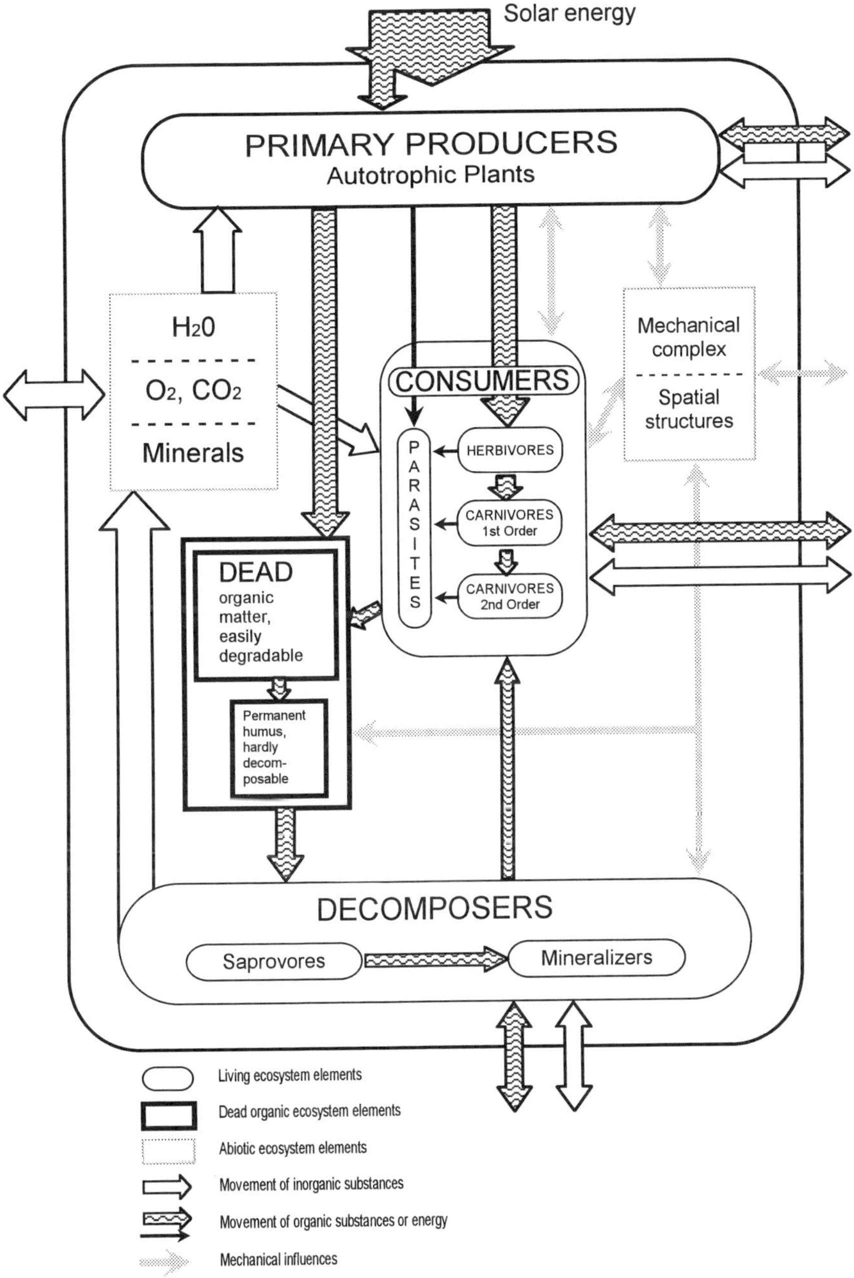

Figure 4.1 Metabolic cycles and energy flow in an idealized ecosystem (from SCHUBERT, 1991b).

Aside from local accidents or spills, environmental chemicals usually enter the environment with effluents, through the air (as gas or dust, or as aerosol) or with sludge (HARMS & SAUERBECK, 1984). Active ingredients in pesticides, on the other hand, are usually sprayed in dissolved or emulsified form. A detailed discussion of these substances is provided in the following.

Pesticide Inputs

Table 4.4 provides an overview tracing the path of pesticide application. The basic components used in production may already be toxic, and formulation aids can have an appreciable influence on the efficacy of the end product.

Accidents have occurred in connection with production on several occasions at various sites (Seveso, Bhopal, Basle), in which high doses of pesticides, their combustion products or primary ingredients have entered the environment and done serious, large-scale damage. Accidents of this type can also occur during transport from the producer to the seller (e.g. the loss of containers holding a seed treatment substance in the North Sea).

Based on the intended application, a relatively precise indication of the environmental concentration of a pesticide can be given, since the quantity applied as well as the size of the area treated are known. However, there have been a growing number of instances in which such substances were found to occur even at long distances from the application sites (e.g. DDT in the fatty tissue of Arctic animal species or Atrazine in rainwater). As for environmental chemicals, a very broad range of concentration is possible in practically all media or compartments in a way which is difficult to predict. In addition, the situation in the environment is complicated by the fact that secondary deposits, metabolic processes or build-ups may occur.

Ecotoxicological studies for pesticide registration simulate conditions based on the assumption that a substance will be applied appropriately and as directed — i.e. in the permissible concentrations, at the right time and on the designated crops. The studies ultimately serve to assess the fate and side effects of a quantity that is applied to achieve the desired effect and, if necessary, prescribe additional requirements for use. In spite of precise instructions for use and residue controls after harvesting, it cannot always be assumed that a substance is applied as presumed in the registration process, even in industrialized countries (MICKLITZ, 1991; TREIDLER & WINTER, 1991). Given this fact, states of critical exposure are inevitable in pesticide use, because mistakes are bound to occur in usage and application which can lead to high levels of contamination (e.g. from the rinsing of application devices).

To a certain extent, the agents applied in crop protection or pest control are used in other sectors of industry as well (HEINISCH & KLEIN, 1989). They may enter the environment through landfills, for example, thus magnifying the effect of pesticides that have already been applied for a specific purpose.

Table 4.4 Stages in the Life Cycle of a Pesticide: Overview of Production Aspects, Risks of Pesticide Application, and Disposal

Life cycle stages	Possible sources of hazard to the environment and man
PRODUCTION Active ingredients and formulation aids	Accidents, explosions, spills, fires, etc. Not specific to pesticide production, but rather a general issue of chemical industry
TRANSPORT	Road and transport accidents
Export and retail	Illegal exports and imports Wrong labeling Missing advice Wrong storage
APPLICATION Exposure and uptake	Over or under dosage Misuse Choice of the wrong product Unfit spraying equipment and improper rinsing Wrong timing Drift and run-off
EFFECTS and side effects	Insufficient effect on the target organism Development of resistance to chemicals Unexpected synergisms of several pesticides Changes and impoverishment of the biocoenoses Damage to beneficial organisms
FATE Translocation, sinks	Accumulation and remobilization Secondary effects by metabolites and residues
DISPOSAL	Unsafe package disposal Reuse of packages for food purposes Long-term and unsafe deposition of unwanted and expired remnants

The application of a pesticide forms the decisive link between a compound and the target organism. The formulation, technique and time of application, the spraying schedule (e.g. in a single high quantity or several low quantities, RUNGE et al., 1989) and the environmental conditions (e.g. climate, vegetation cover) are the parameters which largely determine the scope of actual effects and possible side effects. Overdoses can put a disproportionate amount of stress on the environment, whereas underdoses may make additional spraying necessary and also promote resistance formation. Areas which have not been treated with pesticides may nonetheless be subject to stress from spray drift, for example, or irrigation with contaminated water (EAGLE, 1990).

In temperate latitudes, pesticides are only rarely applied in aquatic ecosystems (with the exception of anti-fouling additives in paints for ship's hulls). However, pesticide application in the aquatic medium is more common in the tropics (e.g. to combat mosquitoes or water hyacinths). Aside from inadvertent discharges due to improper use or accidents, pesticides still enter the surface water, even if properly applied, as a result of spray drift or flow-off. Depending on a substance's characteristics, the application devices and the weather conditions (even if "Good Farming Practice" is observed!), over 10% of the quantity applied may fall on water surfaces which are 10 m away from a treated canopy crop (GANZELMEIER, 1986). Losses from flow-off may be as high as 5% of the applied quantity, depending on the slope, type of soil and weather conditions. Drainage water or groundwater are other paths of entry.

An increasing amount of attention has been devoted to the volatility of pesticides from soil and plant surfaces in recent years, ever since it was determined that, under field conditions, between 10% and up to 90% of the applied volume moved into the air, even in the case of low-volatility substances (BÖHNKE et al., 1989). Once in the atmosphere, they may be transported in the form of gas (as aerosol) or bound to dust particles over long distances. In addition to the chemical structure of a substance, the vapor pressure and the rate of conversion with the ambient air play a critical role in volatility; the prevailing weather conditions are less important. The fact is a state of saturation is practically never reached in the atmosphere due to its high volume. In the case of quantitative data, however, it is important to bear in mind that, in many cases, volatility rates are only indirectly determined (i.e. without taking degradation processes into account).

Whereas relatively few data are available on the volatility from plant surfaces, general conclusions can be drawn on the basis of tests with soil due to the better quality of the data (BÖHNKE et al., 1989):

- the evaporation rate rises in direct proportion to the sandy content of the soil (i.e. with a declining adsorption coefficient);
- the higher the water content of the soil, the higher the volatility.

Because soil has a higher capacity for adsorption, the evaporation from plant surfaces is always higher than it is from soils. The variability of natural surfaces is so great that it cannot be simulated with an artificial surface. For the same reason, indicators such as the log P_{ow} or K_{oc} are not very suitable for a correct prognosis of volatility and, if anything, are only adequate for an initial assessment.

4.4 EXPOSURE

4.4.1 Environmental Fate

Exposure is understood to mean the concentration of foreign substance to which humans, animals, plants or ecosystems are exposed at a certain time

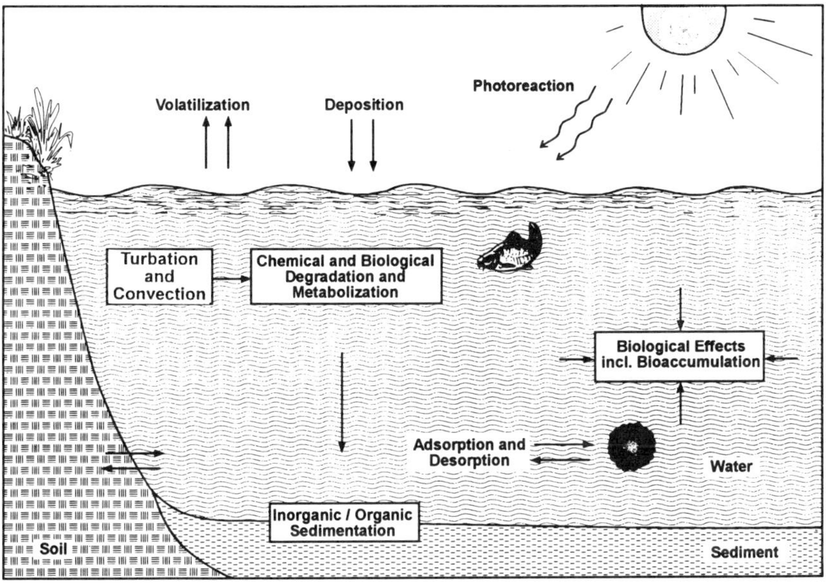

Figure 4.2 The fate and possible dispersion of a chemical in the aquatic medium.

(KORTE, 1987). For the purpose of investigation, a distinction is made between media as well as between compartments (Figures 4.2 and 4.3). The concentration per medium or compartment can be assessed using fate models (cf. Chapter 8.1), but it can also be determined (which is less frequently the case) by means of monitoring data (residue analysis at respective sites).

The exposure of a foreign substance primarily depends on its fate in the environment. A distinction can be made between the following three partial processes, which, however, are usually difficult to distinguish clearly in reality:

- transport/translocation (= mobility)
- metabolism (= transformation)
- mineralization (= degradation)

In the case of transport or translocation, substances spread from the site at which they entered the environment to other places within a compartment or move to a new compartment, where they may continue to have an effect. New substances (metabolites) are created by transformation processes — either through photolysis, in the body of an animal or a plant, or through microorganisms. Their fate and effect are often very different from those of the initial substance (= parent compound). In other words, metabolites may be more toxic than the original chemical; but their damaging effect may also be neutralized (e.g. methylation of heavy metals). Heavy metals function as atoms. Conse-

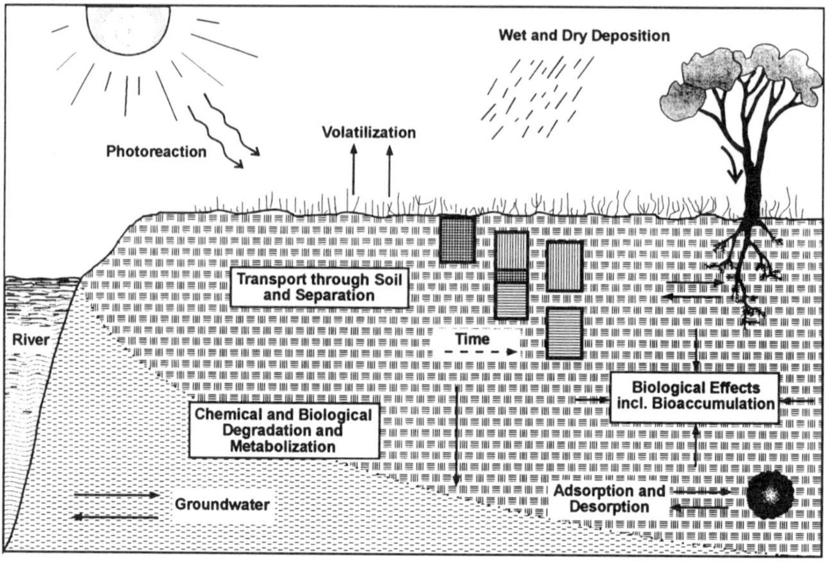

Figure 4.3 The fate and possible dispersion of a chemical in the terrestrial medium.

quently, once they are in the environment, they are not degradable like today's pesticides, whose molecules are completely broken down into water, CO_2 or other small molecules. Heavy metals and some pesticides, however, can be bound almost irreversibly in soils or sediments ("bound residues", HEITEFUSS, 1987), thus eliminating their bioavailability.

Sinks

Sinks are understood to mean a compartment or part of a compartment in which a substance temporarily or permanently loses the capacity for re-entry in a different compartment (STREIT, 1991). This can occur through adsorption (e.g. in the sediment of a lake) or deposits (e.g. in the fatty tissue or bone matter of an organism). Degradation processes can also take place — usually slowly — in sinks. The final sink is the ocean sediment, but many foreign substances are broken down beforehand (KLÖPFFER, 1989).

Soil is the critical sink in the terrestrial medium. However, its role is still not being given sufficient consideration (VON LERSNER, 1982; DUNGER & FIEDLER, 1989). Unlike other compartments in the media water and air, soils which have been contaminated can never be completely decontaminated. This is also true to the same degree of underwater soils (= sediments) (CHAPMAN, 1989).

4.4.2 Surface Water Compartment

Due to the mobility of water and the good conversion conditions, natural factors as well as stress factors in bodies of water are relatively uniform over relatively great distances (in relation to the size of the respective body of water). This is especially true in a horizontal direction, but less so for vertical stratifications (e.g. discontinuity layer in lakes). In view of this characteristic, ecotoxicological tests for the aquatic medium often start out with a very simplistic test substrate: A body of water without sediment or suspended particles is employed with a homogenous distribution of the test substance. In contrast to this simple test situation, the "micromilieus" in the transitional areas of a body of water, and even more so of the sediment, are especially important for degradation and metabolism, due to their special characteristics (e.g. large surface area, good oxygen supply).

The organisms' chances of surviving in water depend on chemical factors (e.g. the concentration of dissolved substances such as oxygen, nutrient salts or organic molecules), on the one hand, and on physical factors (e.g. temperature, water movement and light penetration), on the other. The number of species living in the aquatic medium is very high, but the number of different "basic types" (taxonomic classes) is also much higher than in the terrestrial medium. As far as the organisms to be studied are concerned (regardless of whether they are limnic or marine), there is largely a consensus on the following *minimal* program:

- destruents, e.g. bacteria
- primary producers, e.g. single-cell algae
- primary consumers, e.g. water fleas (*Daphnia*)
- secondary consumers, e.g. fish

The exposure situation of a chemical in the aquatic medium is determined primarily by the following factors (GDCh, 1993):

- the presence of reactive groups of chemicals (e.g. complexing agent);
- the kind and quantity of suspended matter (especially organic particles);
- grainsize distribution of the solid substances;
- microbial activity;
- temperature and pH;
- hydrological factors such as water quantity or flow rate.

4.4.3 Soil Compartment

Soils are a part of terrestrial ecosystems. The bedrock, the regional climatic conditions and the influence of biological factors determine whether and what

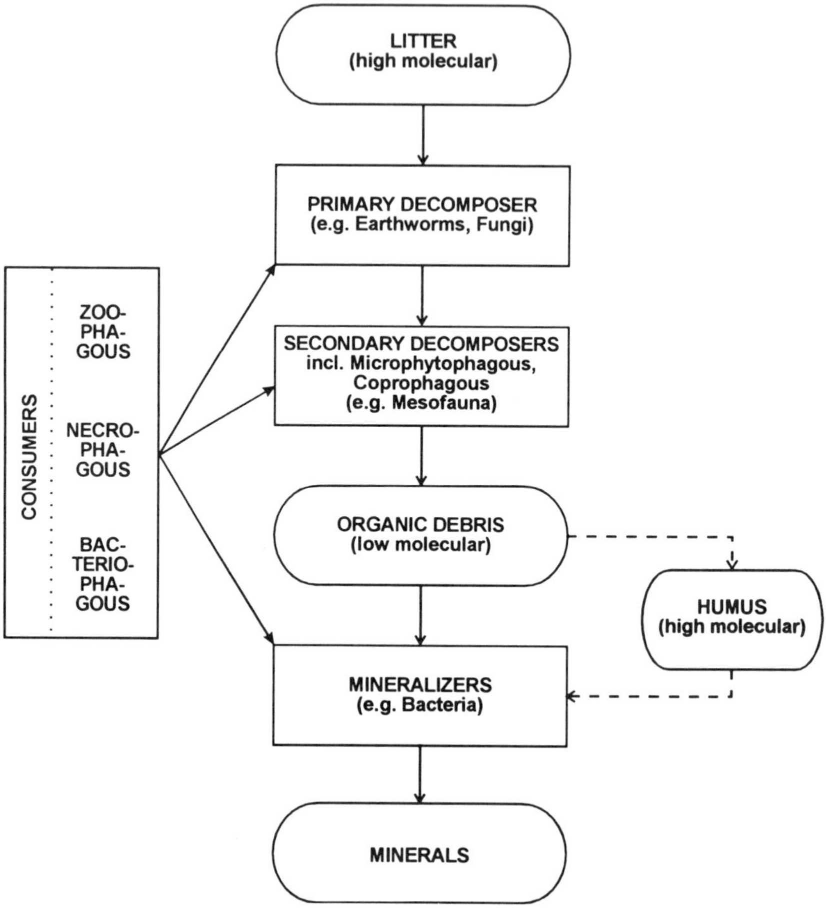

Figure 4.4 Schematic overview on litter degradation and the role of soil organisms (adapted from BECK, 1993).

type of soil evolves. Consequently, there is an extremely wide variety of soil types with significant differences in composition and characteristics, which are constantly changing, even without human influence (ECETOC, 1990).

The most important cycles of a terrestrial ecosystem converge in the soil, namely, the production of biomass by plants as well as the decomposition and mineralization of this organic matter (BECK, 1993). Hundreds to thousands of animal and microbe species are involved in this process (Figure 4.4 and Figure 4.5).

Microbes react more flexibly than animals to stress situations. This capacity for adaptation is also manifested by the fact that they can degrade and mineralize a large number of chemicals, including man-made substances such as pesticides. However, it is still unclear to what extent the same functional

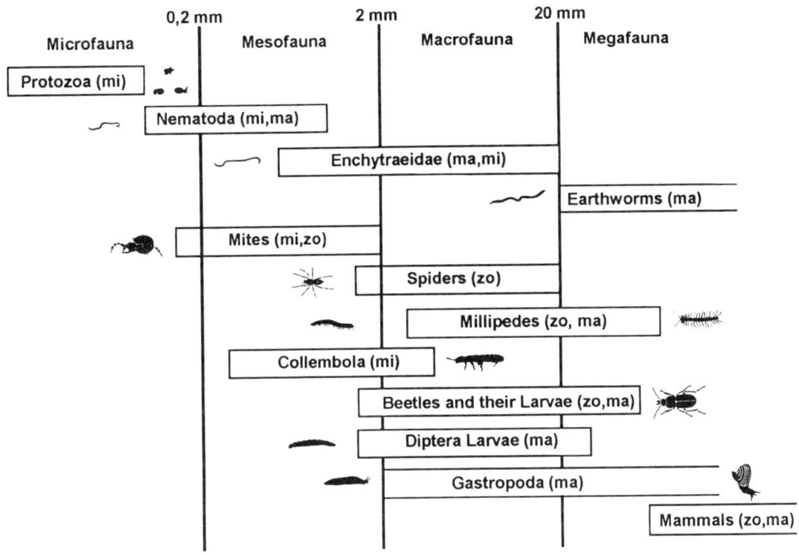

Figure 4.5 Average length of individual animals from different soil animal groups, belonging to three trophic groups: ma = macrosapro- and macrophytophagous, mi = microphytophagous, zo = zoophagous (including necrophagous) (BECK, 1993).

result can also be achieved by different structures in another composition of species, i.e. how high the functional redundancy is (BELOTTI & BABEL, 1993). Up to now, there has been little if any evidence that animal species in the soil were ever totally eliminated due to chemicals; they have many different means of retreat, and cases of large-scale exposure are rare (KLEIN, 1991).

Because ecotoxicological studies have been conducted in aquatic systems for a few decades now, there is a relatively sizeable database available on them. The situation in the terrestrial medium is significantly worse. Moreover, there are a growing number of indications of a dramatic increase in the level of stress from chemicals and radio nuclides on soils as well as other anthropogenic stress factors (especially surface sealing through land development, WINTERINGHAM, 1988).

Soils differ from surface waters due above all to the following aspects:

- Soils are more complex and unhomogenous than bodies of water, i.e. exposure is much more difficult to predict.
- The soil compartment is extremely difficult to distinguish from other compartments such as soil atmosphere or soil pore water. Depending on the species, age, season, etc., many organisms come

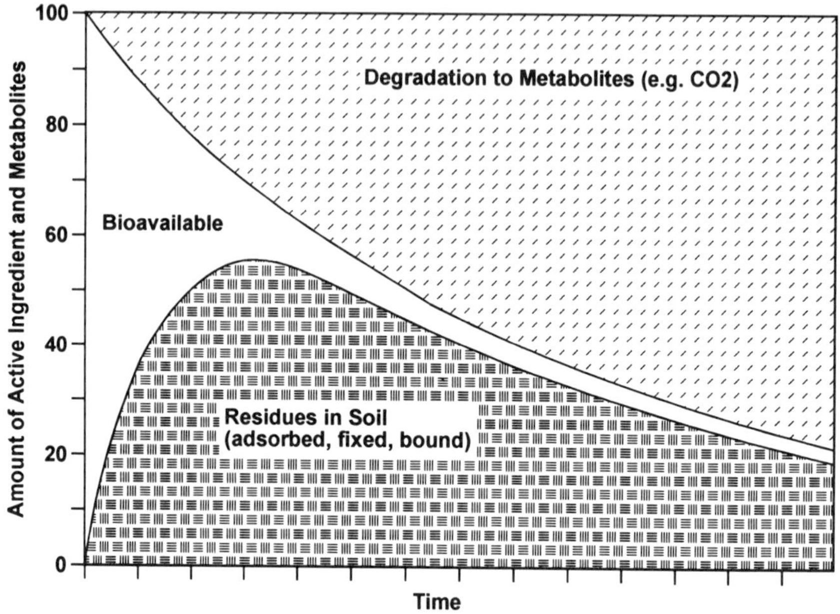

Figure 4.6 Bioavailability of a pesticide in the soil (from KLOSKOWSKI & FÜHR, 1988).

into contact with these various compartments with varying degrees of intensity.

- The chemical reactions in soils are very different, depending on the soil type and its characteristics.

This element of complexity makes prognoses on the fate of chemicals in the soil much more difficult than in surface waters, especially over the course of time. In general, however, a gradient of stress is observable, with the concentration of chemicals decreasing from top to bottom (i.e. there is an accumulation in the upper centimeters). The situation is further complicated by adsorption, bonding or incorporation of the substance to soil components as well as by climatic influences (Figure 4.6).

Organic chemicals can be bound to humic substances as well as to clay minerals. The bonding processes are determined by a large number of soil and substance characteristics, the most important of which are the following (GDCh, 1993):

- type and relative quantity of the clay content
- particle size and porousness of the soil
- capacity for ion exchange
- pH value and conductivity of the soil solution
- redox potential
- composition and quantity of the organic material

A precise knowledge of these factors can be used to estimate the potential for geoaccumulation, i.e. the tendency of a substance to persist in the soil over the long term, with a simultaneously low bioavailability. This potential can be determined more directly by measuring the adsorption behavior (cf. Chapter 8.1).

Bound Residues of Chemicals

Tests on the adsorption behavior, however, do not offer any indication of a possible remobilization of bound residues. What is more, KUBIAK et al. (1990) were able to show that laboratory tests on a small scale are not suitable to arrive at a realistic prediction of their formation under field conditions. The hazard potential of these non-extractable residues can be very high, as shown, for example, by the release of aluminum ions when soils or water are acidified, which has an effect on fish (FÖRSTNER, 1990). On the other hand, CALDERBANK (1989) puts the potential hazard from bound pesticide residues on the low side, as the following line of his argumentation will show.

The use of radiolabeled substances in fate studies clearly showed that a considerable portion of substances in all classes of chemicals are so thoroughly bound to soil particles (usually humic substance or clay particles) that they cannot be extracted with organic solvents. In the case of pesticides, this non-extractable portion may be as high as 90% (the average range is 20% to 70%) of the applied quantity, depending on the substance and the soil type (CALDERBANK, 1989). With time, these residues also undergo microbial or abiotic degradation, albeit in a very slow process (KLOSKOWSKI & FÜHR, 1988), but the minimal quantities of metabolites produced as a result should have little if any ecotoxicological effects. This is also true in cases where the active ingredient may be slowly released as the soil's organic content ages. Sudden releases of larger quantities are a different matter if, as in the case of aluminum, they lead to a dramatic change in the environmental conditions. However, there is hardly any evidence of such cases with pesticides.

4.4.4 Sediment Compartment

The formation of sediments is about as complex as that of soils. They are especially important as sinks, i.e. where a large number of chemicals are deposited (LUOMA & HO, 1993). Despite the possibility of remobilization, sediments can "document" states of stress that date far back into the past from accumulations in certain strata.

Most of the information already provided for soils is also valid for sediments, which are often referred to as "underwater soils". Thus, here as well, the content and composition of the organic matter is probably the single most important factor in the fate of a chemical, and for bonding in particular. As far as the state of exposure is concerned, however, several peculiarities should be borne in mind:

- There is a constant exchange of matter between the sediment and the open water, although substances move from the water to the sediment in much higher quantities than they do in reverse. As a result, the stress gradient in sediment usually runs from top to bottom.
- Due to the activity of animals (e.g. the feeding activity of tubificids), sizeable quantities of chemicals which would have otherwise been found in the sediment enter the water again (remobilization, KARICKHOFF & MORRIS, 1985). Abiotic processes such as fluctuations in pH value may also trigger remobilization.
- Sediments may have extremely different thicknesses. In many cases, there may be a shortage of oxygen, even at only a few centimeters depth, i.e. there are anaerobic conditions (in some places up to the sediment surface, e.g. in the Black Sea). This has a considerable influence on the metabolism and degradation of chemicals, since such processes take a completely different course, qualitatively and quantitatively, in the absence of oxygen.

4.5 BIOAVAILABILITY

It is more important to know what percentage of a chemical concentration in a medium or compartment reaches the organisms living there (both directly and through the nutrients) than it is to know the absolute concentration of a chemical, since this available amount determines the toxic effects of that compound. Various definitions of bioavailability are given in the literature:

- the fraction of the total chemical that is available for uptake by biota from the encompassing environment such as water, sediments, soil, suspended solids (RAMAMOORTHY & BADDALOO, 1991);
- the percentage of external exposure (absorbed or resorbed by the organism) which enters the metabolism and is thus available at a possible site of effect (KETTRUP et al., 1991).

The bioavailability depends on a number of factors (GDCh, 1993):

- the physico-chemical characteristics of the substance
- the composition or characteristics of the surrounding medium
- the absolute quantity of chemicals in the medium
- the possibility of uptake or the contact between substance and organism
- the physiological status of the organism

This list clearly shows that bioavailability is not a substance-specific parameter which could be defined with a single test. Instead, the chemical concentration at a target site (e.g. in a test substrate) as a whole can be measured by residue analysis. However, no distinction is made between bound and available quantities of a substance. No procedure exists to date which could be used to assess the available portion based on the absolute measured concentration. For this reason, the concentration measured in the open body of water (i.e. without suspended solids or sediment) is equated with the available quantity in a strong simplification in the aquatic medium. This approach cannot be applied to the soil, however, because of the high spatial heterogeneity of this compartment and the often high percentage of bound residues. Consequently, an accompanying residue analysis is dispensed with here. An exception is made in the case of heavy metals, where correlation factors between the extraction efficiency of certain solvents as well as the bioaccumulation in plants or earthworms can be determined for certain types of soil (e.g. SHEPPARD & EVENDEN, 1992).

4.6 EFFECTS

The effect of foreign substances can be observed on very different levels of organization of biological systems. The transition from the individual (the object of classical toxicology) to the population opens up entirely new dimensions of interpretation (MORIARTY, 1988):

- Sublethal effects which lower the birth rate over the longer term, for example, can have a much stronger influence on the development of a population than acute mortality, even of a considerable number of individuals in the existing population. In the case of species with a mortality which is naturally high, the foreign substance is only an added mortality factor; in some cases, it may take the place of others. For species which take longer to grow, however, lower birth rates pose a much greater threat. A high mortality, on the other hand, can increase the chances for reproduction of the remaining individuals by reducing the intra-specific competition (HARWELL & HARWELL, 1989).
- Foreign substances have an effect on individual organisms, regardless of their number. In other words, like most physico-chemical factors, they are an environmental factor which does not depend on the density of individuals affected by them. However, an indirect impact — e.g. an alteration in the competitive or nutrient relations — may indeed be density-dependent.
- Environmental conditions influence the effects of foreign substances. For example, climatic factors alter the availability of foreign substances for a population or parts of a population.
- Just as the individuals of a species react in different ways to foreign substances, populations of the same species also react in different ways to a foreign substance (a consequence of the genetic variability). This makes experimental results more difficult to replicate.

It must always be borne in mind that the effects of foreign substances on a population are induced by their direct or indirect effects on the individuals of the population. The stated target levels of ecotoxicology, however, are populations and ecosystems. The study of the effect of pesticides and environmental chemicals at various levels is described in Chapters 7 to 10.

4.6.1 Uptake

A substance is taken up either directly (i.e. orally, through inhalation or dermal contact) or indirectly through the food and drinking water. Foreign substances must be absorbed in order to take full effect. In animals, this generally happens through the skin (integument), the stomach or the respiratory system. Many agents for plants are made to stick to the leaf surface, but so-called systemic agents are taken up via the roots and transported to the leaves.

Pesticides are not very specific in most cases. Insecticides, for example, occasionally have adverse effects on plants, and fungicides may be extremely toxic for soil organisms. Even abiotic compartments may be altered directly by entries of foreign substances (e.g. release of aluminum through acid rain), but such changes are more often a result of changes in the biocoenosis. As an example, the litter layer in forests may grow thicker if litter-degrading organisms are eliminated due to the impact of chemicals (BECK et al., 1988) or heavy metals (COUGHTREY et al., 1979).

Organisms can accumulate foreign substances in the body up to a level at which there is an observable effect. Accumulations occur, for example, from consuming a large quantity of contaminated prey. Although there have been several cases, especially in the 1950s and 1960s, in which persistent pesticides (especially DDT) accumulated in food webs, no steady increase has been observed from one trophic level to another. The reasons are that many organisms occupy several trophic levels and that specific metabolism and elimination rates have a significant effect on the quantities in the body (MORIARTY, 1988; HOFFMAN et al., 1990). Accumulation factors (i.e. the ratio between the concentration in the surrounding medium and the body tissue) of up to tens of thousands are possible (e.g. in mussels or fish; PHILLIPS, 1993).

4.6.2 Types of Effects

Pesticide application is aimed at specific targets. However, because most pesticides are non-specific in action, they not only affect the target organisms but also a number of other, beneficial or neutral species (non-targets). In hundreds of cases, massive numbers of deaths in birds or mammals have been attributable to pesticides; however, it is not known in how many of these cases gross negligence played a role (HOFFMAN et al., 1990). Indeed, deliberate use is often made of extremely toxic organophosphorous pesticides to destroy alleged seed consumers (HILBIG et al., 1994).

The most direct and obvious effect on organisms is the acute poisoning of affected individuals and, over an extended period of time, diminished reproduction from sublethal effects. The prognosis or interpretation of direct biological effects is difficult due to the following aspects, among others (HARWELL & HARWELL, 1989):

- differences in the sensitivity of target organisms to chemicals (inter- and intra-specific);
- the relative significance of the directly affected species with regard to their ecological role in the ecosystem or their aesthetic, economic or other value for human beings;
- the influence of physico-chemical environmental conditions on the direct dose-response relationships;
- the interactions with other chemical stressors in the target organism, including the possibility of additive, synergistic or antagonistic reactions.

There are relatively few data available on the simultaneous effect of several pesticides or a chemical with other stress factors (synergisms). Results from acute tests with birds have confirmed the presence of synergistic effects in pesticides (e.g. DDE and Parathion) as well as in environmental chemicals (HOFFMAN et al., 1990). In field studies with birds, it was shown that natural factors (e.g. cold temperatures) as well as the form of application (e.g. granulates) can influence the effect considerably. The concurrent effect of various heavy metals on water fleas proved to be higher than the effect of individual metals alone (ENSERINK et al., 1991). This aspect is particularly relevant because concentrations considered to be unharmful in the framework of a hazard assessment (cf. Chapter 12) are related to *one* metal only, and do not take the effect of metal compounds or combinations into consideration.

The direct effects at the individual level can have a vast range of *indirect effects* on the level of populations, biocoenoses and ecosystems (HARWELL & HARWELL, 1989), such as:

- trophic effects, i.e. changes or interventions in the food webs and resulting shifts in the species composition;
- destructive impact on the habitat, often resulting in a reduction of the spatial heterogeneity and the number of ecological niches;
- changes in competitive relationships, such as a decrease in competitive pressure due to a decline in one species which shared a resource with another; at the level of ecosystems, this mechanism is a means of compensating for effects at population level;
- effects on populations which play a role in the relationships between two species, e.g. as parasites, symbionts, pollinators, etc.;

- resistance formation as a frequently observable consequence of selection by the pesticide as an environmental factor. Natural genetic variability in a population opens up the prospects for an evolutionary adaptation to continuing chronic exposure to chemicals, especially in species with a short succession of generations.

To take the last point one step further, the individuals which are then naturally resistant are benefited by the death of other members of the species. Natural resistances arise spontaneously through mutation. If the toxic effect of a pesticide is based on a simple biochemical reaction which, for example, is linked to the specific structure of the membrane, resistances are relatively easy to generate with a minor mutation of the receptor. Major mutations usually result in the death of the cell, but the formation of a resistance to complicated active mechanisms induced in many cases by natural insecticides, occur much less frequently.

In much the same way as direct biological effects, indirect effects may be influenced to a critical degree by the presence of several chemicals and interactions with the physico-chemical environmental conditions. Moreover, indirect biological interactions and ecosystem processes may be subject to considerable time lags between exposure, the direct or primary biological reaction and, finally, the indirect effect.

The direct and indirect biological reactions to a chemical are integrated at ecosystem level. This combination of mechanisms leads to qualitatively new reactions to chemical stress:

- direct or indirect effects on key species: those which play a decisive role in the structure or function of the ecosystem;
- changes in the community structure: these are produced by the change in the relative abundance of species involved and changes in the interspecific links;
- impact on ecosystem processes: especially effects on primary production, decomposition and the nutrient cycle. These process-related effects are frequently characterized by changes in the photosynthesis rate of the dominant plants or by changes in microbial activity which control the decomposition. Effects on ecosystem level are strongly influenced by the kind of nutrient cycle and the physico-chemical environmental conditions.

4.6.3 Genetic Variability of the Effect

From an evolutionary standpoint, the lives of individuals are very short; only the species exist for longer periods of time (SCHUBERT, 1991b). This continuity is the result of genetic mechanisms which transmit the genetic species-specific program from one generation to the next. The totality of the hereditary factors of all individuals in a species is referred to as the gene pool of the species. These characteristics also ultimately determine the sensitivity to foreign substances in the environment. Each of the various populations of a species is a reproductive community unto itself and thus contains its own gene pool, which represents a segment of the gene pool of the species. Every population of a species can thus have a differing degree of sensitivity to foreign substances. There is a weaker or stronger flow of genes to other populations in the same species. The isolation of populations is the prerequisite for species formation.

The taxonomic categories above the species (genus, family, order, class) are artificial and have been revised several times throughout history. The species can also only be established in the temporal cross-section, since the characteristics of a species are obviously subject to change in the course of time.

The genes which are responsible for a certain characteristic may exist in various states (alleles). Consequently, the characteristic (e.g. clutch size) may occur in two or more forms (polymorphism). This genetic variability, together with environmentally conditioned modifications, is the reason why, generally speaking, no two individuals in a population are exactly the same. The variability is expressed through the relative frequency of the alleles of a gene in the population (allele frequency). All forms of genetic variability originate in mutations which occur as a result of defects during replication of the chromosomes. The genetic variability is preserved through continuous recombination of the genes during sexual reproduction, the previous processes of chromosome splitting, new mutations and the migration of individuals of the same species from other populations.

The species-specific characteristics of an individual (the phenotype) are the expression of its genotype. Given the large number of organisms with a double (diploid) set of chromosomes, however, the genes from one set of chromosomes may predominate over those of the other (dominant genes). The phenotype of an individual as a unit is affected by environmental factors which determine a higher or lower capacity for survival and reproduction (fitness) of the organism relative to other members of the species. Consequently, the subsequent generation will have a larger number of offspring which are fitter. There will also be a shift in allele frequencies in the population, but only if a genetic variability and selection pressure from an environmental factor are given. The examination of this effect of environmental factors is ecology in the above sense and represents the causal link between physiology and genetics.

Natural selection is ecology in action (KREBS, 1985). This ongoing evolutionary process is shaped by the biological and spatial environment in which an organism lives. The current distribution and frequency of animals and plants was determined by evolutionary processes of the past, which continues to have an effect on the environment of the present. The current gene pool is the result of adaptation that has taken place in the past. Adaptation may take place in individuals with a flexible reaction capacity (physiologically or ethologically), or in the population, through the plasticity of the gene pool as a whole. As an example, the development of resistance to insecticides is a major form of adaptation to effects of foreign substances in the evolutionary sense and one which has also been extensively investigated.

5 PESTICIDES AND ENVIRONMENTAL CHEMICALS

5.1 PESTICIDES

5.1.1 Classification

Definition and Description

Pesticides — also known as plant protectant agents or crop protection products — are (primarily chemical) substances designed to combat and control pests on cultivated plants or carriers of viral or bacterial diseases. They also include substances to combat vegetational competitors (weeds) as well as fungal plant diseases. So-called growth regulators are a special type of agent. By influencing the vegetative growth of plants, they support the stability of grain stalks or simplify harvesting by inducing the simultaneous abscission of ripe olives. Chemicals are also utilized to combat disease vectors which are harmful to human beings and domestic animals, both in areas out of doors and in the home. Such agents are sometimes referred to as biocides.

"Agrochemicals", which is the common term in English-speaking areas, is often used to mean all of the synthetic substances — including mineral fertilizers — which are applied for specific purposes in agriculture. In addition to the active ingredient, many pesticides contain a large share of formulation aids, which serve as protective colloids, for example, or help to improve the application characteristics. Aside from fertilizers, agents for crop treatment and disease vector control are the only chemicals which are intentionally released in the environment. This is the reason why, unlike other environmental chemicals, they were among the first substances to be thoroughly tested for their fate and effect.

The German Agency for Technical Co-operation (GTZ) has made an initial attempt to trace and assess the various stages in the "life" of a pesticide in a given developing country, in the form of an ecological information system (KNIRSCH, 1993). This approach is oriented to the normative framework of the local pesticide legislation in the respective country, the FAO Code and the guidelines of the International Association of Pesticide Producers, GIFAP. It

Table 5.1 Classification of Pesticides by Target Organism

Pesticide classes	Target organisms
Acaricides	Mites
Algicides	Algae (seaweeds)
Bactericides	Bacteria
Fungicides	Fungi
Herbicides	Plants, usually weeds
Insecticides	Insects
Molluscicides	Molluscs, especially snails and slugs
Nematicides	Nematodes
Rodenticides	Rodents, especially rats and mice
Termiticides	Termites

Note: Additional classes (e.g. avicides against birds or piscicides against fish) are hardly used today.

Adapted from WARE, 1986.

essentially amounts to an ecotoxicological approach. Up to now, however, only partial aspects of this so-called life-cycle analysis have been realized. As the name implies, it will ultimately serve to describe and assess the fate and effect of a pesticide from the "cradle" (production site) to the "grave" (degradation and waste treatment).

SCHÜEPP et al. (1990) provide an historical overview of the development of pesticides in the industrialized countries. A summary of key terms in the field of pesticides can be found in WARE (1986); they are divided into categories based on the target organism and mode of action. The applications are outlined by SCHÄFER & TISCHLER (1983).

Groups of Active Ingredients

Pesticides can be classified in various ways — by target organism, mode of action, type of application, or class of chemical, for example — as shown in Tables 5.1 to 5.4. Our format here does not allow for a more detailed treatment of individual active ingredients.

Status of Registrations

Roughly 700 active ingredients in pesticides are currently registered for distribution and use on the world market (WORTHING & WALKER, 1987; PERKOW, 1989; THE AGROCHEMICALS HANDBOOK, 1992). The number of registered products runs into the tens of thousands. After a period of breakneck development in the 1970s and 1980s, in which the discovery of new classes of substances led to the production of a large number of active ingre-

Table 5.2 Classification of Pesticides by Mode of Action and Effect Type

Cell toxicants	Inhibit important steps of the cell metabolism; e.g. photosynthesis in the case of herbicides or phosphorylation of ADP to ATP in some acaricides.
Neurotoxic substances	Affect the function of membrane permeability of neurons or their metabolism, especially in the case of insects.
Growth regulators	Are artificial or natural hormones used against undesired plants (rarely insects) by accelerating or inhibiting growth in an unnatural manner.
Defoliants	Remove the leaves of plants.
Attractants and pheromones	Attract animals, especially insects, even over large distances, into traps; very often, they are sexual pheromones.
Repellents	Are usually gaseous substances which deter insects or other organisms from feeding but do not necessarily kill them.
Chemosterilants	Sterilize males of insects or pest vertebrates; classical method of biological pest control.
Disinfectants	Destroy or inactivate harmful microorganisms.
Contact effects	Uptake of the active ingredient at the site of application by direct contact; substances that cause localized injury to the plant or animal tissue on contact.
Systemic effects	Compounds that are adsorbed by the crop and translocated mostly upwards into the stem and the leaves, where they have contact with the target organisms.
Selectivity	Whereas insecticides are more or less selective, herbicides are often designed to be non-selective for preplanting applications or selective only against either grasses or dicotyledons.
Preplanting, preemergence or postemergence application	The timing for the application of a herbicide determines the possible side-effects to the crop: prophylactic (and maybe unnecessary) application against weeds prior to planting or emergence of the crop versus postemergence application according to a weed threshold level in a herbicide-tolerant crop.
Protective or eradicative application	Especially concerning fungicides, bactericides and viricides: prophylactic protection against the respective microorganisms versus treatment after infection.

Note: Growth regulators are also used for stalk shortening, especially in cereal plants. This reduces the danger of lodging.

Adapted from WARE, 1986.

Table 5.3 Classification of Pesticides by Type of Formulation

Sprays	Insecticides, herbicides or fungicides as emulsifiable concentrates, water-miscible liquids, water-soluble powders, oil solutions, etc.
Dusts	Undiluted toxic agents, aerosol dust, etc., mainly insecticides and fungicides.
Granulars	Insecticides, herbicides and algicides in the form of an inert carrier impregnated with a pesticide.
Coatings	Insecticides and fungicides for seed treatment.
Aerosols	Indoor application of insecticides, repellents and disinfectants in the form of a spray.
Fumigants	Vaporous liquids or gases for soil treatment with mainly nematicide and insecticide effect; also widely used in storage pest control.
Baits	Practically no release of the active ingredient into the environment; the toxic substance is ingested by the target animals (insects, molluscs and rodents) after they have been attracted to the bait.
Slow-release formulations	Microencapsulated insecticide compounds, paint-in lacquers or resin strips for indoor use.

Note: This list contains the most common formulations for pesticides. There are several other specialized types.

Adapted from WARE, 1986.

dients, registrations are on the decline today. For one thing, a whole series of old active ingredients are being taken off the market for toxicological and ecotoxicological reasons. For another, it has become very costly to develop new substances (approx. DM 150 to 250 million according to industry figures; ZURHAKE, 1989). As it now stands, the probability of newly synthesized chemicals making it from the screening stage to the status of marketable compounds is only on the order of about 40,000 to 1 (IVA, 1993).

Figure 5.1 shows the status of registrations in Germany to date (IVA, 1993). Just over 200 active ingredients in more than 700 formulations are registered at present. The declining numbers of active ingredients and compounds on the market has both positive and negative implications. There is certainly nothing wrong with eliminating or excluding problem substances — in other words, substances with an undesirable behavior or side-effects in the environment. On the other hand, it is good to have a broad range of active ingredients in order to avoid the development of resistances and meet the specific requirements of certain applications. In fact, ecotoxicological problems are more likely to arise when a small number of low-cost active ingredients are applied in large quantities for many different indications.

Only slightly less than 6% of the pesticides that were registered in Germany in 1992 are classified as poisonous or very poisonous, and approximately

Table 5.4 Classification of the Main Chemical Classes of Pesticides

Insecticides	Herbicides	Fungicides
Organochlorines (= chlorinated hydrocarbons)	Criteria often used for classification: • in relation to the sowing or planting date of the crop to be protected	Inorganic fungicides (based on sulfur, copper or mercury)
Organophosphates (= phosphoric acid esters)	• in relation to the exposure pathway (e.g. as contact herbicide)	Organic fungicides (e.g. dithiocarbamates)
Carbamates	• in relation to the mode of action (e.g. as contact herbicide	Systemic fungicides (e.g. benzimidazoles)
Natural and synthetic pyrethroids	Phenoxyaliphatic acids	Antibiotics (produced by microorganisms)
Dinitrophenols	Substituted amides	The chemical heterogeneity of fungicides is even larger than that of herbicides. In addition to substances applied in the field or greenhouses, a large number of fungicide seed treatments are of great importance.
Organotins	Nitroanilines	
Microbials (disease microorganisms and their metabolic products)	Substituted ureas	
Botanicals	Carbamates and thiocarbamates	
	Heterocyclic nitrogens (= Triazines)	
	Dinitrophenols and phenol derivates	

Adapted from WARE, 1986.

23% as less poisonous or irritating. Roughly 20% of all compounds are subject to water conservation requirements. Some 10% are hazardous to bees (information on active ingredients and formulations is published regularly on floppy disk by the BBA).

5.1.2 Pesticide Market and Application Quantities

Worldwide sales of pesticides in 1992 were estimated at DM 36.5 billion (IVA, 1993). Volumes have remained more or less unchanged since the mid-1980s, but with a declining tendency. Most agents are used in the industrialized countries. Africa, Latin America and Asia (excluding Japan) together make up only about 30% of the world market. The ranking of individual product groups in the various regions (as a percentage of the world market) is shown in Table 5.5. There is a marked predominance of herbicides in Western Europe and North America whereas, in Africa and Asia, insecticides play an important role in crop protection (FONDS, 1992). The hotter climates of Asia and Africa favor the development of harmful insects to a greater degree than in the cooler

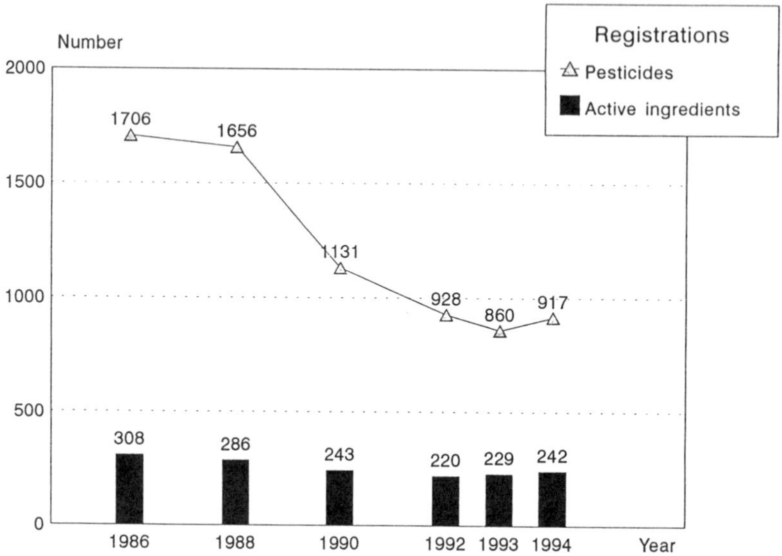

(according to BBA)

Figure 5.1 Status of registrations in Germany up to the beginning of 1993 (figures from BBA).

climates of Western Europe and North America. In the warmer developing countries, weeds are combatted simply by mechanical procedures.

The 10 most important crops in terms of global output as well as the requisite crop protection costs are ranked in Table 5.6 by their share of the world market. They are followed by coffee, potatoes, tobacco, banana and peanut crops (FONDS, 1992).

The German pesticide industry exports some 80% of its domestic production. Exported agents are worth only about half as much per metric ton on the international market as on the domestic market. Such disparities exist world-

Table 5.5 World Crop Protection Market (in Percentages of DM 36.5 Billion) by Region and Product Category

Region	Insecticides	Fungicides	Herbicides	Others	Share of the world market
Western Europe	5.58	9.92	13.95	1.55	31
Eastern Europe	0.80	1.84	1.36	0.00	4
North America	5.46	1.56	18.98	0.00	26
Latin America	4.29	1.98	4.62	0.11	11
Africa	2.08	0.76	1.04	0.12	4
Asia + Japan	11.04	1.08	7.92	0.24	24
World total	29.25	20.86	47.87	2.02	100

FONDS, 1993.

Table 5.6 Level of Pesticide Use for the Ten Most Important Crops in 1990 (in %)

Crop	Share of the world crop protection market	Type of product used (%)			
		Insecticides	Fungicides	Herbicides	Others
Cereals	17.0%	6	38	56	—
Rice	9.6%	34	22	44	—
Corn	9.3%	26	1	73	—
Soy	8.7%	10	1	89	—
Vegetables	6.8%	52	27	220	1
Cotton	6.0%	72	2	26	—
Grapes	4.3%	17	63	18	2
Pome	3.7%	48	39	11	2
Citrus	3.5%	62	16	19	3
Sugar beet	3.3%	23	9	68	—
TOTAL	72.2%	29	22	49	—

FONDS, 1993.

wide. Prices in Germany, for example, are high, but the overall amount is relatively low here because many modern pesticides are used with low application quantities (in the range of 100 g/ha). In the developing countries, many older and cheaper agents are still being used in applications amounting to several kg active ingredient/ha. This makes it very difficult to correlate the world market sales figures to shipped tonnage.

The comparison of data on the quantities of pesticides applied in Germany from 1977 to 1979 to those in 1987 indicate the following (HILDEBRANDT et al., 1991), which is more or less true for other industrialized countries as well:

- The frequency of treatment per unit of surface area rose from 2 to 3.6 applications per year.
- The total application quantity rose from 4.8 to 6.3 kg formulation/ha, whereas that of a single application declined from 2.4 to 1.8 kg/ha (by comparison, cotton crops in Egypt receive about 30 kg/ha per year (1990; GHABBOUR, personal communication)).
- The distribution of the higher total cost over more and smaller treatments is one reason why pesticide applications were more target-oriented in 1987 than at the end of the 1970s.
- On the whole, domestic sales since 1975 have increased 24%, but have remained more or less constant at 30,000 metric tons since the mid-1980s. Sales only recently declined to about 25,000 metric tons (IVA, 1993).

Generally speaking, there has been a noticeable reduction in pesticide use in Europe. There are various reasons for this:

- phasing out of agricultural production for surplus crops;
- as a result, reduction of arable land (set-aside programs);
- ongoing structural change in European agriculture with a tendency towards fewer but larger operations, which as a rule make more rational use of pesticides;
- government-promoted reduction programs and the broad-scale dissemination of Integrated Crop Production procedures;
- increase in biological pest control strategies (FRANZ & KRIEG, 1982);
- changes in the product range in favor of newer agents with lower application quantities (e.g. by separating isomers; CARL, 1985).

5.1.3 Pesticide Use and Yield Losses in Various Crops

In order to assess the role of pesticides in securing the production of food and raw materials, and thus to provide the basis for a risk-benefit estimate, a brief look at the losses in the yields for various crops is provided below.

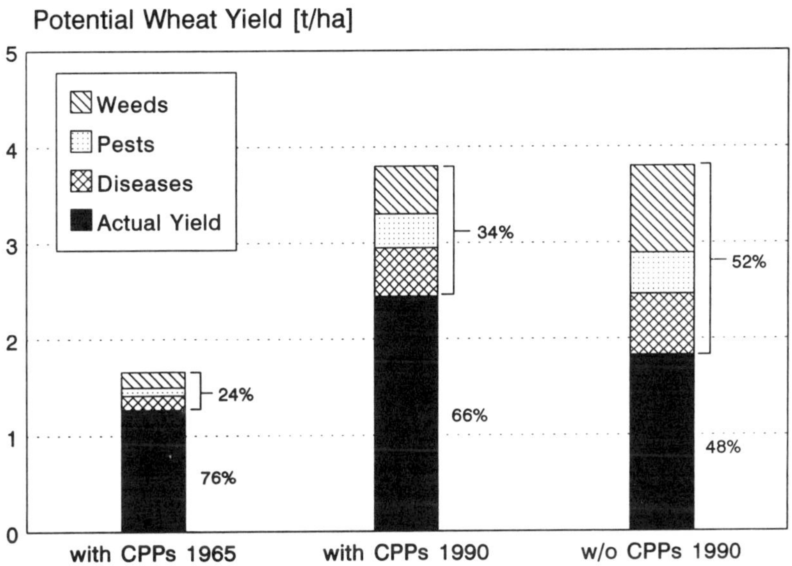

Figure 5.2 Yield and crop losses for wheat on a global scale (adapted from OERKE et al., 1994).

CRAMER (1967) prepared an initial assessment of crop losses due to pests, diseases and weeds. In the meantime, these data have been revised and updated by OERKE et al. (1994), who presents the following picture:

Wheat — The Number One Crop

Wheat is the most widely cultivated crop. The major wheat-growing regions are located in the former Soviet Union and North America. Yields have doubled on average since 1965, accompanied by an average increase in the rate of crop loss, from 24% to 34% (Figure 5.2). Without crop protection agents, the global losses would be as high as 50% today, with sharp regional differences.

Due to climate, fertilization and the selection of varieties, the potential yield of wheat — in other words, the maximum levels of production which can theoretically be achieved — is about two and a half times higher in northern and western Europe than in North America and the former Soviet Union (Figure 5.3); at more than 6 t/ha, the real yield in Europe is in fact three times higher. The risk of damage from disease and pests, as well as the pressure from weeds, is also higher due to intensive cultivation, but it is effectively combatted with crop protection strategies. As a result, crop losses are limited to 20% of the potential yield.

In North America, on the other hand, where the actual yield is 2.2 t/ha, crop losses are on the order of 35% and, in the former Soviet Union, upwards of 40%, with a yield potential that is roughly equal to that of North America

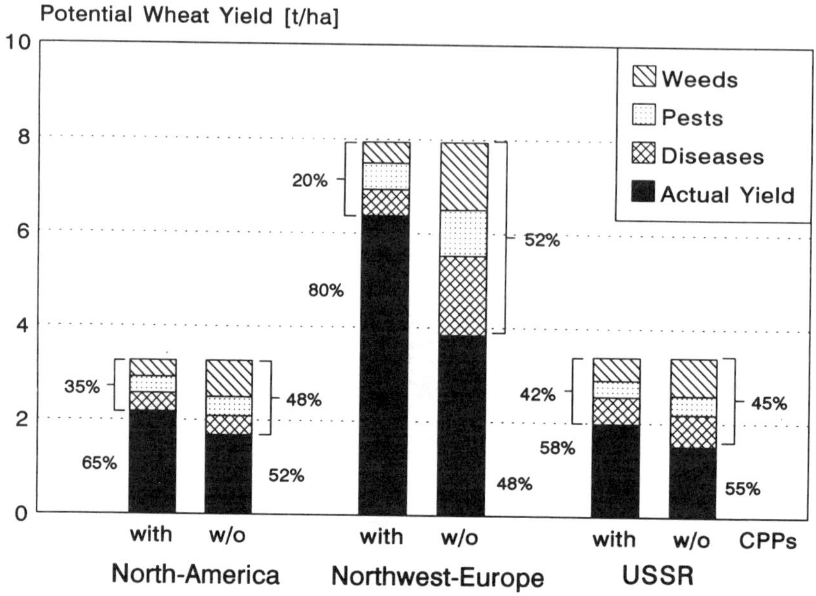

Figure 5.3 Yield and crop losses for wheat on a regional scale (adapted from OERKE et al., 1994).

(Figure 5.3). Crop protection in both regions is largely restricted to weed control.

Rice — The Staple Crop in Asia

Rice is a staple for nearly half of the world's population. Over 90% of the rice harvested worldwide is grown in Asia; India and China are the leading rice-growing countries. The yield per unit of surface area on a worldwide average today is 3.6 t/ha (Figure 5.4). This corresponds to roughly 46% of the potential yield; 1.9 t/ha were harvested in 1965 — 53% of the potential yield at that time.

In the case of rice, as in the case of wheat, the effect of crop protection only becomes evident if the regional differences, especially the state of technological development in the individual sectors of agriculture, are taken into account. Rice is exposed to a greater risk than wheat in all categories of potential damage. In both India and Japan, roughly 85% of the yield would be lost due to harmful organisms if no form of crop protection were applied (Figure 5.5).

In Japan, for example, crop losses are being reduced 23% due to the very intensive crop protection, resulting in yields of 6 t/ha today. Farmers in India, on the other hand, usually cannot afford to purchase and apply chemical agents to fight disease and pests. In that country, crop protection essentially means manual weed control. Consequently, yields in India are no higher than 2.6 t/ha, with a 60% rate of loss in relation to the possible yield.

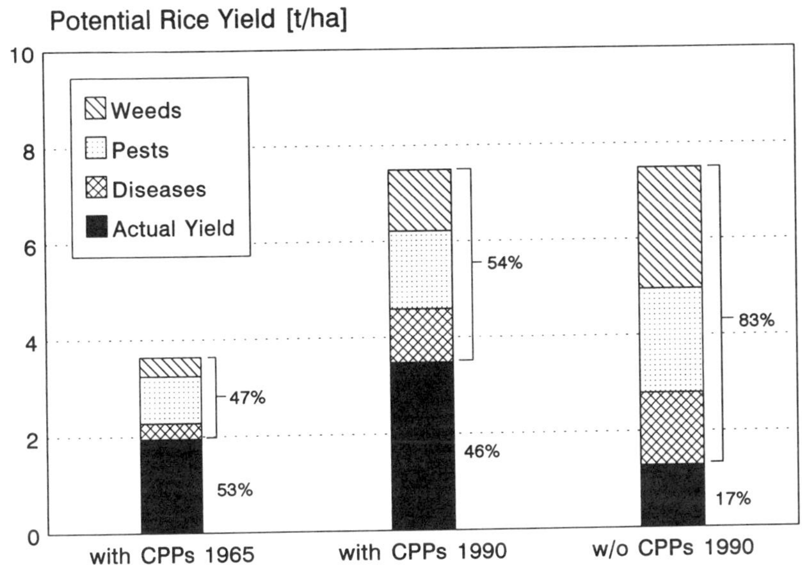

Figure 5.4 Yield and crop losses for rice on a global scale (adapted from OERKE et al., 1994).

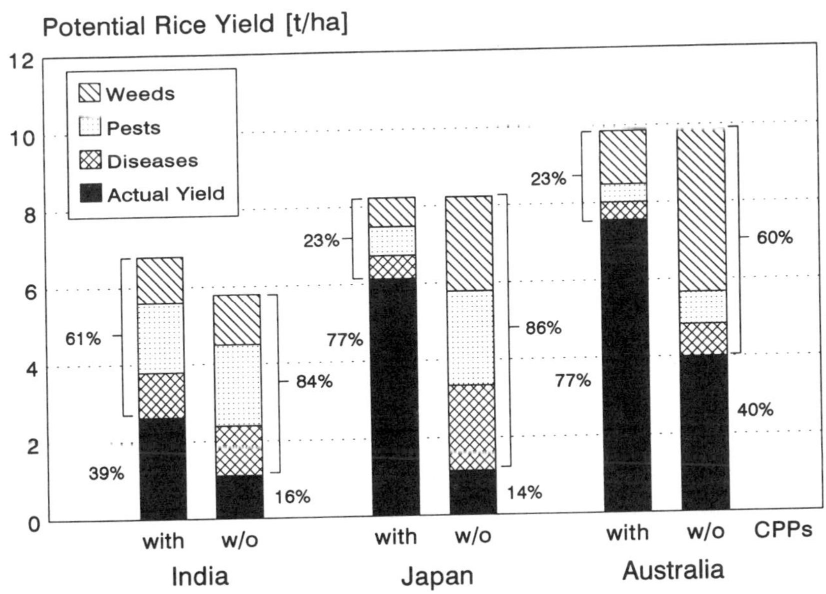

Figure 5.5 Yield and crop losses for rice on a regional scale (adapted from OERKE et al., 1994).

Australia represents a special case in rice cultivation, which was only very recently initiated there. For this reason, there is not much loss from disease and pests. The extremely effective approach to weed control is reducing the otherwise considerable potential for crop losses to 23%. Therefore, rice yields are as high as about 7 t/ha.

5.1.4 Non-Chemical Techniques and Quasi-Natural Procedures

The use of non-chemical techniques and procedures which imitate nature are possible alternatives to chemical pesticides for crop protection and pest control (FRANZ & KRIEG, 1982; KLINGAUF, 1987). They are often lumped under the category of biological pest control agents in the broadest sense of the term (e.g. CROFT, 1990). Alternative agents and procedures, like chemical plant protectants, represent an intervention in natural processes. Therefore, they, too, are subject to hazard and risk assessments in accordance with EU, FAO, OECD, and EPA standards (see Chapter 11). Due to the broad heterogeneous nature of such alternatives, the parameters and criteria of assessment may differ from those of chemicals in many cases. In particular, the use of bacterial active ingredients as crop protection agents may lead to the formation of resistances which, in turn, can cause human or veterinary health problems (SCHÜEPP et al., 1990).

Nearly all alternative approaches to the use of chemical pesticides are characteristically difficult if not impossible to assess for their toxicological effects. The efficacy and the side-effects of such methods do not become manifest until the level of populations or ecosystems. In most cases, moreover, they are not aimed at completely eradicating a particular pest in a given area. This conflicts with their mode of action, which is density-dependent in a variety of respects. The target objective is achieved if the pest is kept below the economic threshold level. Therefore, biological approaches cannot be expected to have the dramatic “knock-down” effect of chemical pesticides (MOLTMANN, 1989).

The economic threshold level is the pest density at which the pest control measures should be initiated to prevent any further increase in the pest population, which would result in crop losses that could no longer be offset (STERN, 1973). The calculation of the economic threshold level is based on the costs of control, the efficacy of control, the yield and proceeds from the harvest, the percentage reduction in yield per pest, the survival rate of the starting population up to the damage density, and a factor which takes into account the long-term socio-economic costs of the pest control strategy (exploitation of raw materials, environmental pollution, damage of predators, etc.) (CHIANG, 1982). The last-mentioned parameter is not the only one which is difficult to appraise; the other parameters may depend on climatic changes, world market prices, consumer tolerance, etc.

One problem requiring intensive evaluation, however, is the precautionary forecast of side-effects, as well as monitoring for early catastrophe detection. Chemical and non-chemical pest control strategies are alike in this respect. A forecast is particularly necessary with regard to microbial pathogens, which can hardly be kept from spreading to non-targets if the latter are susceptible to them.

5.2 ENVIRONMENTAL CHEMICALS

Environmental chemicals are "chemicals in the environment" (HUTZINGER, 1991), be they natural or man-made. The fact is we live in a chemical environment. Environmental chemicals are only one of many environmental factors (temperature, humidity, competitors, etc.) with which organisms have to contend.

In ecotoxicology, environmental chemicals are understood to mean those substances which either enter the environment as a result of human activity; occur in the environment as a consequence of human activity in the environment; or, as the case may be, occur in much higher concentrations than they would in nature (e.g. heavy metals). The number of environmental chemicals is estimated at 100,000 active ingredients and many different compounds or formulations (approx. 1,000,000) (CONNELL, 1986; KOCH & WAGNER, 1989; STREIT, 1991). About 2,000 to 3,000 chemicals are produced in quantities of more than 1,000 metric tons per year, and another 8,000 in quantities higher than 10 metric tons per year (MURPHY, personal communication, 1994).

In the past, a distinction was commonly made between natural and foreign substances (xenobiotics). These are substances which are foreign to the biosphere in terms of their structure and properties — in other words, they are strictly synthetic. However, this distinction is not very revealing as far as the potential hazard of a substance is concerned, especially since natural substances may also be accumulated through human activity (e.g. nitrate).

Independently of these definitions, a distinction has become established in common usage between substances which enter the environment inadvertently (environmental chemicals in the literal sense) and substances which are deliberately applied (i.e. pesticides and fertilizers). Somewhat reliable quantitative data on the environmental concentrations are only available for substances in the second category. As a result of the studies conducted during the research and development phase, extensive data are also available on the environmental fate, and especially the biological efficacy, of these chemicals.

In the case of all other substances, estimates of the environmental fate (exposure) must be made on the basis of random, incomplete measurements. These forecasts are based on the amount of a substance produced as well as the application pattern (local sources such as factories or diffuse entry as in the case of tire abrasion). However, due to the regional and temporal differences (e.g. after accidents), the actual exposure is not quantifiable in absolute quantities for the overwhelming number of environmentally relevant substances (KOCH, 1989). Only the relative dispersion between various compartments can be described fairly well with the help of mathematical models (cf. Chapter 8.1).

In the German Chemicals Act (ChemG, 1990), environmental chemicals in the literal sense are defined in negative terms — i.e. to the exclusion of certain categories of substances (foods and feeds, fertilizers, cosmetics, pharmaceuticals, various wastes and explosives); there are special laws applying to each of these. The remaining, extremely heterogeneous group is described with a total of 16 properties, ranging from "explosive" to "cancerogenic" and "environmentally hazardous". The latter category, in turn, is defined in the Chemicals Act as follows:

> Substances and compounds or their derivatives which are inclined to alter the quality of the natural balance, of the water, soil or air, climate, animals, plants or microorganisms in such a way that they may cause risks for the environment, either immediately or at some later time.

In this definition, the term "natural balance", which can be described as the "nexus of relationships amongst the ecosystems and with their environment" (HEIDLER, 1987), is especially problematic. In the German Plant Protection Law, it is defined as: "Soil, water, air, animal and plant species as well as the interaction between them." The term cannot be clearly defined in scientific terms (ROTHERT et al., 1990). "Environment", on the other hand, is understood to mean the totality of all factors (both abiotic and biotic) which have a direct, active effect on the external world (BICK et al., 1984).

In order to label "environmentally hazardous" substances, standardized descriptions of special hazards (including selection criteria for the aquatic medium, but not for the terrestrial) were published by the EC in 1991, as well as safety recommendations (e.g. "Do not discharge in the sewer system") (AHLERS, 1991; also see Chapter 11.2).

In the discussion of ecotoxicological hazard potential, it should be borne in mind that many chemicals lead to the production of metabolites (e.g. chlorine-substituted anilins or phenoles) which are much more stable than the original substance (OTTOW, 1985) and whose fate and effect may be distinctly different from those of the original substances.

There is not much knowledge available on the impact of substances for special applications, which have a high potential toxicity, such as organo-tin compounds. They are employed in paints for ocean-going vessels to ward off colonizers on the hulls. With time, they eventually peel off with the paint and can thus lead to considerable build-ups of tin in the harbor basin. High persistence characteristics are desirable for hull impregnations, which thus makes them different from biocides used in agriculture (HEINISCH & KLEIN, 1989).

The impact of so-called mass pollutants such as SO_2, CO_2 or ozone, which can have a severe ecotoxicological effect especially on plants, will not be discussed in any further detail here. As a matter of principle, an ecotoxicological assessment should also be required for wastes (mixtures of the most diverse substances in concentrations and compositions which are extremely disparate from one location to another; BAHADIR & FREITAG, 1989). At present, however, there is no uniform approach available to deal with this problem.

6 MEASUREMENT PARAMETERS

6.1 PARAMETER CHARACTERISTICS

In order to assess the ecotoxicological hazard potential of a foreign substance, the characteristic features of the substance must be determined, if at all possible in quantitative terms, as well as its environmental fate and possible effect. Various measurement parameters are employed for this purpose, which are defined as follows:

- Physico-chemical parameters describe the features of the substance itself, which, depending on the prevailing environmental conditions, largely determine the fate and effect.
- Fate parameters describe the occurrence of a substance in the various environmental compartments; the transition from the physico-chemical parameters is flexible.
- Effect parameters are biological measurement variables which describe the effects on organisms. These effects may be directly or indirectly manifested at the various levels of organization of biological systems.

In the context of our discussion here and in the following chapters, a distinction is made between test parameters and test procedures, primarily for practical reasons. Most of the tests commonly employed today are based on only a very small number of measurement variables which are selected above all because they are easy to measure. Consequently, the measurement parameters outlined in the following discussion represent more or less a maximum selection of all such parameters.

The values identified for a given parameter form the initial basis for assessing the effect of a foreign substance at the level of organization from which the parameter was derived. The results obtained at this level must be extrapolated to predict the effects at other levels, which invariably gives rise to uncertainties. The fate parameters are subject to the same principle as effect parameters in this respect, making it necessary to organize the ecotoxicological test procedures in the form of a tiered test program (see Chapter 7.1). Generally

Table 6.1 List of Potential Assessment and Measurement Endpoints Used in Ecotoxicology

Assessment endpoint	Measurement endpoint
Individual	
(Generally not used in ecotoxicological assessments except in limited cases, e.g. with rare or endangered species)	Death Growth Fecundity Overt symptomology Biomarkers Tissue concentrations
Population	
Extinction Abundance Yield/production Age/size class structures Massive mortality	Occurrence Abundance Yield/production Age/size class structure Frequency of mass mortality Frequency of mass morbidity Reproduction
Community	
Market value Recreational quality Usefulness/desired type	Number of species Species evenness/dominance Species diversity Pollution indices Community quality indices Community type
Ecosystem	
Productive quality Productivity	Biomass Nutrient dynamics

Adapted from SUTER (1989) and KAPUSTKA & REPORTER (1993).

speaking, there is no need to make any distinction between pesticides and environmental chemicals in our discussion of measurement parameters, although the issue of pesticides has often led the way in efforts to define and implement measurement variables and methods.

There is a new approach which distinguishes between measurement and assessment endpoints in connection with hazard and risk assessment procedures (Table 6.1; SUTER, 1989). Measurement endpoints refer to the measurement parameters in laboratory tests (e.g. mortality) or monitoring studies in the field (e.g. abundance). Assessment endpoints, on the other hand, are ecosystem parameters, such as the disappearance of populations or negative changes in the spectrum of species, which are often extremely difficult to measure. No distinction is made between a measurement endpoint and an assessment endpoint if the latter can be measured directly (SUTER, 1990). A link between the two endpoint levels can usually only be established through the use of models (BARNTHOUSE, 1992). However, the differentiation between these two endpoints cannot be considered definitive as yet. In addition to parameters which have been clearly defined in scientific terms, assessment endpoints such as "Recreational Quality", which are difficult to define precisely, have also been proposed in the meantime. Furthermore, other authors use the same terms with different connotations: parameters are measuring variables such as mortality or growth, whereas endpoint is understood to mean a numerical value derived from the measurement data (e.g. LC_{50}) or a qualitative assessment (OECD, 1989a).

The following chapters provide a definition of the most important parameters which are necessary to assess the ecotoxicological hazard potential of chemicals (RIPPEN, 1987). Generally speaking, parameters which are measured in the laboratory, e.g. to determine the acute effect, are easier to quantify than field parameters. Of those which come under the category of laboratory tests, standardized physico-chemical parameters can, in turn, be defined much more precisely than biological parameters. The biological parameters must then be differentiated according to whether they are toxicological parameters at the level of individuals or system parameters at higher levels of organization (populations, ecosystems). Toxicological parameters are often used to assess the effects on human beings, although many of the actual tests are conducted on laboratory animals. The degree to which a parameter lends itself to immediate interpretation declines as the complexity of the level of organization increases, making it necessary to introduce additional, e.g. socio-economic, criteria to assess the environmental risk.

6.2 PHYSICO-CHEMICAL PARAMETERS

There is currently no general consensus as to which physico-chemical parameters are necessary for the ecotoxicological assessment of a substance. One model favored by the EPA, for example, recommends the following list of 13 measurement endpoints, which can be arranged in various groups depending on the environmental chemical and medium involved (DONALDSON, 1992):

- ionization (pK_a)
- Henry's law constant (H_c)
- vapor pressure (P_v)
- water solubility (S_w)
- soil/water partition coefficient (K_p)
- soil/water partition coefficient, organic carbon normalized (K_{oc})
- octanol/water coefficient (P_{ow})
- oxidation rate (k_{ox})

direct (aqueous) photolysis rate (k_d)
aqueous photolysis reaction quantum yield (O_x)
molar absorptivity (E_l)
hydrolysis rate (k_h)
biotransformation rate (k_{bio}).

Similar standards are also required for the notification or registration of chemicals and pesticides in Europe, usually based on OECD guidelines (1984). As one can see from this list, the transition to fate parameters is flexible, e.g. in the case of vapor pressure and the partition (adsorption) coefficient (K_{oc}).

Water solubility, hydrolysis, vapor pressure (volatility), the Henry's law constant, octanol/water and adsorption coefficients (see Chapter 8.1) are most often employed to predict the environmental fate. These parameters will now be examined somewhat more closely.

With regard to water solubility, which is given in g/l, the biggest problem is that, under the artificial conditions of a physico-chemical test, the value measured often proves to be much different from that which occurs in laboratory effect tests or even in the field. Normally, pesticide formulations are highly soluble, whereas active ingredients and environmental chemicals may be either highly soluble or extremely insoluble (<1g/l).

Hydrolysis is understood to mean the splitting of a chemical molecule through the uptake of water. Depending on the polarity of a chemical bonded, hydrolysis requires a high temperature and the interaction of protons. It can be catalyzed by enzymes. Hydrolysis is characterized by the speed reaction rate (s^{-1}).

Nor is it all that simple to conclusively determine the vapor pressure (measured in Pa), i.e. the tendency of a substance to evaporate. Despite the use of standardized methods, the data reported in the literature may diverge by as much as a factor of two for the same pesticide (BÖHNKE et al., 1989). This is one reason why volatility tests under more realistic conditions (i.e. from soil and plant surfaces) have been required for pesticides in Germany for several years now. Although the data cited in the literature are not conclusive, a value of >0.1 Pa is a reasonable indication of a volatility which cannot be considered negligible.

The Henry coefficient H_c is of critical importance in assessing the rate at which a chemical passes from water to air. This coefficient is defined as the ratio between the concentrations of a substance in the gas and liquid phase. It is often given as the ratio between the saturation vapor pressure and the maximum water solubility [kPa $\times m^3 \times mol^{-1}$]. The H value can be gauged on the basis of the molecular structure of the respective chemical using the Molecular Connectivity Index (= MCI; SABLJIC & PIVER, 1992). Volatile substances have H values in the range of >0.1-1.

The octanol/water partition coefficient is the ratio of the concentrations of a substance in the two-phase system n-octanol and water. It is an indication of their polarity. The higher the numerical value, which is usually given as a logarithm, the stronger the tendency of the chemical will be to accumulate in the fatty tissue of organisms. In various guidelines, a log P_{ow} of approximately 2.7 to 3 is regarded as the threshold value for this tendency.

The adsorption coefficient describes the partitioning of the substance between the soil and the aqueous phase, normalized to the organic carbon (= o.c.) content of the soil. The K_{oc} value is measured in practical terms, e.g. according to OECD guideline 106 (OECD, 1981b). However, it can also be estimated on an order of magnitude from the partition coefficient log P_{ow} as well as the water solubility (LYMAN, 1982; VON OEPEN, 1990). Substances with low adsorptivity have very low K_{oc} values, whereas extremely high-bond chemicals such as pyrethroids or dioxins achieve values of several hundred thousand. K_{oc} values of more than 10,000 suggest a low anticipated mobility in soil, in which case there is little or no probability of a hazard to ground water (ESSER et al., 1988).

It must be borne in mind that physico-chemistry data are invariably based on measurements taken under artificial conditions. A substance may behave much differently under the conditions of fate or effect tests conducted in the laboratory (and to an even greater degree, of course, under field conditions), and thus also yield values which deviate significantly from the physico-chemical values.

6.3 FATE PARAMETERS

The transition between physico-chemical parameters and fate parameters is flexible since the physico-chemical characteristics of a substance determine their fate in the environment, depending on the prevailing environmental conditions. In the following chapters, persistence and mobility will be discussed in more detail. Volatility will be dealt with in Chapter 8.1.

6.3.1 Persistence

Persistence refers to the degree to which a substance in an environmental medium remains in its original form. This factor depends on the mobility as well as the degree of possible transformation processes. The less a chemical leaches, for example, the more time there is for microbial degradation. With respect to degradation, a distinction is made between primary degradation, in which the test substance is chemically altered, and mineralization, in which the substance is completely broken down into small molecules such as water or carbon dioxide. The single most important process is aerobic degradation (biochemical oxidation). A distinction must also be made between the persistence of biological activity and the persistence of all residues. The Paraquat herbicide, for example, may be biologically active in the soil for no more than one day or so, whereas its residues have a virtually unlimited persistence, due to irreversible sorption (HEITEFUSS, 1987). The residual effect of pesticides in target organisms is also referred to as remanence.

Persistence can be quantified in a number of different ways. The degradation time is the most widespread quantification of persistence in the terrestrial medium; it is the amount of time it takes for 50% (DT_{50}) or 90% (DT_{90}) of the applied concentration (day 0 = 100%) of a substance to be broken down (Figure 6.1). The DT values are calculated from the residue-analysis measurements using empirically determined first or second order functions. The reference to functions as "orders" is intended to convey a formal link to the laws of reaction kinetics. The relative deceleration rate is only constant in the case of first-order reactions, where the DT value does not depend on the original concentration of a substance. In the case of all other functions, the deceleration curve tapers off over the course of the degradation process. Root functions in which deceleration slows down most are especially well-suited for so-called "bent" degradation curves, i.e. degradation processes in which more than one factor (e.g. abiotic or microbial degradation or two different physiological groups of microorganisms) probably had an influence. Experiments with pesticides have shown that the DT_{50} value is rapidly achieved in the case of many agents, but degradation slows down considerably ("bent" curves) thereafter. There are four main reasons for this:

- simultaneous or successive abiotic or microbial degradation processes;
- the abundance of microorganisms decreases during the test due to a deficiency of nutrients;
- the test substance is increasingly bonded to organic soil components and is thus no longer available for the microflora;
- depending on the concentration, the test substance is more or less attractive for various microbe populations.

Due to this great complexity of factors which have an influence on degradation, the question of degradation kinetics and its mathematical condition cannot be considered clarified.

Especially if degradation does not follow a first-order function, the DT_{50} value is slightly overestimated, i.e. degradation appears to be more rapid than it really is. For this reason, the BBA (1986a) also required that the DT_{90} value be determined for the assessment of degradability. This parameter, although relevant, is hardly to be found in the literature, which instead often contains estimates of the half-life (= DT_{50}) or total degradability (variously defined in the range of 75% and 100% of the applied volume).

OTTOW (1985) proposed a division into four groups as a rough basis of orientation for assessing degradation times. In this model, a substance with a degradation time of 12 weeks or less is not persistent; if the value for this period is longer than one year, the substance is considered to be highly persistent.

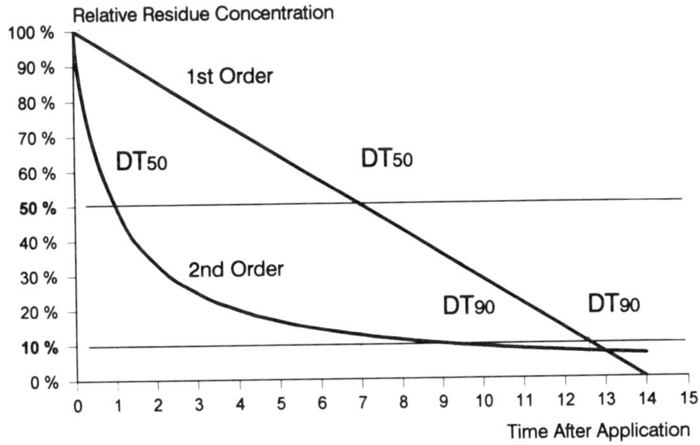

Figure 6.1 Schematic degradation curves. The time it takes for 50% or 90% of the active ingredient to be metabolized is often used as a measurement parameter (DT_{50} and DT_{90}). A "more realistic" 2nd order degradation curve results in different DT_{50} and DT_{90} points in time.

SCHINKEL (1985), on the other hand, developed a pesticide classification system based on the DT_{50}: not very persistent (DT_{50} < 20 days), moderately persistent (DT_{50} 21-90 days) and very persistent (DT_{50} > 90 days). However, the DT value required under the German BBA guideline (1986a) is still considered to be the most useful persistence criterion for a hazard assessment: on this scale, substances with a DT_{90} value >100 days are persistent. This value is taken as a preventive standard to safeguard subsequent crops (BBA, 1988).

In the case of these data, it should be noted that a decrease in the concentration of a chemical is not the same as complete degradation (= mineralization; BBA, 1986a). Even slight changes in the molecule can make it inaccessible to the analytical method employed. The test results may also be distorted or may lead to erroneous conclusions due to the volatility or a strong bonding of the test substance to soil components (= bound residues), especially humic substances or other organic compounds (FÜHR et al., 1985). This gives the false impression of degradation. Studies on metabolism offer the only means of determining whether degradation is involved or not, especially if radiolabeled substances are used; the use of non-labeled substances would be extremely costly (SCHINKEL, 1985; FÜHR et al., 1991). As a matter of principle, it is more important to clarify the question of bioavailability in such bound residues than it is to establish its chemical identity (ROBERTS, 1984).

A few authors, such as KLÖPFFER (1989), consider the parameter of persistence to be the key (negative) criterion for assessing the environmental fate of chemicals (or, expressed positively, their degradability). It is argued that a combination of persistence and accumulation capacity can lead to high concentrations in the environment or in certain organs of organisms (for example, in the case of polychlorinated biphenyls). Even if no acute toxic effects are found whatsoever, this is no guarantee that persistent substances will not cause global damage to the environment on a massive scale (e.g. chlorofluorocarbons in the ozone layer).

6.3.2 Mobility

Mobility is understood to mean the migratory behavior of a substance within or between the media soil, water, air and the respective compartments, primarily as leaching of a chemical from the soil surface down to the ground water. An estimation of mobility can also be conducted with the help of physico-chemical parameters, which, in soil analyses, may also include water solubility (S_w > 30 mg/l), the adsorption coefficient (K_{oc} < 300-500) and the degradation time (DT_{90} > 100 days) in the soil (BBA, 1990a; RÖMBKE et al., 1993). In very simple form, such values may also be calculated in a "hazard index" (e.g. the GUS index, GUSTAFSSON, 1989) or in models of varying complexity.

Since there is no single measurement parameter for mobility in soil, in practical tests (both in the laboratory and in semi-field tests, e.g. lysimeters) the percentage of the applied volume collected in the leachate is used for the assessment. In any case, there is no scientifically sound approach in a leaching test that would allow a certain percentage of a chemical found to exist in the leachate to be defined as a threshold value. Consequently, the results of such tests only allow for rough classifications (HERZEL, 1987). Major advances in residue analysis have made it possible to detect even the most minute traces anywhere. Therefore, it cannot automatically be assumed that mere evidence of a substance warrants an unfavorable assessment; otherwise, too many substances with a false positive rating would be tested unnecessarily in further steps. Due to the vital importance of drinking water resources, the EU specified a threshold value of 0.1 g/l per substance and 0.5 g/l for the sum total of all pesticides. In the sense of precautionary health measures, the values are to be regarded as quasi-zero values (= close to the detection limit). Problems arise when these values are used as a "cutoff criterion" in the registration process for substances detected in the ground or surface water that is used to produce drinking water. In other words, when such a low value is also prescribed for surface waters, it makes routine use of pesticides in agriculture virtually impossible since there is no way to keep the substances from entering the aquatic medium.

6.4 EFFECT PARAMETERS AT THE VARIOUS LEVELS OF ORGANIZATION

Foreign substances have a direct effect on individual organisms or affect them indirectly through changes in the surrounding environment. The direct effects

Table 6.2 Definition of Effect Tests Used in the Laboratory

Test type	Main measurement parameter	Duration
Acute	Mortality	Short-term (a few days to approximately 2 weeks); usually without feeding
Chronic	Sublethal parameters (e.g. reproduction, growth)	Long; in small organisms most of the individual life cycle; in larger animals, up to one-third of the mature life cycle; in mammals, 1.5 to 2 years

are studied by toxicology. Although the individual organism may be a target of nature conservation and monitoring, ecotoxicological activities focus on higher levels of organization such as populations and ecosystems. The presence or absence of individuals or the change in their behavior (in structure or function at the level of organisms) induces a change in structural and functional parameters which are characteristic of higher levels of organization. The degree to which the results of effect tests on one level of organization are relevant for higher levels is discussed at some length in Chapter 7.

The following chapters outline the measurement parameters most commonly employed in tests conducted at the levels of individuals, populations and ecosystems, respectively. In standardized ecotoxicological studies, especially in the laboratory, animals are used far more frequently than plants. For methodological reasons, animals are easier to breed and manage, and they seem to be more sensitive to foreign substances. The use of algae in aquatic laboratory tests (MADSEN, 1984), the composition of plant communities in the field (SORTKJAER, 1984) or the use of accumulation indicators for atmospheric pollutants (WEIGEL & JÄGER, 1988) tend to be the exception. Examples of tests for every level are provided in Chapters 8 to 10.

As far as duration is concerned, test procedures, as well as the measurement parameters employed in them, are generally subdivided into "acute" and "chronic" (Table 6.2). More refined subdivisions (e.g. "subchronic" as an intermediate stage; SOLBE, 1993) have been considered but have never become established in general practice because they are defined in so many different ways, depending on which procedure is used.

The two terms "acute" and "mortality" are often used synonymously, to mean "lethal effect". In chronic tests, on the other hand, various kinds of sublethal measurement parameters may be employed. There is definitely a certain amount of overlap or transition between the two types of test. Morphological changes, for example, can be measured in both acute and chronic tests. Moreover, the two test groups also differ in terms of their orientation: acute tests are usually employed as a "screening tool" with high concentrations over a broad range, whereas chronic procedures tend to be applied for in-depth investigations, i.e. concentrations are used which did not have any acute effects, or only a low mortality, in the preceding acute toxicity test (HAMBURGER, 1983).

6.5 BIOLOGICAL PARAMETERS AT THE LEVEL OF INDIVIDUALS (= TOXICOLOGICAL PARAMETERS)

In the case of toxicological parameters, a distinction is made between those with a direct (acute or chronic) damaging effect (Chapter 6.5.1) and those used to ascertain accumulation (Chapter 6.5.3). A separate chapter (6.5.2) will be devoted to a few of the statistical problems involved in evaluating the data. Data on the concentration of a foreign substance in the tissue are discussed in connection with bioaccumulation (Chapter 6.5.3), although such measurements can also be taken in other tests. Toxic effects are either reversible (i.e. they go away after the foreign substance ceases to be active) or irreversible (causing permanent damage or death). At this level, no distinction is made between effects on plants or animals, even if the methods used to measure the parameter differ greatly from one case to another.

6.5.1 Parameters for Direct Damaging Effect (Table 6.3)

The mortality or acute toxicity in individuals is a standard parameter for the assessment of a foreign substance. In mammals (including human beings), a distinction is made between oral or dermal application (concentration indicated in mg/kg body weight) and application through inhalation (concentration indicated in mg/l inhaled air). In tests with other animals or plants, on the other hand, exposure primarily takes place via the surrounding substrate (water or soil). In such acute tests, five concentrations are usually selected after a preliminary ("range-finding") test in order to determine a dose-response ratio of mortality over several levels of concentration. The results are generally indicated as the LD_{50} (lethal dose) or LC_{50} (lethal concentration). These quantitative dose-response ratios are defined as the mean lethal volume or concentration of a substance or its formulation which causes death in half of the test animals after exposure through the stomach, skin or respiratory pathways within a specified period of time.

Until recently, sublethal effects were not included in standardized tests due to the greater complexity of the procedures required to measure them and the more difficult interpretation (PERKINS, 1979). In some of these tests (e.g. plant growth tests), sublethal parameters such as weight development are determined to ascertain concentrations in which an effect occurs in 50% of the test individuals (EC_{50}).

Since stress under conditions of exposure in the field can be expected to be lower but of longer duration than in the laboratory, the length of exposure and analysis of sublethal measurement parameters is being assigned increasing im-

Table 6.3 Key Measurement Parameters in Laboratory Effect Tests

Term	Description
Acute Parameter	
Mortality	Number of dead individuals. Corresponds to immobility in some invertebrates and to growth inhibition in plants. Indicated as LC_{50} or EC_{50}.
Chronic Parameters	
Growth	Change in biomass or body length during the test, measured in relation to the control or to the starting value. Usually given as NOEC, and less often as EC_{50}.
Reproduction	Indicated differently, depending on type, e.g. number of eggs or young animals, clutch size or hatching rate. Usually given as NOEC.
Juvenile development	Included in some reproduction tests, otherwise as a separate measurement; always in relation to the control.
Morphological effects*	External damages such as wounds in animals or chloroses in plants; difficult to quantify.
Biochemical or physiological effects*	e.g. changes in enzyme synthesis or activity or in respiration. Indicated as NOEC if quantifiable.
Behavioral changes*	Observations of non-typical behavior; e.g. uncoordinated swimming of water fleas or fish.
Teratogenity	Inducement of defects in embryos.
Mutagenity	Irreversible changes of the genome.
Carcinogenity	Formation of tumors through transformation of a differentiated body cell to a cancer cell which multiplies uncontrollably.

* These effects may occur in both acute and chronic tests. They are generally difficult to quantify or evaluate statistically.

portance in relation to the data obtained in acute toxicity tests (e.g. VAN STRAALEN & DENNEMAN, 1989). The levels of concentration should be selected, by means of a preliminary test, for example, to determine the concentration at which no effect is observed (NOEC) compared with a control value. The lowest concentration at which effects are observed is indicated as the LOEC.

Theoretically, modes of behavior (e.g. running or breathing activity), productivity (e.g. clutch size and hatching rate; McARTHUR et al., 1983), morphological characteristics (e.g. shell thickness; RATCLIFFE, 1980) or physiological processes (e.g. membrane permeability of lysosomes; BAYNE et al., 1985), the cholinesterase activity or neurotoxic effects (HAQUE & PFLUG-MACHER, 1985) are suitable for the analysis of sublethal toxicity. In basic research on quantitative structure-activity relationships (QSAR; KÖNEMANN, 1986; HERMENS, 1989), similar parameters, especially physiological or biochemical functions, are chosen as an indicator for the scope of effect. Generally speaking, tests with biochemical parameters have little practical value because they are more elaborate and also more difficult to interpret. This is why they do not form a routine part of the registration process for pesticides. If anything, they have more of a diagnostic function in determining the cause of damaging effects.

Growth is usually measured by the change in relation to the starting weight and in relation to a control. It is important to bear in mind, however, that both positive and negative weight changes can have very different causes and meanings. Weight loss, for example, can be a sign of a toxic or deterring effect of the test substance, but it may have also been caused by reproduction processes (e.g. egg-laying). The conclusions to be drawn from the measurement of weight loss in these two cases would be diametrically opposed.

Depending on what organism is involved, reproduction can be measured in very different ways, e.g. as the number of eggs, hatching rate or average size of the eggs laid. Since it is only important whether reproduction is secured for the next generation or not, and not when or how it might potentially be broken off, reproduction can best be indicated as the reproduction success — which is to say, the total number of survivors which are left to reproduce again. It is the same reason why tests in which the early life-stage development of an organism serves as a sensitive parameter (e.g. Early Life-Stage Test with fish) should eventually be replaced by Full Life Cycle Tests.

Until only a few years ago, the parameters of teratogenity, carcinogenity and mutagenity were regarded exclusively from the standpoint of assessing hazards to human beings. Bacteria (e.g. Ames tests; RINKUS & LEGATOR, 1979) or vertebrates such as rats are usually employed in standardized procedures. The mechanisms involved in producing cancer, some of which are very complicated, have only been inadequately clarified to date. For this reason, and because these effects were also observed in organisms in the field, various higher animals such as sea urchins, fish or amphibians are being used successfully in terms of their suitability as test organisms (DUMPERT & ZIETZ, 1984; NACCI et al., 1986; HAWKINS et al., 1988; FRANCIS & HABER, 1990).

These test procedures are based on the genetic variability of the individuals which may be manifested at various levels of organization (e.g. physiological, morphological or behavioral). Damages are understood to mean a disturbance in the interaction of nervous and humoral mechanisms which impair the biochemical and physiological processes in the organism (homeostasis; MORIARTY, 1988). Because the individuals have differing degrees of sensitivity to the foreign substance, some of the test organisms survive, grow, or reproduce, etc. in a certain time period. However, this also implicitly restricts the relevance of the results outside the laboratory. Sensitivities to a foreign substance not only differ from one individual to another but also from one stage of development to another, depending on the external and internal conditions, and of course from one species to another as well.

6.5.2 Excursus on Statistical Evaluation of Tests

Tests which are conducted to assess the ecotoxicological hazard potential generate data of varying quality. The transferability of the data depends on the extent to which the tested species, with its mode of life, physiological and morphological characteristics, is representative of the organisms exposed to the substance in the field. The quality of the results of the individual tests is also determined by the characteristics of the test procedure itself, such as the number of replicates and test organisms and especially the statistical methods used to evaluate the raw data.

The following types of data are generated primarily by biological test procedures:

Mortality data (usually from acute tests):
Discrete data, with values which may range between 0% and 100%.

Weight data (usually from chronic tests):
Continuous data within a range which is specific to the species in each case. Excessive weight loss results in death. A stimulation of weight development is often observed with low chemical concentrations.

Reproduction data (usually from chronic tests):
Discrete data within a range which is specific to the species in each case. Again, a stimulation of reproduction relative to the control for low chemical concentrations cannot be ruled out.

In order to compute the LC (= Lethal Concentration), EC (= Effect Concentration), NOEC (= No Observed Effect Concentration) and LOEC (= Lowest Observed Effect Concentration) values from the raw data obtained in the experiments, either the deviation from the control is checked for its significance (e.g. ANOVA) or dose-response relationships can be described and used to extrapolate the concentrations at which a certain effect occurs (e.g. regression analysis). Not only such values as LC_{50} or NOEC but also the shape of the dose-response curve can give valuable information on the effects of the test substance.

Determination of LC or EC Values

The determination of LC and EC values is very often followed up with a probit analysis, in which strict prerequisites must be met; for example, a linear dose-response relationship must exist in the probit probability network after the transformation (FINNEY, 1971). For this reason, a probit analysis can only be carried out after a careful check and, if necessary, a pre-selection of raw data. In acute tests with usually only five different and often relatively roughly gradated concentrations, the prerequisites for conducting a probit analysis to determine the LC_{50} are often not met. In this case, the data can be transformed into probits and a regression line may be calculated (so-called probit transformation), resulting in an estimate of the LC_{50}.

If two immediately consecutive concentrations exhibit 0% or 100% effect, the geometric mean of the two concentrations can be used to determine the LC_{50}. In a subsequent procedure, an arcsinus transformation is carried out instead of a probit transformation of the data (cf. STEPHAN, 1977). If no linear dose-response relationship exists, various procedures are indicated to determine EC and LC values (e.g. KOOIJMAN et al., 1983), which are not explained here in detail.

A special problem involved in determining EC values is posed by the frequently observed stimulation in growth parameters (e.g. in the plant and earthworm test) at low concentrations, which can influence the results of the probit analysis. A regression model, on the other hand, may show an increment of the reaction in low concentrations (BRIAN & COUSENS, 1989).

In many cases, acute tests are also required to indicate the concentration at which no effect is observed (NOEC) or the lowest concentration at which an effect can be measured (LOEC). Due to the inherent limitations of acute tests (e.g. short duration), these NOEC or LOEC are not really “measured” but estimated.

Determination of NOEC or LOEC Values

In the case of sublethal parameters, the simple analysis of variance (ANOVA) is usually employed to determine LOEC or NOEC values. This method is one of the standard procedures used to examine issues of this kind and it has met with widespread acceptance for the evaluation of ecotoxicity tests (LÖKKE & VAN GESTEL, 1993). In the ANOVA, the variance of measured values within the individual groups (concentrations) is compared with the variance between the groups. The ANOVA only provides information about whether there are significant differences between the individual groups. To analyze where differences exist between which of these groups, so-called multiple comparison tests such as the Bonferroni test, Scheffe test or Duncan test are subsequently conducted. The highest concentration without significant differences is defined as the NOEC and the lowest concentration with significant differences as the LOEC (separately in each case for every parameter).

If the prerequisites for conducting an ANOVA are not given, parameter-free procedures, such as the H test according to Kruskal-Wallis, with a subsequent test for contrasts (e.g. Nemenyi comparisons), can be applied instead.

However, these tests are not as refined as the ANOVA. This is why differences which exist in fact may not be recognized and the estimated NOEC values are too high.

The NOEC value determined by an analysis of variance may vary depending on the technical givens of the test, such as the number and volume of the tested concentrations, the number of replicates (PESTEMER & GÜNTHER, 1989) and the variability within the control. In the process, smaller and less precise experiments may lead to higher NOEC values; in other words, possible effects of a substance are underestimated (HOEKSTRA & VAN EWIJK, 1993). Another point of criticism is that the level of the detected NOEC value depends on the gradations employed for the concentration. Last but not least, it is not possible to indicate any confidence interval for LOEC or NOEC values.

Under these circumstances, an assessment of NOEC values with empirical biological data would also appear to be acceptable if, for example, an inhibition of weight development by more than 10% in comparison to the control is taken as a threshold value for the estimate of an NOEC value. It is important to note that the empirical value employed is *not* derived from the respective test (e.g. from the standard deviation of the control) but is based on a large number of tests previously performed. The requirements for tests with reference compounds (negative controls) in the framework of the BBA/IOBC beneficial arthropod tests are based on similar considerations (e.g. BBA, 1991a).

In order to avoid unacceptable generalizations, the results of an ANOVA should always be indicated with the measurement parameter to which they refer ($NOEC_{Biomass}$, $NOEC_{Reproduction}$, etc.). This not only serves to distinguish them from NOEC values which integrate several measurement parameters, such as those arrived at by the end of chronic studies on ecotoxicity. It also keeps them from being confused with NOEC estimates yielded by acute tests with relatively short test periods, which are usually not designed to analyze sublethal measurement parameters but to determine a median mortality (LC_{50}).

In view of the described deficiencies of the methods used to determine the NOEC, a number of suggestions for improvements have been discussed in more recent literature. In addition to deriving NOECs directly from tests ("observed"), a value of this kind can also be statistically determined. In view of the fact that a higher priority will be assigned to NOECs and the exposure period in the future, PESTEMER & AUSPURG (1986) and PESTEMER & GÜNTHER (1989) suggest determining a threshold concentration (EC_{10}, EC_5 or EC_1) for plant tests which can be calculated with the help of a regression analysis. HOEKSTRA & VAN EWIJK (1993), on the other hand, prefer to calculate a so-called safe concentration, the effect of which is approximately 25%. This value is used to perform a linear extrapolation to a concentration which "causes an acceptably small deviation to the control."

In many cases, the design of ecotoxicological tests is not optimal for subsequent statistical evaluation (e.g. in relation to the number of concentration stages or replicates employed). In addition, the limits on the variability of the

Table 6.4 Definition of Bioaccumulation Parameters

Term	Description
Molecular weight	Characteristic of substance; substances < 700 have a propensity for accumulation.
Partition coefficient log P_{ow}	Relationship of concentration in the two media n-octanol and water; the higher the value (especially as of approx. 2.7 to 3.0), the greater the tendency for bioaccumulation. The value can be assessed on the basis of water solubility.
Bioconcentration factor (BCF)	Value for the relationship of the concentration of a substance in an organism to the concentration in the ambient medium at steady-state. It can be estimated roughly from the P_{ow}.

test organisms are often poorly defined. Therefore, special attention should be devoted to precisely this aspect of a test strategy since, given a high variance in the control and/or the small n, only sharp discrepancies will be recognized as statistically significant. For this reason, it is advisable to conduct routine evaluations of the test organisms' sensitivity with a reference substance, an internal laboratory standard or a general standard for the variability of the results in the control. Generally speaking, however, it must be said that the interpretation of results derived by statistical methods must always be based on a biological understanding of the subject and revised if necessary.

6.5.3 Parameters on Bioaccumulation

Bioaccumulation refers to the deposits of a foreign substance in the organism (especially in the fatty tissue) in a concentration exceeding that which is found in the organism's environment. A distinction has to be made between the bioconcentration (uptake from the ambient medium) and the biomagnification (input via the food supply). The extent of this input strongly depends on the biochemical-physiological processes which differ from one species to another as well as on the biotransformation (= metabolism) and elimination processes.

It is extremely difficult to quantify the amount of foreign substance taken up in the organism's body as well as the rates of partition, metabolism and elimination (Table 6.4). In particular, the residue analysis accompanying this evaluation must be highly sophisticated because of the often extremely low concentrations; consequently, the use of radiolabeled substances is recommended. The characteristic features of substances, such as water solubility or molecule size, can have such a decisive influence on accumulation processes that they can be estimated with QSARs (STREIT, 1992).

The bioconcentration factor (BCF) is the standard of measurement for accumulation. In experiments, it is usually determined for aquatic organisms (e.g. fish or mussels); standardized procedures do not exist for terrestrial organisms (cf. Chapter 8). BCF values higher than 100 are considered ecotoxicologically problematic, especially if other risk factors are also involved (GDCh, 1993). In any case, the degree of accumulation strongly depends on the conditions of exposure, in addition to characteristics of the organism (e.g. volume of fatty tissue). If the log P_{ow} value is higher than 2.7 to 3, this is regarded by various registration authorities as a strong indication of a higher bioaccumulation potential in the case of both pesticides and environmental chemicals, which means that laboratory tests, e.g. with fish (OECD, 1981c), can then be required. It is important to bear in mind that accumulation will again begin to decline if the log P_{ow} values are extremely high (> 6).

The mathematical concept of compartment models was developed for such persistent substances. In the strict sense of the term, compartments in this context refer to biological structures in which foreign substances exhibit a uniform transformation and transport kinetics. The key compartment in animals, for example, is the blood, the hemolymph, the extracellular fluid or another transport system. Environmental compartments can be defined in an analogous manner (e.g. sediment, open water, biota and suspended matter as compartments of the aquatic medium). Thus, for example, the plant biomass is proposed as another compartment for "real world" models (MacKAY, 1991) due to its capacity for the bioaccumulation of chemicals (CALAMARI et al., 1991).

Changes in typical metabolic function, measured and explained with the help of compartment models, can serve as parameters to determine the effects of foreign substances. On the other hand, they may also be responsible for masking simple dose-response relationships (MORIARTY, 1988). Laboratory tests of this type are conducted on individuals or small groups ultimately for the purpose of extrapolating the results to other populations of the same species or to other species. At the same time, the relevance of these results is severely limited by major species-related differences in the uptake, accumulation, fate and elimination of foreign substances.

6.6 BIOLOGICAL PARAMETERS AT SYSTEM LEVEL

Before the individual parameters at the three levels of organization — population, biocoenosis and ecosystem — are discussed in more detail, a few basic problems related to the parameters used in this connection will be touched on here. The main problem is the contradiction between the number of samples or replicates which are necessary for a statistical evaluation of the natural variability and the high methodological cost of determining the species. More than 1,000 species of invertebrates were found in the soil of a very simply structured beechwood forest in southern Germany, for example, and by no means were all of them known to science (BECK et al., 1988). As will be discussed in connection with the example of the diversity index (Chapter 6.6.2), the already narrow database in such cases should not be reduced any further by mathematical-statistical procedures. In other words, if the number of replicates is not high enough for an evaluation for methodological reasons, then raw data should be employed instead of mean values. Due to the great variability of ecological measurement data, only major or long-lasting deviations can be statistically secured (MATHES et al., 1991).

6.6.1 Parameters at Population Level

To a greater degree than at the level of individuals, the problem that occurs at the level of populations is how the concentration of a foreign substance is correlated with the effect on a population parameter. The parameters become more abstract in relation to those measured at the level of individuals. This is true to an even greater degree of the transition from the population to the biocoenosis (community) and, with the ambient environmental factors, ultimately, to the ecosystem.

The parameters at the level of populations are derived from characteristics of the population dynamics (CALAMARI & PACCHETTI, 1987). The population dynamics is influenced on the genetic variability of the organisms. Consequently, a shift in the gene frequency could serve as a parameter for the effect of environmental factors, including foreign substances, but not to the exclusion of other factors. However, because the genetic composition of a population is difficult to measure directly, primarily the following measurements are taken to describe a population:

- number or biomass
- growth rate
- age structure
- distribution of size classes

Otherwise, relatively little use is made of parameters which describe a population's overall state of health, such as the degree of parasitization or the proportion of deformations. Generally speaking, studies limited purely to the population level are rare; such measurements are usually taken in the framework of ecosystem studies.

In any case, the maximum period of stress must be determined in addition to the maximum extent of the effect, as this is the only way to judge the recovery of a population. It is especially important to determine whether or not there are any compensatory reactions (surplus reactions) after the stress factor has ceased to have an influence (Figure 6.2; OTTOW, 1985; BECK et al., 1988).

6.6.2 Parameters at the Level of Biocoenoses

The number of species as well as changes in the dominance structure are examples of parameters applicable to biocoenoses (and higher levels; MATHES

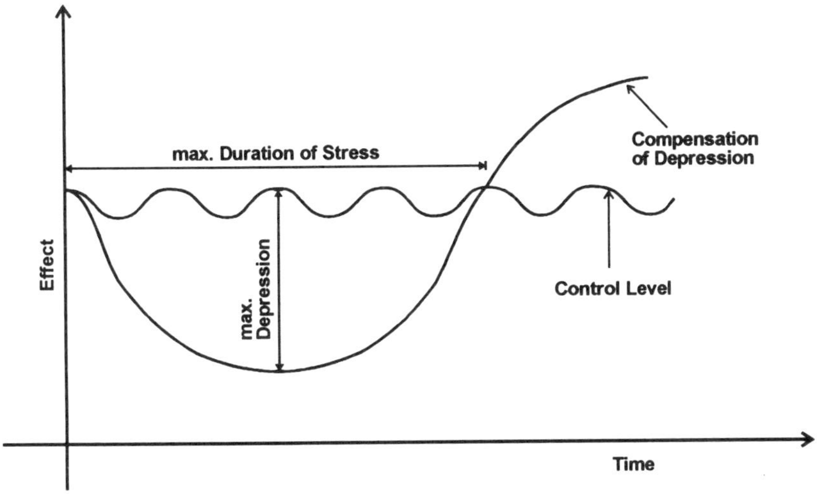

Figure 6.2 Schematic graph of soil flora reacting to the input of a pesticide. When assessing the depression of production capacity and population level, one has to keep in mind that the duration of the stress impact is of more importance than the maximum intensity (OTTOW, 1985).

et al., 1991). Since these data are easy to quantify or calculate, attempts were already made long ago to derive biotic pollution indicators from the dominance structure or the distribution of organisms and species (SHEEHAN, 1984a; MALTBY & CALOW, 1989). In the 1970s and to a lesser degree in the 1980s, many attempts were made to quantify the results that had been obtained through the use of bioindicators in order to compare the effect of foreign substances in various ecosystems (CAIRNS & PRATT, 1987). This raised the broader question of whether there are any general rules whatsoever which determine the structure of biocoenoses (ARNDT et al., 1987).

The simplest example of this are data based purely on the relationships between certain species (e.g. chironomids in relation to insects), which can be assigned to a certain sensitivity or tolerance (WINNER et al., 1980). These data were developed further into a large number of qualitative or semi-quantitative biotic indices. The relationships which have been empirically identified with regard to the composition of macroinvertebrate fauna in the benthos of a polluted river, for example, have never gained currency because they are overly dependent on the local situation and do not yield any additional information. Diversity indices could be called the culmination of this approach as a whole (review for limnic systems: MALTBY & CALOW, 1989).

There are a large number of diversity indices and formulas derived from them (e.g. evenness) which are based on various theoretical distributions and in which the species are weighted differently in terms of their number and relative frequency (OWESEN, 1976). The Shannon-Wiener index is by far the most widespread. Diversity indices such as biotic indices have a number of drawbacks, aside from the statistical errors which occur in processing the data (MORIARTY, 1988):

- Diversity indices do not change if one species is replaced by another and, for all practical purposes, they are not affected if only a few species, especially rare species, disappear.
- Biotic indices can only be compiled if a great deal is already known about the ecosystem and the foreign substance.
- Biotic as well as diversity indices call for a precise determination of the individual species. Related species can react in very different ways to environmental changes and a specification which goes no further than the genus or family invalidates diversity indices.
- The community on which any index is based takes in only a limited range of species.
- Biocoenoses are not supra-organisms, which makes it difficult to assign any biological significance to a diversity index in the first place.
- Every change in the index must first be explained in terms of the dynamics of the individual populations before it is attributed to an effect of the foreign substance.

A diversity index as such cannot provide any information about the condition of or hazard to a given site, which can only be judged in relative terms by monitoring changes over the course of time and comparing the site with unpolluted sites (KNEITZ, 1983). To summarize, the use of diversity indices cannot be recommended due to the above-mentioned limitations, since a high value may be an indication that the ecosystem is in "good" condition, but may also be a sign of a disturbance (NÄHRIG, 1991). GOODMAN (1975) gives the index a categorically negative rating: "Shannon-Weaver (sic!) diversity is a dubious index. The biological meaning of what it represents is unclear; and there are no available criteria for precisely measuring community diversity under natural conditions."

Similarity coefficients (e.g. the Jaccard or Renkonen index), which describe the similarity between two communities of organisms based on a comparison of species stock and their relative dominance by pairs, is more suitable than diversity indexes (LAMONT & GRANT, 1979). These usually quite simple comparative values are widespread in ecological research, but they are seldom used in ecotoxicological studies.

In projects oriented to conservation and wildlife preservation, the comparative use of values which are calculated in very similar ways is being discussed to an increasing degree in order to plot sites on a scale of "ecologically very valuable" to "ecologically less valuable" (PLACHTER, 1991).

Mathematical models are used to balance out method-specific errors which occur in the species count (ACHTZIGER et al., 1992).

In addition to the systematic-taxonomically oriented parameters at the level of biocoenoses, the state of knowledge is so good for a few groups of organisms that a comparison of functionally defined units is possible (e.g. for species with the same trophic strategy such as micro- or macrophytophages). Species which are not related to one another taxonomically are thus formed into "guilds" (HAWKINS & MACMAHON, 1989), which are in turn more or less finely differentiated, depending on the state of biological knowledge about them. The guilds can be expanded into "leagues" by adding another criterion, e.g. microhabitat (or various strata of soil in the case of soil organisms, FABER, 1991; BELOTTI, 1993). However, it is hardly possible to analyze microflora other than as functional groups (e.g. nitrificants, cellulose decomposers) due to the large number of species and the difficulties in determining them. The rationale behind this "guild" approach is that both absolute and relative changes which may be found to occur in the biocoenosis under observation are easier to extrapolate for the ecosystem level.

One example is the ecological classification of earthworms (annelids) into three main groups (Figure 6.3), as proposed by BOUCHE (1977) and LEE (1959; quoted in LEE, 1985) in their respective models:

Mineral level dwellers (= endogées):
Live in horizontal burrows in the soil, feed on earth and benefit from its organic content; not pigmented; weak burrowing musculature.

Vertical burrowers (= aneciques):
Dig vertical burrows (up to 3 meters deep) with an opening to the surface, take leaves on the surface and consume them deep in the soil; bright red pigmentation, at least dorsally; strong burrowing musculature.

Surface-ground dwellers (= epigées):
Dig no burrows in the soil; sometimes found even on above-ground vegetation; consume litter and the microflora living in them; brightly colored, often as camouflage; very strong musculature for quick movements, not burrowing.

In the meantime, this classification has been refined many times over, primarily due to empirical data on tropical worms (LAVELLE, 1984). The groups mentioned may be comprised of 1 to approximately 10 different species per site which manifest a relatively similar ecological behavior and thus a similar exposure. Due to the dominant role of these animals in many holarctic systems, a toxic effect on them can be expected to have direct effects on the ecosystem as a whole.

Figure 6.3 Three earthworm species as examples of the three main ecological groups (*Lumbricus terrestris* = anecique; *Aporrectodea longa* = endogée, *Eisenia fetida* = epigée).

6.6.3 Parameters at Ecosystem Level

Generally speaking, the following parameters are frequently suggested for the investigation of this level of organization (cf. Table 3.1 and Table 6.1):

- **Structural parameters**:
  Abundance and its dynamics, spectrum of species and dominance structure, diversity as well as spatial distribution.
- **Functional parameters**:
  Degradation of organic substance, nutrient cycle (e.g. of nitrogen) and the biomass or productivity as well as altered food-web relations.

SHEEHAN (1984a, 1984b) provided an initial detailed outline of the structural and functional measurements that need to be distinguished for this research. A discussion of the general reactions by ecosystems to natural or anthropogenic stress factors can be found in RAPPORT et al. (1985).

In Figure 6.4, the structure and functional relationships of a terrestrial ecosystem which can be used to study the effects of chemicals are illustrated in a highly simplified diagram (MATHES et al., 1991).

The structure and functional parameters selected must meet the following criteria (MATHES et al., 1991):

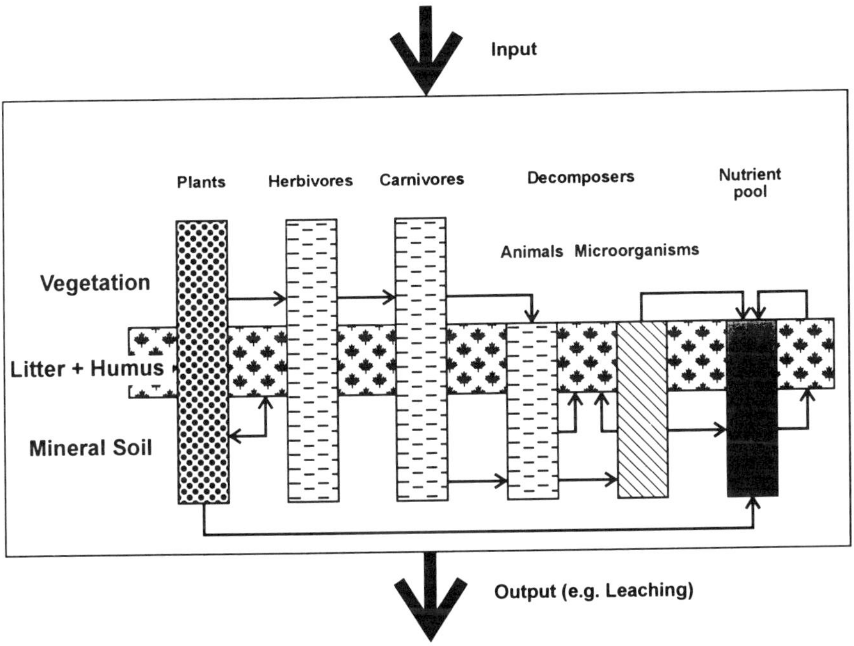

Figure 6.4 Simplified structure and functional relationships of a terrestrial ecosystem (adapted from MATHES et al., 1991). Decomposers are typically restricted to the humus and mineral soil layer as well as the nutrient pool.

- definite relevance for the investigated site
- simple to analyze
- qualitative differentiation is possible (e.g. determination at species level)
- high density and little aggregation of individuals
- comparability of long periods

In any case, ecotoxicological studies should include an unpolluted control plot as well as knowledge about the development of the site that is to be subjected to stress *prior* to application of the test substance.

SCHÜEPP et al. (1990) call for the following measurements as a concrete example of a practical approach to the assessment of a foreign substance's effects on microflora in the soil:

- changes in the microbial biomass
- effects on metabolic cycles, especially of nitrogen
- inhibited degradation of organic substances

- investigation of the vesicular-arbuscular mycorrhiza
- clarification of the interactions of microflora with meso- or macrofauna

All of the investigations should include the time dimension, especially as a function of the period and frequency of application. The assessment of these productivity measurements in soil microflora is not uniform at present. The functional measurements employed represent sum parameters, whose reactions to stress may consist of direct and indirect effects which can cancel each other out. Moreover, the microflora coenosis being studied must exhibit a minimum of activity in order to be analyzed in the first place, e.g. by means of respiration or dehydrogenase activity.

In certain cases, e.g. in soils and sediments, functional parameters for assessing the effect of a foreign substance have been employed and expressed in mathematical terms (TORSTENSSON, 1988; NEBELUNG & BODE, 1988). Primary production, metabolic and energetic conversions, degradation, mineralization and bioretention should be included as integrative functions and capabilities of the biocoenosis at ecosystem level (WEIDEMANN et al., 1988). More specifically, methods for the analysis of soil respiration, the nutrient cycle and the degradation of organic substances, for example, are now available for use in the study of soil ecosystems.

In a large-scale research project at eight different terrestrial sites, the litter decomposition (e.g. measured using litter bags) and the nitrogen level proved to be the most suitable, highly-integrative functional parameters. At the same time, it should be noted that, due to the high degree of integration, structural changes which only become apparent at functional level over the long term may be overlooked unless both groups of parameters are studied concurrently (MATHES et al., 1991).

It was shown that the species and their functional roles are not randomly interchangeable (FABER & VERHOEF, 1991). Nonetheless, functional redundancies can occur in ecosystems, in which case functional parameters may remain unchanged even though serious damage has been done to organisms (BELOTTI & BABEL, 1993). In many ecosystems, however, it has to date proved difficult to arrive at a sufficiently precise description of a state or a course of development by means of the energy flow and metabolic cycles within the food web. Basically, long periods of investigation are necessary to ascertain structural and functional changes in the system as a whole. This makes it possible to distinguish natural fluctuations from induced fluctuations.

System parameters such as the identification and dynamics of species which are critical for the function of an ecosystem (key species) do not become important unless they are considered in the context of broadly designed research programs (ELLENBERG et al., 1986). Due to the related high expense, such studies are not feasible everywhere, nor is there any adequate database as yet which could be used to apply more than common knowledge to other, similar ecosystems (KIMBALL & LEVIN, 1985).

Since it would not appear possible to achieve this depth of investigation in every case, the concept of the reference ecosystem (cf. Chapter 10) was developed as a standard of comparison (MATHES et al., 1991). In other approaches to the solution of this problem, an attempt is made to change the classification systems for ecosystem types (see e.g. SCHUBERT, 1991b) in such a way that each category will cover those ecosystems which react to a particular stress in a similar way and follow similar paths of regeneration (KIMBALL & LEVIN, 1985; HARWELL & HARWELL, 1989). In initial approaches, structural parameters (e.g. number of individuals, number of species and dominance relationships) are linked with functional parameters (e.g. net production and stress) as well as with abiotic factors in a web of relationships. This leads to a distinct characterization of ecosystem types (BUSCH et al., 1989). If a concrete ecosystem deviates from the standard for these parameters, this is a sign of a disturbance. The problem here is that these parameters are examined with varying degrees of thoroughness in the individual ecosystem types.

6.6.4 Biomonitoring

In the framework of monitoring programs (cf. Chapter 10.3), organisms can function as bioindicators for the occurrence and effects of foreign substances (SCHUBERT, 1991a; ARNDT et al., 1987; STREIT, 1989). Bioindicators are generally understood to mean organisms or groups of species whose occurrence, behavior, habit, etc. are so closely correlated with defined ecological factors that they can serve as direct indicators of the state of an ecosystem, stress on the ecosystem or changes in the ecosystem. In ecotoxicology, bioindicators in the strict sense of the term (monitoring organisms) are understood to mean organisms or communities of organisms which (necessarily) respond to stress from foreign substances with characteristic changes in their life functions (reaction indicators), or which accumulate the foreign substance (accumulation indicators). Monitoring organisms are either deliberately introduced into an ecosystem (= exposure monitoring) or, if they already exist in the system, specifically selected (= passive monitoring). The highly standardized organisms for ecotoxicological laboratory tests can also be regarded as bioindicators. The need to determine precisely which organisms are to be used is a major problem (CRANSTON, 1990).

The following parameters are usually employed for the analysis of indicators in monitoring programs (CULLEN, 1990; PAOLETTI et al., 1991):

- change in the composition of species
- presence or absence of key species
- biomass (subdivided at least by ecosystem compartment)
- biochemical stress indicators
- pathology or parasitization of populations
- analytical evaluation of bioaccumulation

Reaction Indicators

Reaction indicators can provide information about the overall state of stress at a site, but not about the substances which have been found there on the basis of chemical analysis. Both acute and chronic effects are monitored, depending on the species or parameter. The species selected should have the following characteristics:

- sensitivity to stress factors
- high density of individuals
- easy to analyze/monitor
- rapid succession of generations
- differentiation according to life forms or consumer type
- confinement to certain geographical region
- if possible, key functional role
- no conflicts with wildlife conservation considerations

Examples of reaction indicators are caddice fly larvae and fish in aquatic systems or lichens or gamasids in terrestrial systems (cf. Chapter 10.3). Exposure monitoring, where at least partially standardized organisms are actively released into the environment, makes almost exclusive use of plants (e.g. genetically uniform clone spruces; ZIMMERMANN & RUDOLPH, 1986).

Accumulation Indicators

Accumulation indicators must meet the following conditions:

- uptake of the foreign substance and concentration of the substance in relation to the ambient medium in due time;
- tolerance of the foreign substance to permit uptake;
- no problems with the residue analysis;
- laboratory keeping and breeding are possible.

Well-known accumulation indicators are plants, mussels or earthworms, but parts of higher organisms (e.g. the fatty tissue of dolphins or the feathers of birds of prey) can also be used.

To an increasing degree, samples from environmental specimen banks are being used as a reference for the current foreign substance values found in monitoring programs (STOEPPLER, 1984; LEWIS, 1985; LEWIS & KLEIN, 1990). Generally speaking, the use of accumulation indicators has shown that, as opposed to what was previously thought (e.g. in the case of DDT), accumulation is less apt to occur along a relatively simple food chain. There is now more of a tendency to regard build-ups as the result of complex food webs which are extremely variable (MORIARTY & WALKER, 1987; STREIT, 1992).

Compared with the chemical analysis of the abiotic compartment, biological accumulation indicators such as sea mammals (VETTER & LUCKAS, 1991) have three advantages in particular:

- A few foreign substances, especially heavy metals and chlorinated hydrocarbons, occur in organisms in significantly higher concentrations than in the surrounding area and thus make the chemical analysis easier.
- When organisms are analyzed, the bioavailability of the foreign substance is measured, which is more important for the biological effects than the total quantity.
- Organisms integrate the volume of foreign substance over time and space.

However, a few problems do occur in connection with the use of accumulation indicators which restricts their usefulness, especially since standardized organisms should be available (NEUMANN, 1982; MORIARTY, 1988):

- Aside from the need to make accumulation indicators easier to analyze, two conditions must be met: First, the concentration of foreign substance in the organism must be a function of the concentration existing in the environment. That is actually only ensured if something is already known about the foreign substance, the species and the habitat involved. Moreover, the results obtained for one species are not necessarily relevant for other species. Secondly, the individuals taken for a sample must be representative of the species as a whole — in other words, individuals which have already died from acute toxicity should not be included in the sample. Probably, it is best to examine the fate of a foreign substance in samples with species which are of immediate economic interest to human beings.
- Due to the high specificity of the way in which the organisms deal with the foreign substance, there is no universal indicator for environmental pollution. For this reason, one should analyze various species which differ in terms of their exposure.
- In some situations, an integrated value is less important than the kind of fluctuation in order to understand the kinetics and the fate of a foreign substance in the environment as well as its biological effects.

Nevertheless, the advantages outweigh the limitations to such an extent that use of bioindicators has even been recommended to monitor disaster situations (especially those caused by human activity, JEFFREY, 1990). In any case, however, the methods used for this purpose should be standardized in much the same way as test procedures are, i.e. in the form of guidelines.

Biomarkers and Bioassays

Morphological, biochemical and other changes with a measurable normal variability which are studied in specific target species (e.g. developmental disturbances in birds) as part of monitoring programs are referred to as "biomarkers" (CAIRNS & MOUNT, 1990; HOFFMAN et al., 1990). The use of such parameters is taking on increasing significance, especially in the United States. However, the term "biomarker" should be avoided because it is sometimes given the overly broad connotation of "biological responses or biological changes" related to the effect on all biological levels of organization, from the cell to the ecosystem. Bioassays are another approach which has come into use, especially to avoid the high costs of residue analysis. Material from sites which may be contaminated by several unknown foreign substances is examined in the screening procedure that forms a part of standardized biological test systems. This procedure is regularly applied for food crops such as maize (SUNDERLAND et al., 1991), especially to study risks to subsequent cultivation. A follow-up chemical analysis is only conducted if toxic effects were found to occur in the bioassay. The use of immunoassays (ELISA tests) is particularly noteworthy in this connection. They were developed to an increasing degree specifically for certain chemicals and pesticides (e.g. for Atrazine (GOH et al., 1991) or for PCP (MAPES et al., 1992)). Consequently, the use of such testkits can save a considerable amount of analytical expense.

7 TESTS AT VARIOUS LEVELS OF INVESTIGATION

7.1 TIERED TEST PROGRAM

Today, many countries have regulations governing the registration of pesticides or the notification of environmental chemicals. Ecotoxicological soundness is one of a wide variety of criteria including those derived from human toxicology which must be met by these substances in order to qualify for notification. These assessments, which must follow a procedure codified in the form of guidelines specifying limitations on the level or duration of exposure to a substance and its effects, are usually conducted on a rising scale of complexity, proceeding from laboratory and semi-field tests to studies in the field (e.g. KIMBALL & LEVIN, 1985; SCHINKEL, 1985; CAIRNS, 1989; BOLLER et al., 1989; TARDIFF, 1992). SHEEHAN et al. (1984) provided an initial detailed overview of this approach.

This tiered test program is designed to minimize the amount of time and expense that has to be invested in tests. Semi-field and field tests are much more elaborate and costly to conduct than laboratory tests. If a simple fate and effect assessment test is sufficient to tell that a substance poses only a relatively low environmental risk, then there is no need to conduct any further tests. The respective registration authorities have established test-specific trigger values marking the transition from one stage to the next higher stage of the procedure. In the laboratory test for acute toxicity in fish, for example, it can be assumed as a rule of thumb that an LC_{50} of 10 mg active ingredient per liter of water is not achieved in agricultural practice, even under unfavorable conditions (HOLLIHN et al., 1991). This approach, in which the individual stages of the test procedure can roughly be assigned to the different levels of organization of biological systems (Table 7.1), was first defined as the standard general procedure by the U.S. National Academy of Sciences (NAS, 1981). At about the same time, the first tests were adapted as part of a tiered test program in the form of guidelines, e.g. the effects of pesticides on bees (MAFF, 1979; BBA, 1980).

Today, the OECD test guidelines (e.g. 1984) are probably the most widely recognized procedure in the world. They are accepted, either partially or

Table 7.1 Scheme of a Tiered Test Program

Test stage	Level of biological organization
Laboratory	Individual, population
Semi-field	Population, biocoenoses
Field	Population, biocoenoses, ecosystem

integrally, even in many countries which have enacted their own standardized procedures (e.g. Germany, Holland, United States; OECD, 1989a). Intensive efforts are currently under way within the OECD to fill gaps which still exist (e.g. in the terrestrial medium) either by revising existing guidelines or by enacting new ones.

Table 7.2 provides an overview of the kind of tiered model that might be prescribed in a notification procedure to test the mobility (leaching) and the effects on a soil animal (earthworms).

The higher one goes on these levels, the closer one comes to ecological reality in the field, but the expense increases at a disproportionately high rate (Figure 7.1; NUSCH, 1991). A tiered procedure of this kind is not only used to test the spectrum of side-effects but also to estimate the efficacy of pesticides on the target organisms (ROBERTSON & WORNER, 1990). In the case of certain tests, e.g. when assessing the potential hazard to ground water (ECETOC,

Table 7.2 Examples of a Tiered Procedure for a Fate Test (Leaching) and an Effect Test (Toxicity in Earthworms)

Test stage	Capsule description of test	Example of trigger values
	Leaching behavior	
Laboratory	Small glass columns filled with standard soils	Amount of applied quantity found in the leachate
Semi-field	Lysimeter studies with large intact soil cores	Amount of applied quantity found in the leachate
Field	Monitoring in the field (e.g. well, ground water)	Amount of substance detected
	Toxicity in earthworms	
Laboratory	Acute mortality for compost-worm in artificial soil	LC_{50} > 1000 mg/kg PNEC/LC_{50} ratio
Semi-field*	Chronic reproduction test with compostworm in artificial soil	NOEC reproduction per soil concentration
Field	One-year sampling of natural populations after application	Discrepancies in abundance and dominance spectrum compared with control area

* Due to the small vessels and artificial composition, it is more of a prolonged laboratory test than a real semi-field test.

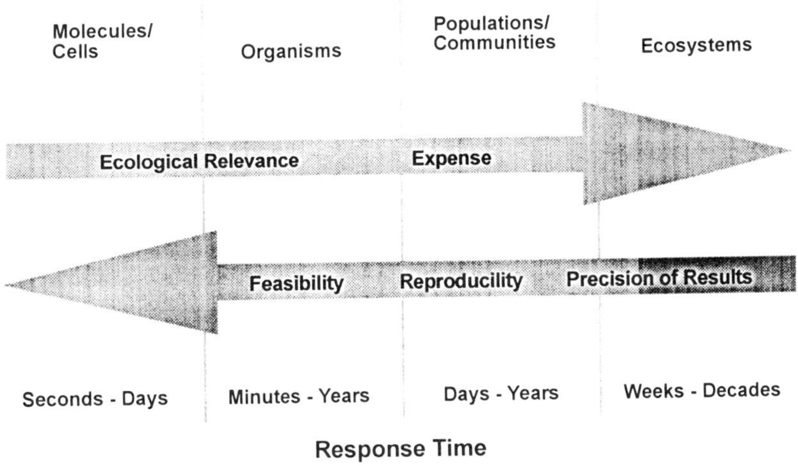

Figure 7.1 General relationship between biological levels, ecosystem relevance, methodological aspects and response time.

1992), mathematical compartmental models can be substituted for practical experiments to some extent. But these models must also be validated, optimally through practical experiments in the field.

The parameters (e.g. mortality) that are employed in most laboratory tests are not very refined (IGLISCH, 1985, 1986). Consequently, for scientific as well as ethical reasons (to minimize the "consumption" of animals), sublethal parameters are deemed preferable and have been incorporated into guidelines since the early 1980s (HAMBURGER, 1983). The LC_{50}, an acute test value which has served as the endpoint of such experiments, has been replaced, for example, by the reproduction NOEC from the chronic tests. It should be emphasized again that laboratory tests cannot predict the variety of ecological effects in the field (CAIRNS, 1986a; PERSOONE et al., 1989); this is also true of fate tests, such as the one used to analyze soil degradation (KÖRDEL, 1992).

Another fact which can make it difficult to apply the results of laboratory data to conditions in the field is that most ecotoxicological studies focus on the "average" reaction of individual members of a species. Because the variability among members of the same species is disregarded, the range of possible reactions in a natural population, with its physiological differences, is hard to predict (DEPLEDGE, 1990; cf. also Chapters 2.4 and 6.6). The effects of behavioral changes (e.g. avoidance reactions) can also not be represented in the laboratory. Moreover, none of the interactions between various species in the system are included in the picture. It is not yet clear how this deficiency in the currently available laboratory methods can be eliminated. CAIRNS (1984) has summarized this dilemma as follows: "No scientifically justifiable evidence exists to indicate the degree of reliability with which one may use single-species tests to predict responses at higher levels of organization."

This dilemma is also not resolved by the fact that, in some cases involving plants (PESTEMER, 1988) or algae, for example (LEWIS, 1986), laboratory tests have revealed a higher toxicity than subsequent field studies. This discrepancy is usually due to different exposure conditions. The sharp divergence in the toxicity of pyrethroids is one example. These substances are highly toxic in aquatic laboratory tests with *Daphnia* and fish, but under field conditions (e.g. in pond studies), they exhibit only very minimal effects due to their high rate of adsorption to particles in the water or in the sediment, thus reducing their bioavailability (HILL, 1989). In a review of the studies conducted up to 1987 on the degree to which laboratory results can be applied to the field or the results of single-species tests are valid for multi-species systems, MALTBY & CALOW (1989) discovered that examples of every constellation can be found in the literature. In other words, sometimes the laboratory test proved to be more sensitive and, at other times, the field study.

Under certain prerequisites, the scale of direct effects from such diverse substances as copper or Cypermethrine in the field can be forecast by means of acute experiments in the laboratory. The indirect effects of these substances, however, have been vastly underestimated, even when chronic laboratory tests were conducted (KIMBALL & LEVIN, 1985). Herbicides, whose acute and sublethal toxicity to animals is usually lower than that of insecticides, are another example of the way in which this effect in the field can be underestimated. These substances induce a change (or, rather, a depletion) of the phytocoenosis which, in turn, alters the food supply in such a way that clear-cut indirect effects also occur in animals such as earthworms (HEINISCH & KLEIN, 1989). In addition, the lack of soil cover and the lower root density of the soil lead to a decrease in organic matter, a reduction in biological activity and a deterioration of the soil structure and crumb stability (SCHÜEPP et al., 1990). Similar examples of indirect effects are also known from the tropics, e.g. when DDT was applied for mosquito control in Borneo and side-effects in predators indirectly led to an outbreak of previously negligible diseases in humans (KIMBALL & LEVIN, 1985). It is remarkable, however, that in spite of these examples and numerous calls for studies to be conducted at higher levels of organization, the actual number of test guidelines which exist for these levels is extremely small (cf. CALOW, 1993).

In spite of these limitations, laboratory tests have obvious practical advantages, especially in view of the large number of substances that have to be studied. If only because of their simplicity, lab tests are indispensable, at least as a "screening" step (KLEIN et al., 1992). Even so, laboratory test procedures must fulfill the following, minimum criteria:

- the procedure used in the test can, for the most part, be standardized;
- the test results can be reproduced and replicated;

- the test results are easy to manage for statistics;
- comparative data are available for the evaluation of test results;
- the test is practicable (e.g. not overly time-consuming or costly).

The world's main standardized laboratory test procedures are outlined in the following chapters and subdivided into fate and effect assessment tests in each case. The overview also includes a list of promising test designs, but, due to the rapid development in this area, the list is necessarily incomplete.

At the lower levels, the physico-chemical properties and, at a more complex level, the environmental fate of a substance, are tested separately from its biological effect. Field or semi-field procedures (e.g. ecosystem sections) focus directly on how a substance affects the system. However, if the potential exposure and the effect of a substance are tested individually, the results will not provide an adequate basis for a hazard and risk assessment. Such assessments must be based on an evaluation of both aspects and, if at all possible, should include monitoring and general field data (from residue analyses as well as ecological studies; see Chapter 12).

7.2 SOURCES OF ECOLOGICAL UNCERTAINTY

The problem faced by those who must convert ecotoxicological insights into policy or legislation is that ecotoxicology can provide a relatively precise indication of the toxic effect at the level of individuals (e.g. single-species tests in the laboratory), but the practical relevance of these data is rather limited, or that it can yield more generally valid assessments (usually based on field tests) which, however, also harbor a great many uncertainties. As one can see from a critical analysis of ecotoxicological assessment data, the number and range of uncertainties increase in proportion to the level of organization. Inadequate methodology is one reason for this, but the increasing complexity of ecological systems as such is also a major factor. The fact is ecological uncertainties are immanent to the systems and cannot be completely eliminated. Generally speaking, they are attributable to the following causes:

- *insufficient data* and an inadequate understanding of ecological processes and systems due to deficient methods;
- the numerous, *potential indirect effects* in ecosystems which lead to unpredictable events (interwoven systems);
- the system's inherent environmental *stochastics*, a combination of chance and an enormously wide range of variability;
- the need to *extrapolate* from special conditions of the respective ecosystem or test substance (= transferability).

The first three categories of uncertainty (insufficient data, indirect effects and stochastics) require no further explanation, but the need for extrapolation must be considered somewhat more closely (Table 7.3). By now it has become obvious that we are not yet in a position to predict the effects of pollutants down to the last detail.

Because the assessment parameters of laboratory tests have the advantages of practicability and precision, such tests are considered preferable in practice over other test approaches in the sequence microcosms → ecosystem sections → mathematical models → manipulations of ecosystems. Consequently, extrapolations are often made from this lower level. The degree to which the results of a laboratory test can be applied to field conditions will be discussed in Chapter 12 in detail. Generally speaking, this depends on (1) the difference between a certain test species and other species; (2) the difference between acute and chronic exposure; and (3) the increase of complexity from laboratory to field level.

7.3 TECHNICAL ASPECTS INVOLVED IN CONDUCTING TESTS

To secure the benefits of standardized tests (especially in the laboratory), all ecotoxicological tests for pesticides and environmental chemicals in industrialized countries must be conducted according to the "Guidelines of Good Laboratory Practice" (GLP). This elaborate system of standardized tests, documentation and monitoring is designed to allow the results of laboratory tests to be reproduced and thus ensure that the laboratory meets a minimum quality standard. The GLP rules became necessary in the late 1970s, when it was no longer possible for the respective national authorities or the producers to carry out the growing number of tests. The GLP system was finally enacted into law after it was discovered that a research institute in the United States had fabricated a set of test results.

Since the significance of test conditions (abiotic factors or the biological variability of test organisms) which influenced the results of ecotoxicological experiments was not documented up to that time, many older studies are only useful to a limited extent (KOLLIG & KITCHENS, 1990). Because the validity of test results is now based on bilateral acceptance and is supported by a framework of international institutions, it is not necessary for a given test to be repeated in every country. This has helped to reduce the number of animal experiments and the overall costs significantly.

Whereas most of the above-listed biological tests are designed to analyze the absolute toxicity of a pesticide or of an environmental chemical (independently of the anticipated concentration in the environment; in terms of LC_{50} or NOEC parameters), the tests on environmental fate as well as those that are used to assess the effects of pesticides on beneficial organisms are based on a different philosophy: only the "highest recommended normal application" of a pesticide is tested, sometimes with a safety margin figured in. The first type of procedure is designed primarily for the testing of environmental chemicals, where exposure usually takes the form of an airborne input in low concentra-

Table 7.3 Extrapolation of Test Results: Inherent Uncertainties at Various Levels of Investigation

Test type/source of data	Examples of reasons for uncertainty in extrapolation
Laboratory tests on single species	The influence of physical environmental conditions on the relationship between dose and response and the exposure regime are not taken into consideration. A particular stage in the life cycle is selected as representative of the whole cycle. A few species which are easy to breed and control are selected as representative of a large number of species in the environment. Interaction with other species is not considered. Effects at the individual level are applied to the population level without taking compensating mechanisms or behavioral adjustments into account.
Semi-field procedures	Artificial relationships between species are created. Important trophic relationships, among others, are excluded. Effects are determined on the basis of discrepancies in scale between the microcosms and the actual environment. Border effects can cause disturbances. Spatial heterogeneousness of the natural ecosystems from which the samples are taken. The approach is geared more to precision and replicability than to a reliable description of real effects. Exclusion of natural, climatic factors in some cases.
Manipulations of intact ecosystems	A specific ecosystem is supposed to be representative of the general ecosystem type. The stress triggered by the tested chemical is supposed to be representative of the stress from non-tested chemicals. Effects of specific climatic and physical environmental conditions as well as of historical events on the behavior of the ecosystem. Insufficient replicability. Difficulties of gathering relevant data yielded by the dose-response relationship on the ecosystem. Duration of the experiment in relation to the time required for the ecosystem to manifest an adequate reaction and recover from the intervention.
Models	Imprecise or incorrect mathematical formulations. Effects based on an uneven distribution of the organisms (aggregation) in the field. Insufficient database for verification of the model. The model is applied on a level which exceeds the scope of its capacity to yield a relevant assessment. Sensitivity of the results to changes in parameters or to the model's design.

tions over an extended period of time. Extremely high local entry levels can only be studied in the event of accidents. The second approach, on the other hand, is oriented more to pesticides, which are applied deliberately in clearly defined quantities. This approach does not allow for the fact that, in practice, it is not at all uncommon for pesticides to be applied in concentrations which are higher than those recommended by the manufacturer (MICKLITZ, 1991; TREIDLER & WINTER, 1991). This test strategy also fails to include the effects of a possible build-up in sinks.

There was a time when only the active ingredients of a pesticide were tested. Today, however, the formulation itself is also examined because it definitely has an influence on the toxicity of the active ingredient; moreover, the active ingredients cannot even enter the environment without auxiliary transport substances. In aquatic tests involving about 300 different pesticides, for example, the formulation was responsible for increasing the toxicity of the active ingredient in 32% of the cases and reducing it in 11% of the cases (MAYER & ELLERSIEK, 1988).

At the same time, an extremely important consideration is the form in which a substance is applied, since this factor largely determines the level of exposure for organisms in the field. In the case of pesticides, an attempt is made to apply the substance in the laboratory in a manner that approximates conditions in the field as closely as possible (e.g. by employing spraying devices with the same kind of nozzle as is used in agriculture). Other chemicals may be applied directly (topically), in the form of gas or through substrate concentrations (e.g. mixed in with the soil). In any event, the "worst case" should be chosen in order to achieve a certain degree of reliability for the situation in the field (BROWN et al., 1990). One example that is currently being discussed is whether the artificial soil employed in some acute soil tests, which has a high content of organic matter (approx. 10% compared to the 2 to 3% average in central European soils), adsorbs the test substances at a much higher rate on the substrate than natural soils (RÖMBKE et al., 1992). The resulting, reduced bioavailability can lead to significantly lower toxicity, an assumption which has been confirmed in enchytraeid and collembola tests (RÖMBKE, 1991).

There have been considerable differences in the time and effort (and thus also in the costs) required to implement biological tests in aquatic and terrestrial media for several years now. Tests with pesticides in the aquatic medium are, for all practical purposes, only recognized by the authorities in industrialized countries if they are accompanied by residue analyses. This makes it possible to correlate observed effects with the real values (instead of the applied nominal values) of the chemical concentrations. In the case of volatile or slightly degradable substances, it is necessary to apply flow-through procedures, which are very costly from a technical standpoint. In the terrestrial medium, on the other hand, it is hardly possible to separate the bioavailable portion from the portion that is bound to the soil. Consequently, these tests rarely need to be accompanied by a parallel residue analysis except in the United States (e.g. in the chronic earthworm test required by the FDA (1988)).

For all of the standardization in procedure as well as the increasingly elaborate documentation (GLP), the results vary on average from one laboratory to another by a factor of 10 or less and, in very extreme cases, by as much as 250, if only because of the biological variability of the test organisms (RUDOLPH & BOJE, 1986). In the meantime, so many data are available for a few laboratory test systems, e.g. acute bird tests (results of 75 species and over 1,000 substances; HOFFMAN et al., 1990) or fish and earthworm tests (BAIER et al., 1985; MAYER & ELLERSIEK, 1988; OOMEN et al., 1991; EDWARDS & BOHLEN, 1992), that the relative toxicity of a substance can be assessed fairly accurately. Due to the quantity of material available for comparison as a result of these tests, a few of the procedures, such as the earthworm test, are still being employed, at least as screening instruments, in spite of well-known deficiencies.

In the case of all laboratory studies, it is always assumed that individual substances have isolated effects, since it is practically impossible to test all of the potential combinations, much less under variable environmental conditions. Initial practical approaches for the testing of combined effects are available for effluents, i.e. for unknown mixtures of environmental chemicals, with bacteria (Microtox), algae or *Daphnia* (VOUK et al., 1987; VASSEUR et al., 1991). Exemplary investigations of pesticides and soil microorganisms (AUSPURG, 1985) have shown that all of the theoretically conceivable interactions between several chemicals, in the sense of both a positive and negative reciprocal influence, can occur in reality. The same studies also demonstrated the presence of indirect effects — on the degradation and thus the persistence of a substance, for example — in addition to the chemicals' direct effects.

8 LABORATORY LEVEL

8.1 TEST METHODS ON ENVIRONMENTAL FATE

The test methods described in this chapter can be divided into two groups:

- "Classical" test methods on degradation, metabolism and mobility in the soil and in the sediment as well as volatility on soil and plant surfaces, including mathematical compartment and degradation models;
- Biological methods on biodegradation in the aquatic medium (including sludge).

The most important test methods on environmental fate are given in Table 8.1.

No really new test methods for the environmental fate of pesticides have been developed in recent years. Instead of developing new tests, work is being done primarily on standardizing test conditions, especially for substrate characteristics, within the framework of existing guidelines (HAIDER, 1985; STRATTON, 1986).

In the case of terrestrial tests with pesticides, the German BBA calls for standard soils which must be taken from a defined location (SCHINKEL, 1985). In the future, the EC will also require that five elaborately selected "regional soils" which are more or less representative of the wide variety of soil types found in Europe be employed in the tests (KUHNT & MUNTAU, 1992). Other authorities such as the EPA in the United States give only rough specifications (e.g. the soil type), but require a very precise characterization of the soil used (EPA, 1982a). In any case, an appropriate sample preserving the natural microflora in the soil must be taken and stored (ISO, 1992a). Very precisely characterized substrates taken from the field are usually required for sediment tests (BBA, 1990b; EPA, 1982a). Artificial soils employed in a few effect assessment tests are not used in these experiments because it is difficult to transfer the results to field conditions.

Table 8.1 Examples of Environmental Fate Test Methods

Term	Brief description	Example of a guideline
Adsorption/desorption	Distribution of a substance between solid and liquid phase (K_{oc})	OECD 106
Soil degradation	Degradation and metabolism of a substance (DT_{50}, DT_{90})	BBA IV, 4-1
Sediment degradation	Degradation and metabolism of a substance (DT_{50}, DT_{90})	BBA IV, 5-1
Leaching	Mobility of the substance in soil columns with or without aging	EPA N §163
Volatility	Dissipation of a substance from soil or plant surfaces	BBA IV, 6-1

Adsorption/Desorption

As far as the test procedures for environmental fate are concerned, the transition from the physico-chemical analysis to the actual distribution tests is flexible. This is particularly clear with regard to the adsorption tests, which are sometimes classified in one group and at other times in another. In the practical adsorption/desorption tests, the fate of a chemical is ascertained between the aqueous and solid phase of a water-soil system (e.g. OECD, 1981b; EPA, 1982b), for which at least three different (standard) soils should be used. The objective of the tests is to determine the adsorption coefficient (K_{oc}) that is normalized to the carbon content of the respective soil. The K_{oc} is an important variable for assessing the fate of a chemical, especially its mobility (VON OEPEN, 1990). As far as this parameter is concerned, it should be borne in mind that, in addition to indicating the organic content of a soil, it may also be influenced by other characteristics of the soil, especially the pH value as well as the composition of clay minerals (HERMOSIN & CORNEJO, 1989). In addition to the use of practical tests, this coefficient can be empirically derived by means of physico-chemical parameters (especially the solubility in water and the octanol-water partition coefficient), in which case different, substance-specific formulas must be employed (KENAGA & GORING, 1980; BRIGGS, 1981; LYMAN, 1982; KANAZAWA, 1989; DI TORO et al., 1991).

Degradation

Degradation is the key variable in assessing the environmental distribution of a chemical: a high persistence value is considered undesirable by anyone's standards (KLÖPFFER, 1989). The analysis of degradation takes the form of laboratory experiments which are codified in various guidelines (e.g. OECD, 1981a; EPA, 1982a; BBA, 1986a; ISO, 1992b). Only in the United States are tests on degradation processes required under anaerobic conditions from time to time (EPA, 1982b). Since the purpose is to examine the fate of metabolites which may be produced in addition to the degradation of the substance being tested, it is advisable to make use of radiolabeled substances as well as several, if possible different, types of soils (OECD, 1981a).

Tests on the degradation of pesticides in the soil have indicated rates of degradation that vary by as much as a factor of 80 (<20 in most cases), depending on the substance and soil type (LASKOWSKI et al., 1983). Even greater differences — more than a factor of 100 — were found to exist in the half-life values for the volatile substance Methylchloride if deeper layers of soil were included in the analysis (DAVIS & MADSEN, 1991). This range of variation under laboratory test conditions suggests that it would be extremely difficult to forecast degradation rates in the "real world" (SCHEUNERT, 1991; HOWARD, 1993). The same is true when it comes to predicting the formation of bound residues (KUBIAK et al., 1990).

In addition to the soil characteristics (e.g. organic content, pH), the quantity (the biomass should be at least >10 mg Bio-C/100g dry weight of the soil) and the composition of the soil microflora, which is also subject to strong fluctuations under natural conditions, are primarily responsible for this potential variability (COOK & GREAVES, 1987). The concentration of the test substance must be high enough (which is generally not the case in the ppb range) to trigger the induction of decomposer-enzymes in the microbes. If it is too high, however, the microorganisms will be damaged as well. In other words, an optimal rate of degradation is only given in the range between the "induction concentration" and the "active concentration", which varies from one chemical to another. DOMSCH (1992) provides an overview of pesticide degradation, and especially the influence of microorganisms on these processes. However, it must be borne in mind that laboratory tests limited to microbial cultures do not have much validity for assessing degradation in soil under real conditions (ECETOC, 1990).

From the standpoint of the methodological approach, similar to the aerobic soil degradation studies, tests on the degradation and metabolization of pesticides are conducted in anaerobic soil samples and in water-sediment systems (e.g. BBA, 1990b; HARLAND et al., 1990). This approach could be developed further through the use of so-called "ecocores" instead of artificially composed water-sediment samples. Ecocores are small glass tubes which are used to take simultaneous field samples of sediment and water (including small invertebrates); they could also be called very simple microcosms (HOUX & DEKKER, 1987).

Tests have repeatedly shown that xenobiotics in sediment systems are broken down at a more rapid rate than in pure water (WALKER et al., 1988). There are two plausible reasons for this (PRITCHARD, 1987):

- the microbial biomass is higher in sediment than in water, especially in the uppermost layer;
- organisms on the surface of suspended and sediment particles are more active than microbes which live in the water.

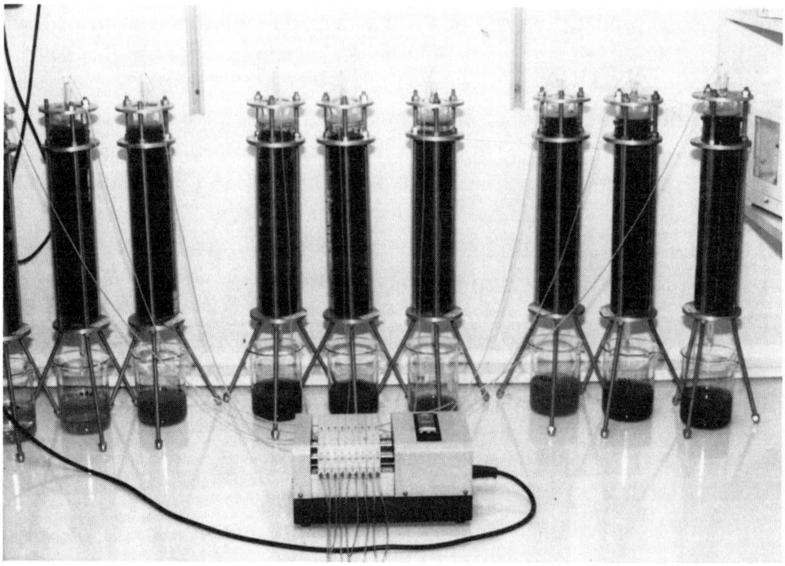

Figure 8.1 Glass columns in laboratory leaching tests.

Leaching

Leaching tests, which are intended to assess the risk to the groundwater from chemicals, are a good example of the inherent limitations of laboratory experiments. In a leaching test, the test substance is applied in droplets to the upper end of small glass columns filled (to the top) with sieved soil (Figure 8.1; BBA, 1986b; EPA, 1982b). Water is precipitated over these columns for a period of varying specified duration (e.g. BBA, 48 hours; EPA, 45 days). The leachate that collects at the lower end of the column is subsequently analyzed for the test substance and any metabolites which may have been produced. Even if not required in all guidelines, the concentration of the test substance must also be analyzed at various levels of the soil column itself in order to identify deposits which have not yet been manifested in the leachate (EPA, 1982b; ESSER et al., 1988). However, it is also necessary to examine the extent to which degradation processes should be included as a mandatory part of the tests for the soil section (PFARL et al., 1990). The entire test can also be conducted with "aged" test substances, i.e. after the chemical has incubated (30 days) in the soil and is subsequently transferred to the glass column — especially in order to take the formation of metabolites into greater account (HERZEL & SCHMIDT, 1991).

Although this test forms a standard part of the registration process, pesticides, especially persistent or water-soluble herbicides such as Atrazine and its metabolites, are nonetheless detected in the groundwater (LUPI et al., 1988;

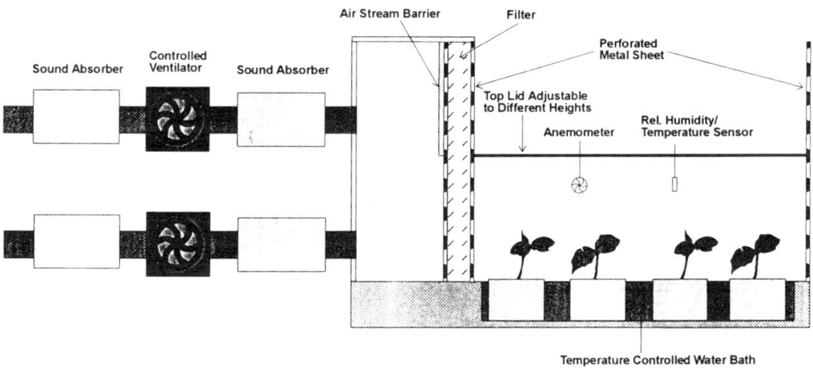

Figure 8.2 Schematic diagram of a volatility chamber.

HALLBERG, 1989). Approximately one-seventh of all registered active ingredients have already been found there (SCHMITZ, 1991). This is most likely due to "site-specific" factors such as macropore systems. These systems cannot be simulated in the laboratory with routine tests employing small, packed soil columns, but a close approximation is possible if large, intact cores of soil (lysimeters) are extracted, representing the transition to the microcosm (KRUGER et al., 1993). Another difficulty is posed by the fact that chemical-analysis methods for the detection of substances in groundwater have only been available for a few years.

Volatility

There has been a call for tests of pesticide volatility on soil or plant surfaces for several years now, ever since monitoring studies showed that physico-chemical laboratory tests failed to yield appropriate results (e.g. low vapor pressure) even though many of these substances had, by and large, entered the atmosphere (SPENCER et al., 1984). The problem is that most of the tests conducted to date are not comparable with each other, especially when it comes to the comparison of results obtained on artificial surfaces such as glass plates with those taken from more realistic substrates such as soils (BÖHNKE et al., 1989). This is related to the difficulty of applying the test results to field conditions: in enclosed test chambers (Figure 8.2) under controlled conditions, the volatility rate can be measured directly with radiolabeled substances (KRASEL et al., 1992), whereas the volatility in field tests is determined indirectly based on the difference between the quantity applied and its recurrence in soil and plants, without taking degradation into account (e.g. MAJEWSKI et al., 1990). As in the case of degradation, volatility data can be presented using simple mathematical models (e.g. first-order functions or root functions).

Modeling Systems

In all of the tests on environmental fate presented thus far, a semi-field or full-fledged field study of the test chemicals is required if the threshold levels, which vary depending on the laws in force, are exceeded (see Chapters 9 and 10). In order, among other things, to avoid the high level of expense related to such tests, model calculations are used to assess the environmental distribution of chemicals, especially pesticides (GYDESEN, 1984; FIGGE et al., 1986; SAMIULLAH, 1990). In order to perform such a calculation, the influence of the test conditions (e.g. substrate, temperature) must first be quantified (LASKOWSKI et al., 1983). Then the environmental fate of these substances can be forecast to a certain extent on the basis of physico-chemical data for the chemical. Simple calculations are available in the form of a hazard index (e.g. the "GUS-index" to assess the mobility of a substance; GUSTAFSSON, 1989), whereas site-specific models, which are designed to predict the fate of a substance under real conditions in absolute proportions (differentiated by environmental medium), are very complicated and require a large number of data on the respective environmental conditions and application quantities (BACCI, 1992).

A comprehensive overview of the currently disseminated models can be found at ECETOC (1992). The models are classified by medium, i.e. as air (ADMS = Atmospheric Dispersion Modeling System or ISC = Industrial Source Complex Dispersion Model), surface water (EXAMS = Exposure Analysis Modeling System or WASP4 = Water Analysis Simulation Programmes) and soil or groundwater models (e.g. PRZM = Pesticide Root Zone Model or SESOIL). Many examples are available for each of these groups. The main problem here is selecting the model that will be most suitable to examine the question involved.

Although there are a growing number of examples in which results of model calculations were verified and substantiated in the field (e.g. CALAMARI & BACCI, 1987; SMITH, 1990; BROOKE & MATHIESEN, 1991), it must be borne in mind that models invariably generate estimates. Thus, for example, the forecast of leaching behavior (an important aspect in assessing the risk to groundwater and, ultimately, to drinking water resources), in which the use of models is common practice in the framework of the pesticide registration process as well, may lead to overestimations as well as underestimations of the real fate, especially over longer periods of observation (MUELLER et al., 1992).

It is important to make a clear-cut distinction between the above-described models, which seek to analyze the fate of a chemical in actual ecosystems in absolute terms, and models which are designed to ascertain the percentage distribution of substances among the various environmental media. These models — the fugacity approach most likely being the best known (MacKAY, 1991) — are used to estimate exposure levels (to specify the "Predicted

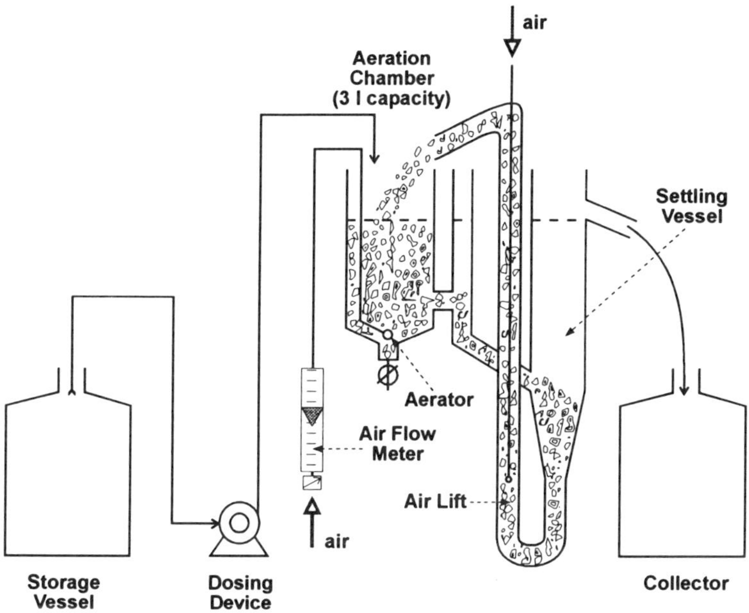

Figure 8.3 Schematic diagram of biodegradation test procedures.

Environmental Concentration" = PEC) when assessing the ecotoxicological hazard potential (cf. Chapter 12).

Biodegradation

Studies on the indirect determination of the degradation of organic substances in water and in sediment form the transition to the purely biological test systems (PAGGA, 1984). Biodegradation, i.e. primarily decomposition through microorganisms, is by far the most important mechanism of irreversible elimination in nature. This is what keeps degradable substances from accumulating in the environment. A distinction is made between primary degradation, in which the test substance is chemically altered (e.g. by modification or loss of its original chemical configuration), and mineralization, in which the substance is completely broken down into small, inorganic molecules such as water or carbon dioxide. The single most important process is aerobic degradation (biochemical oxidation), which is tested in the laboratory with mixed cultures of microorganisms (Figure 8.3). There are a number of methods which vary in cost which can be used in the framework of a tiered test program (NYHOLM, 1991; BIAS, 1993); they are presented in Table 8.2. It must be noted that most of the OECD guidelines listed have also been published by the EC (especially EC, 1984) or as ISO standards (DE HENAU, 1993).

Table 8.2 Overview of the Most Important Biodegradation Tests

Test procedure	Measurement parameter	Example of a guideline
Basic Stage = Ready Biodegradability		
DOC die-away test	DOC decrease	OECD 301 A
CO_2-emission (modified Sturm test)	CO_2 production	OECD 301 B
MITI I test	Oxygen consumption	OECD 301 C
Closed bottle test	Oxygen consumption	OECD 301 D
Modified OECD screening test	DOC decrease	OECD 301 E
Manometric respirometry test	Oxygen consumption	OECD 301 F
Stage I = Inherent Biodegradability		
SCAS test	DOC decrease	OECD 302 A
Zahn-Wellens test	DOC decrease	OECD 302 B
MITI II test	DOC decrease	OECD 302 C
Stage II = Simulation Tests		
Coupled unit test	Chemical analysis	OECD 303 A

Note: DOC = Dissolved Organic Carbon.

Environmental chemicals which are 60-70% degradable in an aerobic aquatic medium within a certain period of time in a screening test, for example, are classified as "readily biodegradable" at the zero-level of the German chemicals law. If this condition is not met, the potential, or inherent, biodegradability of a substance is investigated in Stage I, using the Zahn-Wellens test, for example. An extremely low rate of degradation in this test stage indicates a high probability that the substance will persist in the environment. In the final stage (II, simulation tests), the biodegradability is analyzed under realistic conditions (e.g. laboratory model of a sewage treatment plant); as yet, however, there are very few accepted methods available (with the exception of the OECD Sewage Treatment Test). In much the same way as sediment degradation tests, test approaches referred to as ecocores, which form the transition to the microcosms, have also been described for biodegradation (JENSEN et al., 1988; CARSON et al., 1990; see Chapter 9.2).

The problem with all of the methods currently employed at the first two stages (= Basic Stage and Stage I) is that they do not allow for any distinction to be made between partial degradation and complete mineralization of the substance. Moreover, there is no accompanying chemical analysis; consequently, all that is known is the nominally applied concentration as the starting point (NYHOLM, 1991). Recently, ECETOC (1991) reviewed the literature on the determination of biodegradation rates and their use in predicting the environmental concentrations of chemicals in aerobic freshwater. They concluded that no single test procedure can accurately predict the rate of degradation in a variety of aquatic environments. Therefore, it is very pragmatically suggested that such predictions should be based on first-order (half-life) kinetics for surface waters. This approach does not obviate the need for further refinement of existing test procedures, particularly the elaboration of the influence of the respective test conditions (e.g. illumination, planting, etc.) on the test result as well as the use of radio tracers. Moreover, the values derived in aquatic biodegradation tests are not transferable for terrestrial environmental compartments. Therefore, these tests cannot take the place of soil degradation tests.

8.2 METHODS TO TEST THE EFFECTS ON ORGANISMS

The effect of a substance on various organisms is examined parallel to the investigation of its environmental fate. Biological test systems are expected to meet the following criteria, in addition to the general standards already outlined in Chapter 7 (e.g. practicability; tiered approach; GLP documentation; consideration of uncertainty):

- the organisms are sensitive to chemicals;
- the selected species are representative for the biocoenosis of specific ecosystems;
- test organisms are readily available (e.g. easy to breed);
- the test organisms have a rapid generation cycle (for chronic tests).

In the early years of ecotoxicology, the selection of test organisms with the highest possible degree of sensitivity was a major criterion in addition to that of practicability, which is still an important consideration in testing today. The rationale behind this approach was that, if the active threshold for the most sensitive species of an ecosystem could be determined under laboratory conditions, then all other species would automatically be protected, as it were. However, this assumption has not been borne out, for two reasons (CAIRNS, 1986a; RUDOLPH & BOJE, 1986):

First of all, depending on the foreign substance involved, the ecosystem under observation, the environmental conditions or conditions of application, any one of a number of very different species may exhibit the most sensitive reaction; therefore, it makes no sense to look for *the* most sensitive species ("The myth of the most sensitive species", CAIRNS, 1986a). Individual species may differ in terms of their reactions to chemicals by a factor as high as 100,000 (MAYER & ELLENSIEK, 1988). In particular, the characteristics of the respective environmental compartment (such as the quantity of suspended matter in a body of water or the pH value of a soil) have an extreme influence on the exposure and thus on the kind of effect (HERMOSIN & CORNEJO, 1989). Secondly, if one test species does not reveal any damaging effects, this finding will not necessarily be valid for all of the possible test parameters or at higher levels of the ecosystem.

Today, the degree to which a species can be considered representative of the biocoenosis of a specific ecosystem has become the more important criterion (CAIRNS & MOUNT, 1990; VAN STRAALEN & VAN GESTEL, 1993a) — "representative" in terms of both the species' function in the field (e.g. Can it be considered representative of a specific trophic level?) and its reaction to a test substance (i.e. Does the tested population exhibit effects similar to those of other populations or species, and can its reaction in the laboratory be considered valid for field conditions?). Put a different way, can the species (or more precisely, the population) used in an effect assessment test serve as a surrogate "for the large body of organisms comprising natural environments" (CAIRNS & PRATT, 1989)? This is also understood to include conditions of exposure (e.g. via the surrounding substrate or the food) which come very close to those which occur under natural conditions. It is optimal if the organism employed in the laboratory test can also be tested in a model ecosystem (micro- or mesocosm), in which case the laboratory results of the effect assessment tests can be validated (CAIRNS, 1988; KNACKER et al., 1989).

The individual test procedures have been subjected to a critical review in light of the criteria presented in the introduction, but the conclusions have generally had little effect on the formulation of guidelines due to the overly complicated and tedious process of international coordination (e.g. MALKOMES & WÖHLER, 1983; HEIMBACH, 1984; SCHAEFFER et al., 1987; GERSICH et al., 1986; MALTBY & CALOW, 1989). It is known, for example, that a critical view should be taken of the sublethal test parameter "growth" in the acute earthworm test, given its interaction with reproduction processes, because a weight gain by the animals is negatively correlated with cocoon production. Up to now, however, this observation has not led to any revision of the relevant OECD guideline 207 (BAUER, 1991; VAN GESTEL et al., 1991; KOKTA, 1992). Another problem involved in the evaluation of existing test procedures is the fact that the bulk of this data has been generated in conjunction with the notification or registration process for chemicals and pesticides and are not published due to their confidential nature.

In the following, studies in which the toxicity of a substance is determined by means of direct, topical application will not be discussed in any detail since this exposure pathway is considered to be too artificial. However, experiments of this kind may be justified if they are used in basic research, for example, to investigate the reasons for differences in the effects on various species (SIEGFRIED, 1993).

8.2.1 Aquatic Medium (Limnic Compartment)

The test procedures in the aquatic medium to be discussed here can be divided into the following groups, which have been defined more on the basis of historical requirements (e.g. stipulated in certain laws) than technical considerations:

- quick, simple tests with microorganisms or protozoa (usually several species); available in some cases as a test kit;
- "classical" limnic tests with plants, invertebrates and fish;
- sediment tests with limnic organisms;
- bioaccumulation tests.

Tests in the marine compartment will be presented in the subsequent Chapter 8.2.2.

An overview of limnic test procedures which for the most part is still up-to-date, even today, can be found in MALTBY & CALOW (1989). Relative to the reported predominance of acute mortality tests (approx. 80% of all studies), the share of suborganism tests (e.g. of fish liver cell tests) has increased in recent years, not least of all for animal protection reasons (KETTRUP et al., 1991; CALOW, 1993). Generally speaking, the drawback of these usually quick and simple procedures is that it is difficult to apply the results to higher biological levels. There is a risk that the effect on the whole organism will be underestimated since the test is restricted to examining the effect at only one site (e.g. the liver cells).

Microbe Tests, Microbioassays and Test Kits

Especially in the aquatic medium, there is a tendency to utilize microorganisms and protozoa tests which are not only simple, cheap and easy to reproduce, but also ecologically relevant, if complete single-cell coenoses are used (multi-species tests; CAIRNS, 1984; CAIRNS, 1985; BLAISE et al., 1988; MAYFIELD, 1993). Moreover, so-called "microbioassays" (sometimes also called biosensors; PEICHL et al., 1987) are becoming more popular as screening tests, especially for the routine examination of large numbers of samples: they are cost-effective, not labor-intensive, can be automated or transported in many cases, require only small amounts of substrate as samples and quickly yield replicable and sensitive results which lend themselves very well to statistical interpretation due to the potential for high numbers of replicates (BITTON & KOOPMAN, 1992).

In enzyme tests, the synthesis of the key enzymes (especially of dehydrogenases) is generally a more sensitive parameter than the activity of the enzymes themselves (MAYFIELD, 1993). In bacteria tests, characteristics such as bioluminescence, growth, oxygen uptake, nitrification or heat production are taken as test parameters. A few of these tests are commercially available, and the use of deep-frozen bacteria material ensures that the cultures are standardized. The Microtox test is probably the most widespread; the bacteria's bioluminescence is used as a test parameter for the analysis of water and sediment samples (BULICH, 1979; HERMENS et al., 1985; KANNE, 1991).

When evaluating the results of bacterial toxicity tests, it is important to bear in mind that these relatively simple organisms have an extremely high

Table 8.3 Major Effect Tests for the Limnic Compartment

Test procedure	Measurement parameters	Example of a guideline
	Acute tests	
Daphnia test	Immobilization (EC_{50})	OECD 202 I
Fish test	Mortality (LC_{50})	OECD 203
	Chronic tests	
Algae test	Growth (EC_{50}, NOEC)	OECD 201
Daphnia test	Reproduction (NOEC)	OECD 202 II
Prolonged fish test	Mortality, behavior, possibly growth (NOEC)	OECD 204
Fish, early life stage	Embryonal/larval development (NOEC)	OECD 210
Fish, life cycle	All life stages of the F1-generation and embryonal/larval development of the F2-generation (NOEC)	e.g. BRESCH et al., 1990

Note: Aquatic tests can be carried out under static or semi-static conditions (= single or multiple application) as well as in flow-through systems (= continuous input of the test chemical).

adaptability to various environmental conditions as well as to chemicals. For this reason, bacteria toxicities are for the most part only used in the context of specific activities such as the monitoring of sewage treatment plants (e.g. short-term respiration tests with the species *Pseudomonas putida* or effluents with the luminous bacteria *Photobacterium phosphoreum*; DIN, 1991). However, none of the described methods is equally suitable for all chemicals. Therefore, it is recommended that a choice of appropriate techniques be made from a larger battery of tests, depending on the issue involved (BITTON & KOOPMAN, 1992). An overview of the spectrum of biosensor methods can be found in RAWSON (1993).

Analogous to the microbiological bioassays, test kits with *Artemia* and rotifers, which have in the meantime been validated in elaborate ring tests, have been offered on the commercial market for a few years now. They can be used as a quick screening instrument for all aquatic biotopes (e.g. SNELL & PERSOONE, 1989).

"Classical" Limnic Tests

Table 8.3 provides an overview of the most important limnic toxicity tests. Aside from the examples quoted, a large number of guidelines have also been derived from them (e.g. DIN standards in Germany or EU directives (EC, 1979) which do not diverge very much from the original test recommendations.

Table 8.4 Recommended Test Procedures for the Establishment of Water Quality Criteria in the United States

Selection criterion	Example
One salmonid species (salmon-type fish)	*Oncorhynchus mykiss* = *Salmo gairdneri* (rainbow trout)
One tropical fish species	*Lepomis macrochirus* (bluegill)
A third vertebrate species	*Pimephales promelas* (fathead minnow)
A plankton crustacean species	*Daphnia magna* (water flea)
A benthic crustacean species	*Orconectes immunis* (cray fish)
An insect species	*Tanytarsus dissimilis* (gnat)
A species which does not belong either to the chordata (vertebrates) or arthropods (e.g. annelids, rotifers)	*Aplexa hypnorum* (snail)
Species from an order not yet represented	*Nephelopsis obscura* (leech)

From Stephan et al., 1985.

Most of these tests are performed using a standardized test substrate ("reconstituted water").

Tests with green algae, water fleas (Figure 8.4) and fish are by far the most widely used tests in the aquatic medium. They will continue to form the basis of all aquatic ecotoxicological test strategies in the future, with a tendency to increase the number of species tested. In order to establish "Water Quality Criteria" in the United States, for example, data from tests with eight different families of animals are recommended (Table 8.4; STEPHAN et al., 1985). In order to minimize the cost of such a large number of test procedures and also to make the test conditions even more comparable with each other (especially the same kind of exposure), it is also recommended that several tests be conducted in parallel in a large, compartmentalized tank (EWELL et al., 1986; HOLCOMBE et al., 1987).

Plant Tests

Since the early 1970s, algae species have been regularly required in the limnic compartment for the registration and notification of pesticides and chemicals, whereas higher plant species are rarely used for test purposes, unlike field monitoring studies. Yet, since the mid-1980s, the prevalent view (e.g. KENAGA & MOOLENAR, 1979) which has it that plants are generally less sensitive than animals to chemicals is no longer tenable today. Examples of highly sensitive plant reactions can be found for all relevant categories of chemicals (LEWIS, 1993).

The algae tests commonly employed in the United States and Europe today (e.g. OECD, 1984; ASTM, 1990a), which are similar in most aspects of their implementation such as duration (usually 3-4, sometimes up to 21 days), or species employed (usually *Selenastrum capricornutum*, less often *Scenedesmus subspicatus* or *Chlorella vulgaris*), are used to examine pesticides, environmental chemicals, pharmaceuticals or effluents (NYHOLM & KÄLLQUIST, 1989). Provided prior experience is available, the reproducibility of these tests is very good, which is one of the main reasons why they are so widespread. Depending on the chemical and the species of algae employed, the sensitivity may vary by as much as a factor of 2,000 (LEWIS, 1993). For this reason, it is recommended that more than one species be tested if possible ("test battery"). Moreover, algae tests should always be conducted together with a residue analysis, since the concentration of the test substance may be altered by the algae themselves (extracellular chelates), as well as its own characteristics or those of the substrate (SPENCER & GREEN, 1981).

In addition to the parameter of growth inhibition employed in the commonly known tests, photosynthesis is also used as a measuring variable in short-term tests with algae, whereby field mixtures of various species are often used instead of laboratory-bred cultures. However, since the sensitivity is lower than it is in biomass tests, regardless of which chemical is being tested, this approach never became established (TURBAK et al., 1986). Although there is some practical experience to go on, tests with higher aquatic plant species are not a standard requirement; the only exception is duckweed (*Lemna minor* or *Lemna gibba*), which is required in the United States (especially for effluents, and only rarely even there; ASTM, 1990b). Algae tests which have been modified to approximate ecological reality more closely (e.g. by diluting the standardized medium with river water or using natural algae coenoses) are conducted even less frequently.

Invertebrates

So many different laboratory tests with limnic invertebrates have been developed that they are difficult to keep track of (APHA, 1989; PERSOONE & JANSSEN, 1993). Yet, only a few acute and chronic methods with species of small water flea *Daphnia* (except *D. magna* or *D. pulex* as well as *Ceriodaphnia,* Figure 8.4) have become established in standard test practice. These species are preferred for the following reasons (EPS, 1990):

- *Daphnia* are commonly found in a great many limnic biotopes.
- *Daphnia* play a key role in the food web (they feed on algae and are themselves an important source of food for many fish species).
- *Daphnia* have proved to be sensitive to the widest variety of chemicals.
- *Daphnia* have a short generation cycle and are relatively easy to keep.
- Due to their small size, the test is simple and cheap to carry out.

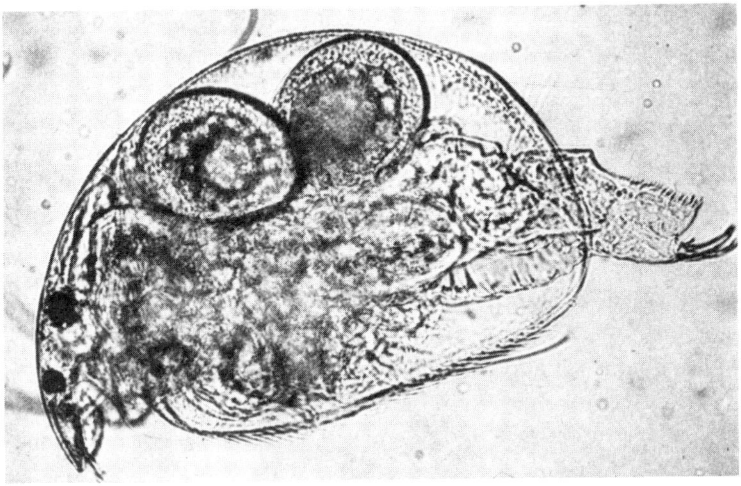

Figure 8.4 The water flea *Daphnia* is the most common test organism in limnic tests.

The test parameters are immobility (mortality is difficult to ascertain in the case of these small animals) and reproduction (juveniles per parent).

Efforts are now under way to improve the existing *Daphnia* tests. Ring tests are being used to focus on the sources of variation (e.g. role of the test medium, influence of the quality of the feed (algae)) in ecotoxicological tests with *Daphnia magna* so as to reduce the variability of the findings. It is now clear, for example, that a synthetic test medium will have to be employed in the future. However, it should be noted that even with this test system, which has been used for at least 30 years and is often called the most standardized aquatic test, considerable breeding problems are still known to occur.

In addition to the *Daphnia* test procedure, a large number of other procedures are described in the literature which have either not been standardized or only in part, such as:

- acute tests with diptera (e.g. chironomids), ephemoptera, crustacea (e.g. gammarids) or annelids (e.g. enchytraeids);
- chronic tests with protozoa (e.g. *Colpidium campylum*), May-fly larvae (e.g. *Stenonema modestum*), diptera (e.g. chironomids), crustacea (e.g. gammarids) or molluscs.

The possibility of using toxkits (e.g. with the rotifer species *Brachionus calyciflorus*) was already mentioned at the beginning of this chapter. Various toxkit tests should always be carried out in parallel, since no one species is the most sensitive to every chemical in a given case (BUIKEMA & CAIRNS,

1980). There are a growing number of indications to suggest that these very simple "toxkits" can be used to obtain results which largely correspond to the data from the more elaborate invertebrate tests (e.g. LC_{50} on the same scale; JANSSEN & PERSOONE, 1992).

However, there are still so many reservations about these additional vertebrate test procedures — limited transferability; the lack of precise knowledge about the ecological role of the organisms employed, which is not completely clear; and the as yet inadequate validation — that it is not very likely they will come into routine use for the chemicals notification process. The main disadvantages of these test designs (excluding the toxkits) are the difficulties involved in breeding as well as the lack of a database. The protozoa growth test is probably the most suitable for routine practice (DIVE et al., 1991). On the other hand, there is probably no ecotoxicological test procedure which has yielded such a vast amount of data for chemicals of all classes as the mortality or reproduction tests with *Daphnia* (BUIKEMA et al., 1980; BAUDO, 1987).

Fish

Fish were one of the first organisms used to examine the effects of chemicals in the laboratory (CARPENTER, 1927). Fish tests are still used today for many different purposes — for example, in the respective notification or registration process for environmental chemicals and pesticides as well as the monitoring of effluents. Fish are highly valued test subjects because of their ecological and economic role in limnic ecosystems. No single species can be representative of the class Pisces as a whole since fish are very heterogeneous in terms of their taxonomy, nutrition and behavior.

SOLBE (1993) differentiates between five test categories with fish (screening tests; tests for establishing Water Quality Criteria; effluent monitoring tests; legal tests and river monitoring tests). In the chemicals tests (= screening tests), a distinction can be made between acute mortality tests, on the one hand, and chronic sublethal tests (behavior, growth, development), on the other, for which a large number of very similar guidelines exist.

The priorities reflected in requirements imposed by the authorities have shifted for animal preservation reasons, among others: instead of acute mortality tests, the prolonged fish test is often required as a minimum. For the time being, chronic tests such as the Early Life Stage test or, even better, the Full Life Cycle test are recommended, which are definitely more valuable as indicators than the methods employed to date (NAGEL, 1994). Such largely standardized bioassays with embryos or early life stages are, moreover, also used in the testing of chemicals and complex compounds such as effluents for teratogenity. The South African clawed toad *Xenopus laevis* is also used for these tests (DUMPERT & ZIETZ, 1984) in addition to various species of fish such as the bluegill sunfish *Lepomis macrochirus* (BIRGE et al., 1985).

Fish tests are normally performed in artificially composed water (= reconstituted water) in three different forms: in the *static* test, the substrate (water) is not refreshed during the test, whereas in the *semi-static* test it is exchanged

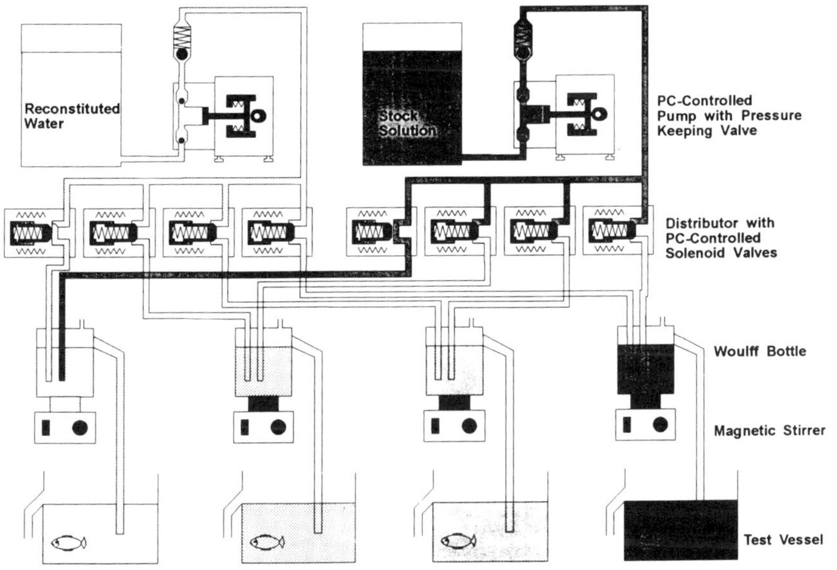

Figure 8.5 Schematic diagram of flow-through test chamber with fish.

in regular intervals (usually every 24 hours). In the very complex *flow-through* test (Figure 8.5), the test substance is continuously added to the water so that the nominal chemical concentration is for the most part maintained at a constant level for the duration of the test (which should be documented with a corresponding residue analysis). The most frequently employed test species are rainbow trout (*Oncorhynchus mykiss = Salmo gairdneri*; Figure 8.6) or carp (*Cyprinus carpio*) as coldwater fish and the bluegill (*Lepomis macrochirus*) or fathead minnow (*Pimephales promelas*) as warm-water fish.

Sediment Tests with Limnic Organisms

Sediments play an important role in aquatic ecosystems; not least of all as (more or less definitive) sinks for many substances which can often be found there in concentrations which are significantly higher than in the bodies of water above them. Physical (e.g. wave motions) as well as biological (e.g. burrowing by worms) activities can cause these substances to re enter the water (ECETOC, 1990). Anthropogenic activities, such as the excavation of waterways, may lead to new exposure to chemicals (REYNOLDSON & DAY, 1993). In spite of this fact and a large number of monitoring studies, especially in the United States, there is no internationally accepted test for the sediment compartment to date. Recommended guidelines (INGERSOLL & NELSON, 1990) only exist for a few sediment procedures such as the chironomid and amphipod test. One reason is that these tests are not routinely required by legislative regulations (if at all, only in the United States when a specific risk

Figure 8.6 Rainbow trout haltering.

is indicated). In any case, the consensus is that more than one test is necessary to assess the hazard potential (AHLF et al., 1989; BURTON, 1991).

Furthermore, existing recommendations for a test strategy applying to limnic sediments often suggest that tests be conducted with organisms in the body of water (e.g. the water flea *D. magna* or fish such as *P. promelas*; NEBEKER et al., 1984; GIESY & HOKE, 1990). Although there are examples of a good correlation between test results from pore water, sediment eluates or reconstituted water and the data derived from whole sediment samples (e.g. WIEDERHOLM & DAVE, 1989), most of the evidence shows that sediment eluates, at least, are not suitable to predict the toxicity of sediment samples (ANKLEY et al., 1991).

In support of earlier recommendations, REYNOLDSON & DAY (1993) therefore propose that a variety of tests with sediment animals and organisms in open water be used to assess the effects of chemicals, drawing on both ecological criteria (e.g. to cover the various trophic levels) and past experience. More specifically, it is a matter of the following procedures, to which another test (with *Tubifex tubifex*) may be added:

- an acute chironomid test (mortality, growth) with the species *Chironomus tentans* or *C. riparius* (Diptera) as actual sediment dwellers (HAMER et al., 1992); currently the most widespread sediment test;

- an amphipod test (mortality, growth) with the species *Hyalella azteca* as a representative of the sediment surface dwellers (chronic reproduction tests are also possible with this species; BORGMANN et al., 1989);
- a chronic tubificid test (REYNOLDSON et al., 1991) with *Tubifex tubifex*; an organism which lives half in sediment, half in the body of water; although these oligochaetes still occur in heavily polluted biotopes, the test is recommended because a relatively large number of data exist (also as an acute test; AMMON, 1985) and because it can be further developed as a bioaccumulation test (EGELER, 1993);
- a growth test with algae as a representative of the primary producers (e.g. MUNAWAR & MUNAWAR, 1987);
- a bacteria bioluminescent test (e.g. the Microtox test);
- an acute fish test, e.g. with the fathead minnow (*P. promelas*).

Moreover, there are a large number of test ideas (especially with oligochaetes and insects), which, however, do not currently exist as standardized guideline proposals (REYNOLDSON & DAY, 1993). Suggestions which make use of sublethal parameters such as the behavior of oligochaetes are especially promising (e.g. WHITE & KEILTY, 1988; COLER et al., 1988). Together with physical and chemical data as well as monitoring results from the field, the above-mentioned tests should be sufficient to assess contaminated sediments (CHAPMAN, 1990a, 1990b). An overview of the current situation with regard to ecotoxicological studies of sediments can be found in BAUDO et al. (1990) and BURTON (1992).

Bioaccumulation in Limnic Systems

To date, only tests with fish, which were standardized by the OECD back in 1981 in various modified versions, exist for the investigation of bioaccumulation in the aquatic medium (NAGEL & LOSKILL, 1991). The aim of such experiments is the specification of a bioconcentration factor (i.e. the relationship between the chemical concentration in fish compared with the concentration in water) as an indication of bioaccumulation (PHILLIPS, 1993). The most complex version (OECD 305E), in which the absorption or desorption of a substance in a flow-through system is examined, is currently being revised (OECD, 1981c, 1988). Very similar tests have also been called for in the United States since the late 1980s (overview in SOLBE, 1993). The bioaccumulation of chemicals in other aquatic organisms, especially invertebrates such as *Daphnia*, has been studied in individual cases, but standardized procedures are only now in the process of being developed (e.g. with chironomids and tubificids).

8.2.2 Aquatic Medium (Marine Compartment)

Outside of the United States, there are relatively few routine requirements for the notification of chemicals and the registration of pesticides in the marine compartment of the environment (including brackish water and estuaries). If anything, such tests are still being called for in connection with waste-water problems as well as the assessment of industrial activities such as oil-drilling (e.g. the Ocean Dumping Regulations passed in the United States in 1977), whereas pesticides are seldom tested due to the low probability of exposure (BURTON, 1992). Nonetheless, the number of standardized test procedures is quite high, although use is made almost exclusively of species from North American waters (primarily brackish or at least neritic species). In Germany, the question of whether marine tests should be called for in the framework of environmental chemicals registration was examined in the mid-1980s. However, no marine tests are required for environmental chemicals or for pesticide registrations (except in the event of an imminent indicated risk). The main reason that is usually given for this is the absence of any differences in sensitivity compared to limnic species, which are tested anyway (UBA, 1986b).

Plants

Over 50 species of marine algae have been employed in laboratory tests to date (overview in WALSH, 1993). The first standardized algae test originated back in the early 1970s (STRAND et al., 1971). In addition, four guidelines or test proposals are currently in the advanced stages of development (EPA, 1974 or WALSH, 1988; PAYNE & HALL, 1979; ISO, 1988; KUSK & NYHOLM, 1991). As in the limnic tests, the measurement parameters usually consist of growth inhibition, less often photosynthesis, respiration or the concentration of certain pigments. Various clones of the same species (e.g. of the diatoms *Skeletonema costatum*), which were taken from the open sea or estuaries, may have varying sensitivities (MURPHY & BELASTOCK, 1980). A new development affects the use of microalgae ("Algal Microtest Battery"; BLANCK & BJORNSATER, 1988), whose small size greatly simplifies the test procedure in practice, similar to the test kits (see Chapter 8.2.1). On the other hand, higher marine plants have hardly been used thus far, although it was possible to demonstrate in the case of red and brown algae, for example, that in many cases they exhibit a sensitivity similar to that of single-cell algae or animals (THURSBY et al., 1993).

Invertebrates

As in the case of the marine plants, many different organisms from the category of marine invertebrates were exposed to chemicals (ERNST, 1984; REISH, 1988; WIDDOWS, 1993). The acute techniques listed briefly in the following were developed for the testing of chemicals and effluents, mainly in the United States. The already mentioned "toxkits" with the rotifers *Brachionis plicatilis* or the brine shrimp *Artemia salina* are attractive for their simplicity and quickness (VANHAECKE & PERSOONE, 1984; SNELL & PERSOONE, 1989). However, both approaches have been criticized for the relatively low sensitivity of the mortality parameter and the low ecological relevance of the test organism (WIDDOWS, 1993). The test with the copepod *Acaratia tonsa*, which is only slightly less costly, is given a similar assessment (EPA, 1976); it is currently being examined by the ISO in a ring test. Because of the possible modification for test purposes, biosensors for the continuous monitoring of waters should also be mentioned, in which the behavior (the shell opening) of mussels is used as an automatically registered measurement parameter ("Mussle Monitor"). The sensitivity of this commercial procedure is so high that there is no need for costly chemical analyses in some cases (Delta Consult, Kapelle, Holland).

An acute test with bivalves (usually the oyster *Crassostrea gigas*, less often *C. virginica*, *Mytilus edulis* or *Mercenaria mercenaria*), which has been available since the late 1960s, is much more widespread and also more standardized. It makes use of the development from the unshelled embryo to the shelled larvae as a test parameter (WOELKE, 1972; ASTM, 1990c; THAIN, 1990). Use of the chronic test parameter shell growth was also proposed at an early stage (e.g. for pesticides: DAVIS & HIDU, 1969). However, this test, which was developed primarily for industrial chemicals as well as waste-water of various pollution levels, has been criticized because the mortality rate is often very high in the controls.

Of the chronic marine tests, the generation test with the mysid crustaceans *Mysidopsis bahia*, in which the young animals' body weight and the mortality are measured in addition to reproduction, is by far the most widespread. The test is usually conducted as a flow-through procedure and is considered very sensitive, e.g. to heavy metals (LUSSIER et al., 1985) or pesticides (CLARK et al., 1989). It is not yet possible at present to assess whether the "SFG" mussle test (e.g. with *M. edulis*), which focuses on growth or other physiological parameters, will become established as a further chronic procedure (WIDDOWS, 1993). This is also true for a reproduction test with copepods (BENGTSSON & BERGSTRÖM, 1987).

Fish

Aside from the kinds of species employed, fish tests in the marine compartment differ very little from those performed in limnic biotopes (ASTM, 1988a; EPA, 1988). Consequently, acute tests can be conducted according to the same guidelines, with appropriate modifications (e.g. OECD, 1984). Since these tests, like other marine tests, are required mainly in the United States (REISH, 1988), experience is available in particular on species of the North Atlantic or

North Pacific (e.g. the sheepshead minnow *Cyprinodon variegatus*, the topsmelt *Atherinops affinis* or the silverside species *Menidia menidia* and *M. beryllina*). However, the need to use regionally relevant species has repeatedly been emphasized (e.g. BELLAN, 1981; MIDDDAUGH et al., 1993).

Since it is assumed that, in the case of fish as well, the juveniles will react most sensitively to chemicals, there is a growing demand (at least in the United States) for Early Life Stage (ELS) tests, analogous to limnic tests (ASTM, 1988b). Such tests were already developed in the early 1970s for the sheepshead minnow *C. variegatus* (SCHIMMEL et al., 1974), which has remained the most frequently employed test species to the present day. In the meantime, embryos and early life stages of many species have been tested, and the individual procedures are very similar for all of them. The main problem involved in conducting these tests is that in some cases it is very costly to maintain the test organisms in a way that is appropriate to the species (especially feeding). The tests last approximately 28-35 days and make use of mortality, growth and, in some cases, the concentration of the substance in the tissue as test parameters (MIDDAUGH et al., 1993).

Sediment Tests with Marine Organisms

LUOMA & HO (1993) provide a very good overview of test procedures that can be used in marine sediments. This review not only summarizes the pros and cons of each test, but also gives an overall assessment of the benefits of biological tests. Generally speaking, a distinction can be made between whole sediment bioassays and tests with various sediment extracts. In the first group, organisms such as amphipods are exposed directly to sediment samples mixed in varying concentrations with unpolluted substrate, e.g. after pollution from sludge (SWARTZ et al., 1990). Alternatively, natural and, less frequently, artificial sediment is tested after the chemical to be studied has been added. The extraction, storage and composition of the substrate (e.g. oxygen saturation, geochemical characteristics) has a significant influence on the test results (LUOMA & HO, 1993).

In the second group, tests with pore water from sediments or (organic) sediment extracts are carried out which are not very different from "normal" aquatic procedures. Oyster larvae (CHAPMAN & MORGAN, 1983) or sea urchin embryos, especially sanddollars like *Dendraster excentricus,* (NACCI et al., 1986, CASILLAS et al., 1992) are used in addition to the already known limnic or marine species. Because the results of these tests are difficult to interpret for the complex sediment compartment, the methods will not be discussed in any further detail here. Microbiological tests such as the Microtox procedures, which are not too different from those applied in the limnic system, were carried out with marine sediment extracts many times, but the results are extremely variable, depending on the kind of extraction method employed. Therefore, these tests cannot be recommended at this time (DUTKA et al., 1990).

As a matter of principle, an attempt is made to have as diverse a battery of test methods as possible for the whole sediment bioassays (DUTKA & KWAN, 1988). For example, brackish hydrophytes such as *Spartina alterniflora* were successfully tested, even in artificial sediments (WALSH et al., 1990). As already mentioned in the algae tests, these methods are used more often to assess contaminated sites or waste-water than in conjunction with the pesticide registration procedure (WALSH, 1993).

When sediment tests came to be required more frequently in the United States during the 1970s, the focus was on acute tests with adult bivalves, polychaetes and fish (SWARTZ, 1987). Since these tests proved to be relatively insensitive, an increasing number of procedures with more sensitive stages of the same species and chronic measurement parameters (behavior or reproduction) were developed. Guidelines in the United States have recently established methods which make use of entomostracans such as amphipods (e.g. *Rhepoxynius abronius* (MEARNS et al., 1986) or *Corophium volutator* (ASTM, 1991)). Tests with polychaetes such as *Ophryotrocha diadema* or *Neanthes arenaceodentata* (HOOFTMAN & VINK, 1980; REISH, 1985) are also in widespread use.

In summary, it is recommended, as proposed by REISH (1988), that a battery of tests also be devised for the marine compartment of the environment. These tests with a selection of species along the following lines would then be regarded as the minimum requirement for assessing the ecotoxicological hazard potential:

- a species living in the sediment, e.g. the polychaete *Neanthes arenaceodentata;*
- a bivalve species due to its great economic role as well as its capacity for bioaccumulation, e.g. the oyster *Crassostrea gigas*;
- a crustacean species, due to its high ecological relevance and the possibility of performing a chronic test, e.g. the mysid *Mysidopsis bahia*;
- at least one, preferably two fish species as representatives of higher trophic stages and due to their high economic importance, e.g. *Cyprinodon variegatus* or *Menidia menidia*;
- a plant species as a representative of the primary producers, e.g. *Skeletonema costatum* or *Spartina alterniflora*.

The selection of suitable species will have to be modified depending on the region and the medium or purpose of the investigation (e.g. brackish water; open water; pesticide registration; waste-water monitoring).

Bioaccumulation in Marine Systems

Despite an enormous number of studies on the bioaccumulation of chemicals, especially of heavy metals, in marine organisms, there is only one test guideline in which oysters (*Crassostrea virginica*) are studied in a flow-through system (EPA, 1989). To an increasing degree, the small species *Macoma nasuta* from the Pacific is also being used (BOESE et al., 1990). Due to their vital filtering function, bivalves are very well-suited for bioaccumulation studies with the widest variety of chemicals, as shown by their frequent use in monitoring studies (e.g. BERGEN et al., 1993).

8.2.3 Terrestrial Medium

For a preliminary overview, the effect assessment tests in the terrestrial medium can be classified as follows, although there is a considerable amount of overlap between the various categories:

- tests with soil microorganisms, usually for the testing of pesticides;
- plant tests;
- tests with soil macroorganisms (mainly saprophagous invertebrates);
- tests of pesticide side effects, employing beneficial arthropods including pollinators such as honey bees;
- tests with higher vertebrates (birds, mammals);
- tests on bioaccumulation (with plants and animals).

The number of standardized test procedures available for the terrestrial medium (Table 8.5) is significantly lower than those which exist for the aquatic medium (BECKER, 1991; VAN STRAALEN & VAN GESTEL, 1993a). However, quite a few ideas for such tests do exist. Biological test procedures (such as that of EPPO) which analyze the efficacy of pesticides on the designated target organisms (e.g. PEDERSEN, 1984) are not included in this context. Test methods or approaches which deal with organisms on the surface of vegetation or in the atmosphere are not discussed in any detail (almost exclusively procedures with ‘beneficial arthropods’; HASSAN, 1985, 1992; SAMSOE-PETERSEN, 1990). In other words, the test procedures just discussed focus on organisms in the soil (defined as mineral soils including the soil surface, e.g. litter layer, including plants; Figure 8.7), whereas vertebrates (birds, small mammals) are dealt with only in brief.

Until very recently, only two sets of soil tests were available in Germany (the acute earthworm test and microflora test for pesticides and the acute earthworm test and acute plant test for environmental chemicals). In other countries such as the United States, even less attention was given to the terrestrial medium. In the meantime, the situation has improved somewhat due to the development of chronic soil animal tests (e.g. with earthworms and collembolans) as well as the inclusion of so-called beneficial arthropod tests (see below). However, because there is a deficit of terrestrial test guidelines (VAN LEEUWEN, 1990), various soil tests are currently in development, especially with arthropods (RÖMBKE et al., 1993; VAN STRAALEN & VAN

Table 8.5 Major Effect Tests for the Terrestrial Medium

Test procedure	Measurement parameters	Example of a guideline
	Acute	
Earthworm test	Mortality (LC_{50})	OECD 207
Plant test	Emergence inhibition (EC_{50})	OECD 208
Carabid test*	Mortality, feeding rate	BBA VI, 23-2.1.8
Avian single dose test	Mortality (LD_{50})	EPA E §71-1
Avian dietary test	Mortality, body weight, behavioral changes, feeding	OECD 205, EPA E §71-2
Bird uptake test	Mortality, body weight, behavioral changes	BBA VI, 25-1
	Chronic	
Microflora test	Dehydrogenase activity, short-term respiration, nitrogen cycle	BBA VI, 1-1
Earthworm test	Reproduction (NOEC)	ISO, 1993
Collembola test	Reproduction (NOEC)	RIEPERT, 1991
Staphylinid test*	Hatching rate, number of eggs	MORETH & NATON, 1992
Avian reproduction test	Mortality, number of eggs, morphological changes of the eggs, effects on the young birds	OECD 206, EPA E §71-4

* Beneficial arthropod tests required in the pesticide registration process.

GESTEL, 1993b). Despite a few initial experiences with a yeast-collembolans-carabid system (GRUTKE et al., 1988), for example, or a collembolans/enchytraeid-gamasid system (SCHLOSSER & RIEPERT, 1992), there are no useful multi-species laboratory tests to date, with the exception of microbiological methods.

Microorganisms

There is a particular lack of soil tests with microorganisms (JAGNOW, 1984; BBA, 1990c; ROSSEL & TARRADELLAS, 1991), whose importance for terrestrial ecosystems (e.g. for the decomposition of organic material or the nitrogen cycle) can hardly be overestimated. For one thing, this has something to do with the fact that the large number of species in these taxonomically very difficult groups (e.g. 5,000-10,000 microorganism species supposedly live in the soil; MAAS, 1987) severely restrict their use. Much the same is true of protozoa (SCHREIBER & BRINK, 1989) and nematodes (OHBA & ISHIBASHI, 1984; VAN KESSEL et al., 1989), which are repeatedly proposed

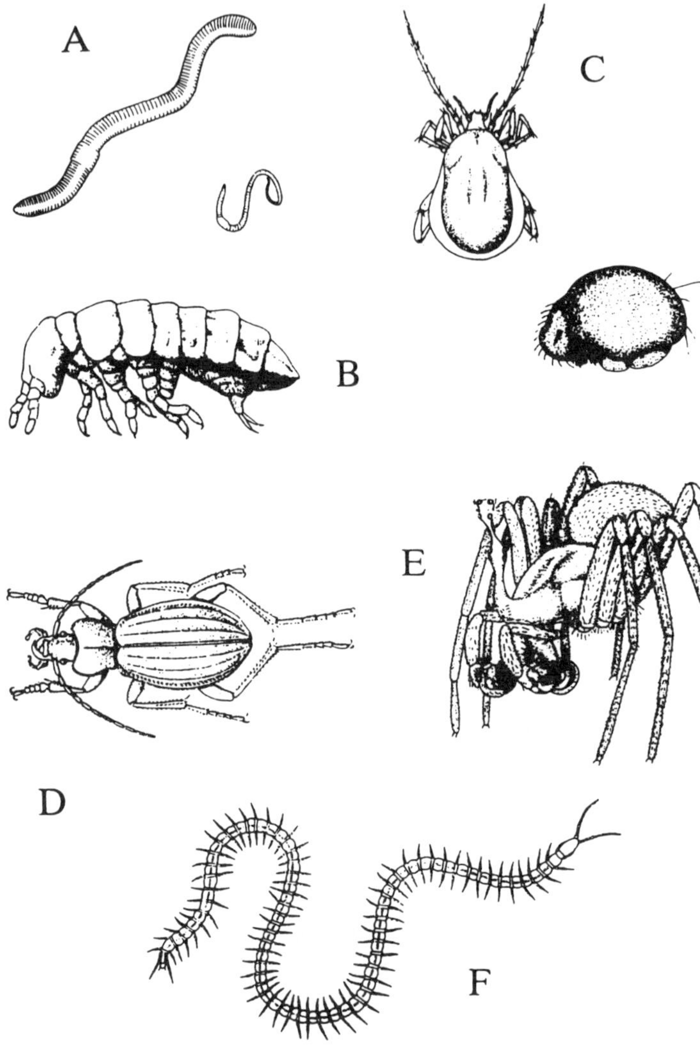

Figure 8.7 Representatives of important soil animal groups: earthworm and enchytraeidworm; gamasid and oribatid mite; springtail (Collembola); carabid beetle; spider; milliped (Chilopoda).

as test organisms due to their major role in the soil ecosystem (LÖKKE & VAN GESTEL, 1993). The main problem is related to the transferability of the results because these tests are usually conducted in artificial media (e.g. nutrient solutions or agar).

Moreover, higher organisms are generally considered more sensitive to pesticides than are microbes with their broad physiological range of reaction and quick succession of generations and thus favorable potential for recovery (OTTOW, 1985; GLEDHILL, 1987). In spite of these difficulties, natural

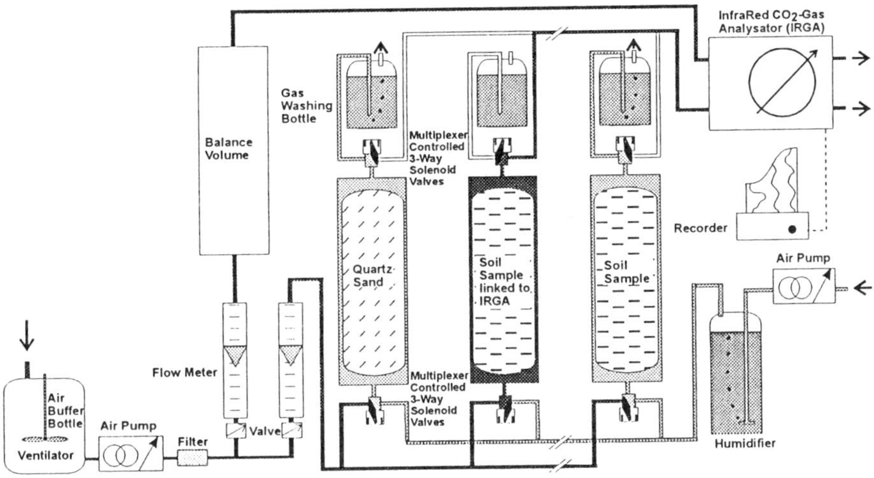

Figure 8.8 Schematic diagram of a measurement device for soil respiration.

microbe populations or physiological groups tend to be preferred for testing. The measurement parameters used in these tests are short-term respiration (Figure 8.8), dehydrogenase activity and also nitrification, which, however is costlier and more difficult to interpret (ANDERSON & DOMSCH, 1978; SOMERVILLE & GREAVES, 1987). A guideline for tests of these parameters is available in Germany for the registration of pesticides (BBA, 1990c). In comparing the first two tests mentioned, MALKOMES (1987) concludes that the dehydrogenase activity shows a sensitivity towards pesticides that is similar or even better than that of short-term respiration. However, BEYER et al. (1993) come to just the opposite conclusions, since the dehydrogenase activity is influenced by too many, mostly unknown, ecological facts.

The results of enzyme tests are generally difficult to assess for two reasons: (1) many of the enzymes may still be active outside the cell, even after the microbes that produced them have already been killed off; (2) enzymes are very sensitive to environmental conditions (SOMERVILLE & GREAVES, 1987). Moreover, reference is made to the fact that small groups of specialists, e.g. nitrificants (as a guideline, NEN, 1988b; BBA, 1990c) often show the greatest sensitivity to pesticides since their function cannot merely be taken over from other groups. For this reason, more than one test method — a battery of tests, in other words — should be available for microorganisms as well (DOMSCH, 1992).

Tests on the effect of pesticides on the symbiotic vesicular-arbuscular mycorrhiza (VAM; a symbiosis between soil fungi and roots of higher plants) are currently in the process of development (SCHÜEPP et al., 1990). They are relevant primarily because of their key role in the soil ecosystem and less so because of their high sensitivity. Other procedures such as dyeing techniques

(PAUL & CLARK, 1989), the ATP method (JENKINSON & OATES, 1979), the Fe(III) reduction method (WELP & BRUMMER, 1985) or the DNA-based hybridization technique (BROCK & MADIGAN, 1991) are less suitable for testing due to their high degree of specificity.

Due to the considerable technical difficulties involved, it has also not been possible to date to develop the important microbial function of the degradation of organic material (litter bag test) as a laboratory test (SIEDENTOP, 1993). Results on the metabolism of microbes usually come from tests with isolated cultures in artificial, often nutrient-rich laboratory media and therefore have only minimal relevance for the real situation in the terrestrial media (ECETOC, 1990).

Plants

Despite their overriding importance as producers, there are relatively few test guidelines for plants (HOLST & ELLWANGER, 1982; GÜNTHER & PESTEMER, 1990; KAPUSTKA & REPORTER, 1993). The physiological active mechanisms of many chemicals in plants are well known (HOCK & ELSTNER, 1988), e.g. from efficacy studies involved in the development of herbicides (KENAGA, 1981). The oldest plant test on the acute effects on growth (OECD, 1984; RIEPERT, 1984; ADEMA & HENZEN, 1989) cannot be recommended. Although the oat (*Avena sativa*) or beet (*Brassica rapa*) plants employed in the test are relatively sensitive to chemicals, the test in its current form is more like a range-finding test (RÖMBKE et al., 1993) due to its very extensive gradation by concentration.

Moreover, it is often criticized because the number of species per test is too low in many cases due to the very specific effect of chemicals on plants (PRICE & HIKINO, 1984); and wild plants in particular are not studied (MARSCHNER, 1992). It is noticeable that none of the plant species suggested for standard tests thus far is a woody species, although — in view of the forest damage problem (Waldsterben), for example — there is a considerable need for such studies (KAPUSTKA & REPORTER, 1993). An analysis of tests conducted with 151 different species indicates that the sensitivity within the vascular plants alone varies on average by a factor of 10.5 and in individual cases by a factor of more than 100 (FLETCHER et al., 1990). Another difficulty is related to the interpretation of these tests, especially with regard to the frequently observed incidence of stimulation effects, on growth, for example, at low chemical concentrations (WIEDMAN & APPLEBY, 1972).

In many cases, a very simple germination test with cress plants is employed for the investigation of contaminated substrate (MOEWUS, 1949; NEURUER, 1975; EPA, 1985). On the other hand, a root elongation test with barley seeds (ISO, 1992c) and a mutagenity test with pea seed to ascertain the relative toxicity of pesticides are still in development (SARBHOY et al., 1991). It is currently difficult to assess the benefit of enzyme tests, especially with bean plants, which were developed primarily for the study of soils containing heavy metals (e.g. VAN ASSCHE & CLUSTERS, 1990). Test procedures on the basis of physiological reactions such as photosynthesis or transparency were proposed but have not been sufficiently standardized to date (cf. RÖMBKE et al., 1993).

KAPUSTKA & REPORTER (1993) point out that plant tests in the past have been limited to two relatively insensitive parameters (seed germination and root elongation) and thus conclude that they are of only limited relevance for ecology. However, they do not give any reasons for drawing this conclusion. In order to compensate for the general lack of relevant plant tests, generation tests (or life-cycle tests) have been developed in recent years for two plant species (*Arabidopsis thaliana* and *Brassica rapa*). This has made it possible to study the sublethal effects of chemicals over the course of an entire generation. In the case of both species, the test lasts only a few weeks (*Arabidopsis* — about eight weeks until the seeds ripen), which means that as many as 10 generations can be studied per year (KAPUSTKA & REPORTER, 1993).

Plant tests can also be used to detect the presence of herbicides in the soil, and the same degree of sensitivity can be achieved with this biotest as in chemical detection procedures (GOH et al., 1991; GARCIA et al., 1992). In tests where nutrient solutions, which present a heavily modified bioavailability, were used as substrates instead of soil, the test plants reacted more sensitively than in soils (ADEMA & HENZEN, 1989; HULZEBOS et al., 1993). The procedure thus has implicit drawbacks: the lack of any generally valid method for the conversion of test substance concentrations from the nutrient solution to the soil solution, the potential for different metabolisms as well as the higher-than-representative sensitivity value.

Invertebrates

The acute earthworm test, which has already been mentioned several times, can be regarded as the most widespread test on animals in the soil (OECD, 1984; NEN, 1988a). This is not only because earthworms play an important role in the soil ecosystem (PETERSEN & LUXTON, 1982; LEE, 1985) but also because the test is easy to perform. The compostworm *Eisenia fetida* is exposed to the test chemical in artificial soil for two weeks. The mortality as well as effects on the behavior and biomass of the animals are the measurement parameters. Due to the negative correlation of weight development with cocoon production (= reproduction), however, use of the biomass parameter is severely limited (VAN GESTEL et al., 1991). A prolonged version of this test is now required for environmental chemicals in the United States (FDA, 1988), where it must be accompanied by a residue analysis, which is also prescribed for aquatic tests.

The acute earthworm test has been harshly criticized, however, on the grounds that *E. fetida* is not a typical soil animal and was only chosen as a test organism because it is easy to breed and handle. The use of artificial soil, which seriously reduces the bioavailability of chemicals in relation to natural soil due to its high share of organic matter, is also considered problematic (CABRIDENC et al., 1984; RÖMBKE et al., 1992). This is why the use of even more artificial substrates such as silica gel (FERRIERE et al., 1981; AFNOR, 1984) cannot be recommended, although positive correlations between the silica tests and those conducted in artificial soil were found to exist in the case of some pesticides (HEIMBACH, 1984). It has also been claimed that *E. fetida* generally react less sensitively to chemicals than other earthworms (VAN GESTEL, 1991), but this criticism is not justified. It is more likely that different species will be the most sensitive species, depending on the chemical, measurement parameter and test conditions involved (VAN STRAALEN & VAN GESTEL, 1993a).

A chronic earthworm test has been developed in the meantime (BBA, 1994a, 1994b). As in the acute test, compostworms and artificial soil are employed because they are easy to manage and compare. However, because reproduction (= number of young animals) serves as a measurement parameter in addition to mortality and biomass, and also because of the longer duration, the potential hazard can be assessed much more accurately (VAN GESTEL et al., 1992). The use of other earthworms (such as *Lumbricus terrestris* or *Aporrectodea caliginosa*) which are more typical of holarctic field conditions has failed because of the longer generation cycle and difficult maintenance conditions. A combined acute/reproduction test with a second family of oligochaetes, the enchytraeids (Figure 8.9), which are widespread particularly in acidic soils (RÖMBKE, 1989, 1991), is very similar to the earthworm reproduction test. In any case, a test which makes use of earth is preferable to procedures in which the animals are kept in agar (WESTHEIDE et al., 1991), where the exposure situation is different, with poor transferability of the results.

Aside from the tests with oligochaetes described up to now, these animals were also used in procedures with suborganismic parameters, e.g. enzymatic tests with earthworms (CHEN et al., 1991) or the expression of stress-proteins in various invertebrates (KÖHLER et al., 1992) as well as ultrastructural effects on colic and chloragogic cells of enchytraeids (HAGENS & WESTHEIDE, 1987). Considerable problems still arise when it comes to extrapolating from the observed changes and distinguishing them from natural fluctuations.

Collembola (springtails) are considered to be one of the most important groups of soil mesofauna. They were selected for test purposes due to the wide variety of evidence showing that they have differentiated reactions to pesticides in the field (e.g. BECK et al., 1988). Acute as well as chronic tests with collembolans have been suggested since the mid-1970s (e.g. THOMPSON & GORE, 1972; EIJSACKERS, 1978; JANCKE, 1989). The collembola species *Folsomia candida* is used in a reproduction test soon to be issued as an ISO

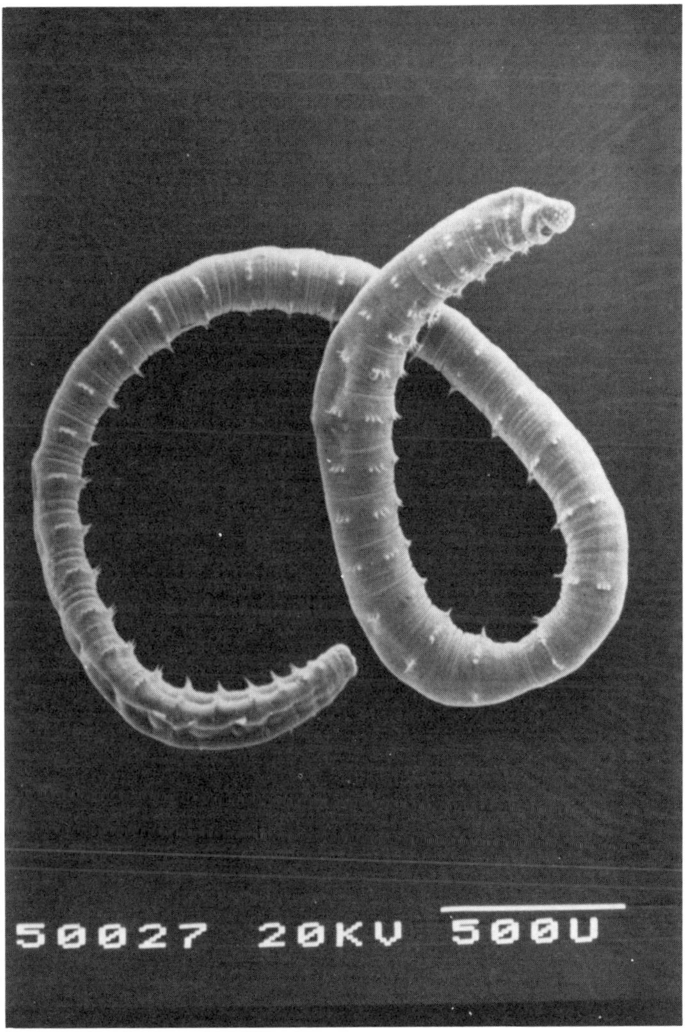

Figure 8.9 ***Cognettia sphagnetorum*** **as a representative of the Enchytraeidae.**

guideline (RIEPERT, 1991). Like *E. fetida*, *F. candida* is a synanthropic species; it can also occur in cultivated habitats, where it plays a role in the regulation of microorganisms through selective grazing (HANLON & ANDERSON, 1979). The test was developed mainly for environmental chemicals, but it can be used just as well for pesticides, although its implementation is still in need of improvement (e.g. introduction of a preliminary test; RÖMBKE et al., 1993). As in the earthworm tests, the use of artificial soil should be reconsidered, although *F. candida*, unlike *E. fetida*, can be kept and bred without any problems, even in sandy soil. Tests with species from other groups of soil organisms, especially arthropods (e.g. isopods, mites or myriapods) are currently being developed (overview in VAN STRAALEN & VAN GESTEL, 1993b).

Beneficial Organism Tests

"Beneficial organisms", a term which comes from the field of pest control, refers to organisms such as spiders or ladybugs which are direct antagonists to harmful organisms (HASSAN, 1985; BIGLER, 1989). The term is often limited to arthropods, the group with the highest number of species, as well as the most well-known beneficial and harmful species (VAN STRAALEN & VAN GESTEL, 1993b). Use of the term "beneficial organisms" has more recently come to include all non-targets which are considered to have a positive ecological function (e.g. earthworms, collembolans) and are thus of some benefit to crops (BOLLER et al., 1989). However, this definition has never gained currency (e.g. HÖGGER et al., 1992; HUANG, 1992).

Tests employing the honey-bee (*Apis mellifica*), which were developed back in the 1970s, are the predecessors of all other tests with beneficial arthropods. In many industrial countries, such as the United States or Germany, bee tests are required for pesticide registration (BBA, 1980, 1991b; EPA, 1985). The reason for this early consideration is the high economic importance of bees as pollinators for many crop cultures. Moreover, extensive knowledge was available on the breeding and keeping of these animals as a result of their use as honey-producers. There are four different tests which can be conducted with bees. They vary primarily in terms of the exposure pathway:

- inhalation, primarily for injected, sprayed or dusting agents;
- contact;
- through netting or dusting;
- ingestion.

The selection of the appropriate laboratory test as well as further testing in semi-field or field tests is specified in a very complex flow scheme of procedure (FELTON et al., 1986).

In the meantime, the working group on "Pesticides and Beneficial Arthropods" initiated in 1974 by the International Organization for Biological and Integrated Control of Noxious Animals and Plants (IOBC) has prepared about 30 laboratory methods in which a decline in "beneficial" characteristics such as feeding or parasitic activity is studied in addition to mortality. All IOBC procedures are explicitly tied into a three-tier test scheme comprising different stages which are referred to in terms which deviate from the usual definitions (the procedures referred to as "semi-field method" by the IOBC are in fact somewhat more elaborate laboratory tests). A few of the methods

Figure 8.10 Wolf-spider (*Pardosa sp.*) as a representative of beneficial organisms (courtesy of A. Hof).

developed by the IOBC have in the meantime been published as guidelines in national (e.g. German) or international (e.g. EPPO) contexts.

With the exception of bee studies, several tests with beneficial arthropods in Germany can be required for pesticide registration (BRASSE & ROTHERT, 1992). The registration authority selects individual methods from this battery of tests, depending on the pesticide to be registered, the crop, the application quantity and risk indication (BRASSE, 1990). Several of these tests are now in the process of being standardized, analogous to other guidelines, by various groups or institutions.

The goal is to specify a manageable number of tests that will be independent of the respective application culture/crop, which will make them easier to compare. The criteria for selecting the test species would be breeding/manageability, taxonomic diversity, consumer type (e.g. prey, parasite, etc.) and estimated degree of sensitivity on the basis of past experience. BARRETT (1992) and JEPSON (1993) have recommended four and seven species, respectively, including the carabid beetle *Poecilus cupreus* as a soil surface dweller and the wolf-spider *Pardosa sp.* (Figure 8.10). Gamasids (several species occur in specific strata of the soil or in vegetation) have also been mentioned because of their high "beneficial" effect, e.g. on spider mites, as well as their high sensitivity to pesticides (KARG, 1993).

The carabid test is the most well-known example of tests with beneficial arthropods involving predatory soil organisms (BBA, 1991a; HEIMBACH, 1992).

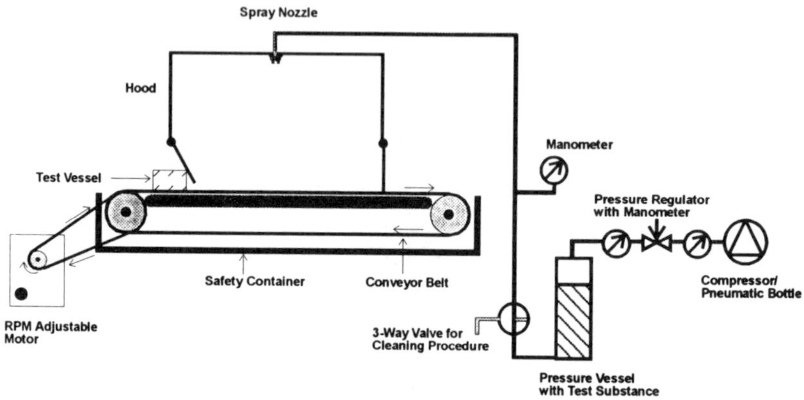

Figure 8.11 Schematic diagram of a test application device.

Adult carabids of the species *Poecilus cupreus* (very similar test recommendations exist for *Pterostichus melanarius* and *Bembidion lampros*, e.g. EDWARDS et al., 1984) on quartz sand are sprayed with the pesticide (Figure 8.11), after which the animals are monitored for two weeks with regard to mortality and changes in behavior and feeding capacity. Since the beetle larvae living in the soil are probably more likely to come into contact with the test substance, KEGEL (1989) developed a test for this stage of development, but it is still in the validation phase. A copepod test for the groundwater compartment is also still in the development stage (NOTENBOOM & BOESSENKOLL, 1993).

A reproduction test with beetles which, like the rove beetle *Aleochara bilineata* (Staphylinidae), live both as larvae and as adult animals in the soil for the most part yield more transferable information than the procedure described for carabids. Guideline proposals for tests with the *A. bilineata* species have been in the final stages for about two years now; they make use of mortality, egg production and hatching rate (SAMSOE-PETERSEN, 1987) or an entire generation cycle (NATON, 1989) as test parameters.

Nothing more will be said in this context about test methods involving organisms in the high canopy and the shrub layer or species employed as biological control agents in greenhouses, for example (see JEPSON, 1993). Methods involving large arthropods such as locusts or earwigs have already been proposed, but have not gained any acceptance to date (SCHMIDT, 1986; WALTON, 1989; SAUPHANOR et al., 1992). THEILING & CROFT (1988) provide an overview of the results in beneficial arthropod tests with pesticides.

Tests with Higher Vertebrates (Birds, Mammals)

Vertebrates are only rarely employed in tests to assess the ecotoxicological hazard potential of chemicals in the terrestrial medium. When environmental chemicals are approved (EC, 1979) or when seeking registration for pesticides in Germany (subject to OECD guidelines; BBA, 1990d), the results of toxicological laboratory experiments with rats and other mammals (mice, rabbits, guinea pigs) are primarily regarded as models for any possible hazard to human beings (SMITH, 1993). This is true with regard to both acute and chronic effects as well as mutagenic, teratogenic or cancerogenic effects. In the United States, tests with wild mammals may additionally be required (acute LD_{50} tests in most cases; SMITH, 1993) if laboratory experiments have indicated a need for further research, although such testing is subject to sharp restrictions based on animal-protection laws and wildlife preservation. The use of data obtained from experiments with rats has already been recommended for assessing the risk to large animals above ground (e.g. FREITAG et al., 1985), but this has never become established practice. Consequently, laboratory tests with mammals will not be discussed any further in the context of ecotoxicology.

Tests with birds, especially various species of quail (such as *Coturnix coturnix japonica*) as well as ducks (e.g. *Anas platyrhynchius*), are required in the various OECD, BBA and EPA guidelines. In the United States in particular, several acute (single-dose oral and dietary) and chronic (reproduction) tests with birds may be required under the pesticide law (FIFRA; EPA, 1982c), whereas in Germany only the acute toxicity (OECD, 1984) and, in the case of granulates and insect-attractants, uptake tests are necessary (GRAU et al., 1992; BBA, 1993). Acute and chronic bird tests are also required for the notification of environmental chemicals in Germany and the EC in the second stage (RUDOLPH, 1987); however, this is rarely ever the case in fact.

The repelling effect of chemicals can generally be used as a basis for a behavior test with an objective that is similar to the uptake test (KONONEN et al., 1986; BENNETT & SCHAFER, 1988). Since the performance of mortality tests is coming under a growing amount of criticism for animal protection reasons (especially if concentrations are employed which far exceed the exposure that can actually be anticipated), at least blood or tissue samples of dead animals should also be used to ascertain any pathological changes. For this reason, there is a strong tendency to employ sublethal parameters for standard tests (WALKER, 1993). These may take the form of biochemical methods such as the inhibition of the cholinesterase enzyme (MINEAU, 1991) or changes in the vitamin A content of the blood (BROUWER et al., 1990).

Generally speaking, none of the bird tests, which are being conducted in large numbers in the meantime (SCHAFER et al., 1983), focuses so much on the ecosystem effect of chemicals. They are carried out more in the interest of protecting endangered (animal) species, especially in view of the dramatic effects which chloro-organic insecticides have had on predatory birds (WALKER, 1993) and the possible accumulation of substances in animals used or consumed by humans.

Bioaccumulation Test

Despite their high value in all ecosystems as well as their far-reaching effects in food webs, there are no guidelines to date for the accumulation of chemicals in plants (OECD, 1989a; PHILLIPS, 1993). Although the literature contains a sizeable amount of individual data, most of it is not suitable for comparison, and the emphasis is on pesticides whereas environmental chemicals are largely ignored (NELLESSEN & FLETCHER, 1993). These data definitely indicate that pesticides may be taken up by plants and, in isolated cases, even provide evidence for the mobilization of bound residues (e.g. EBING et al., 1984). Guidelines for the evaluation of the residue behavior of pesticides in subsequent crops can serve as a point of departure for the development of such standard tests (BBA, 1988). At the same time, uptake rates, accumulation tissue in the plant as well as resulting effects can vary greatly from one species to another (SAUERBECK, 1983).

In contrast to the aquatic medium, there are no test guidelines for the bioaccumulation of pesticides or environmental chemicals in terrestrial organisms. They enter the organism primarily through the food supply and not from the surrounding medium (STREIT, 1992). Based on experiences from the establishment of environmental specimen banks as well as individual studies with pesticides, earthworms are probably the most suitable animals for a standard test due to their size, widespread distribution and easy handling (EDWARDS, 1984; EBING et al., 1984; CONNELL & MARKWELL, 1990). However, in deviation from the general rule, bioaccumulation in soft-skinned oligochaetes may largely depend on the concentration of a chemical substance in pore water (VAN GESTEL & MA, 1988). In this case, one would expect to find bioaccumulation factors similar to those which have been determined in fish tests. Then, however, a worst-case value of 10, as currently assumed in some countries for the assessment of hazard potential for worm-eating vertebrates (OECD, 1989a), seems too low. Moreover, large arthropods such as diplopods or isopods would be suitable for such tests, especially as these animals can concentrate huge amounts of heavy metals in their tissue (HOPKIN, 1989; DONKER, 1992).

9 SEMI-FIELD LEVEL

9.1 CLASSIFICATION

Semi-field tests cover a large number of methodological approaches (see Table 9.1) which combine the advantages of laboratory testing (scope for standardization and reproduction) and field tests (real conditions) in one form or another (METCALF, 1977; LEVIN et al., 1989b; HOFFMAN et al., 1990). In addition to the small number of methods designed specifically for testing purposes, there are many different semi-field studies which focus on isolated issues (MORGAN & KNACKER, 1994). One main problem involved with semi-field approaches is that the results of pesticide evaluations are difficult to generalize if the test design was sharply defined by local factors. Another — and even more frequent — problem to date has arisen if a test was, for example, especially designed for site evaluation to determine whether or to what extent the site should be protected. In contrast to Chapter 8, where the fate tests are classified by parameters, and the effect tests are divided by medium and organism group, the tests in this chapter are outlined in the order of the target level in question.

The classification proposed in Table 9.1 makes a distinction, initially without regard for size, between tests which are conducted under controlled environmental conditions (e.g. in a greenhouse) and tests which are carried out in the field, albeit in delimited areas. The procedures employed in the first group, which are often referred to as microcosms, tend to resemble laboratory tests. The procedures of the second group are, in turn, set off from the first by their size or scale. They are thus referred to as mesocosms, representing the transition to pure field studies (ODUM, 1984). Each of the two groups is subdivided further — into specialized techniques (i.e. focusing on one measurement parameter, be it for fate or effect) and integrated techniques, in which fate as well as effect parameters are studied. Another distinction can be made between closed systems (no interaction with surrounding environment possible) and open systems (exposed to the environmental conditions).

Tests under controlled conditions can be carried out with ecosystem segments which are either taken directly from the field (= undisturbed) or artificially

Table 9.1 Classification of Various Semi-Field Tests

1. Model ecosystem segments (= "microcosms") *Natural or artificially assembled units; a few centimeters in size — up to approx. 1 m^3 (contents up to a few hundred liters); closed and open systems are both possible.*	
Specialized techniques:	e.g. the plant metabolism box of the NATEC (FIGGE, 1992) or small "artificial streams" (CLEMENTS et al., 1989).
Integrated techniques:	e.g. the Terrestrial Model Ecosystem (TME) (VAN VORIS et al., 1984; KNACKER et al., 1990, 1991) or the Standardized Aquatic Microcosm (SAM) (TAUB et al., 1986; EPA, 1987).
2. Ecosystem segments in the field (= "mesocosms") *Field segments which remain exposed to normal environmental conditions; various sizes ranging between 1 m^3 and several hundred m^3.*	
Specialized techniques:	Partial enclosures in lakes or rivers, e.g. plastic bags with algae coenoses (EIDE et al., 1979). Lysimeter (usually about 1 m3 in size), e.g. tests on the mobility of pesticides in natural soil cores (e.g. BBA, 1990a). Semi-field tests (usually tests with beneficial organisms), e.g. effects of pesticides on ground beetles (carabids) in cultivated soil system segments (ABEL & HEIMBACH, 1992).
Integrated techniques (very rarely in the terrestrial medium):	Artificial testing systems, e.g. "artificial streams" — reconstructions of real streams including sediment (EATON et al., 1985). Natural enclosures, e.g. "Bremerhaven-Caissons" in wadden seas (FARKE et al., 1984).

assembled (= gnotobiotic). Based on studies of plants in multi-species systems, ELLENBERG (1984) called for semi-field tests, especially microcosms, *not* to be put together artificially but instead to be taken from nature intact. NIEDERLEHNER et al. (1990), on the other hand, found a positive correlation between an artificially constructed aquatic microcosm and a field test. In other words, which type of test should be recommended in a given case will most likely and largely depend on the question to be investigated.

In addition to the approach presented here, which is based on measurement parameters, a very large number of studies on the aquatic medium can also be classified by size (SETAC, 1992):

Outdoor Microcosms:

Tanks or ponds < 15 m^3 of water volume or artificial streams < 15 meters long, e.g. large plastic bags which separate a body of water with its sediment from its environment in a natural body of water (SOLOMON et al., 1986; BRAZNER et al., 1989).

Mesocosms:

Tanks or ponds > 15 m^3 water volume or artificial streams > 15 meters long, e.g. whole streams which are channeled into a laboratory building or are "reconstructed" at such a site (e.g. PRITCHARD, 1982; WEVERS et al., 1988).

A detailed presentation of this classification and an in-depth discussion of the methodological aspects involved in carrying them out as well as the interpretation of aquatic semi-field tests can be found in HILL et al. (1993).

The data from various aquatic semi-field systems have been shown to be comparable with data on bodies of water in the field (e.g. GUHL, 1991). Most recently, examples have been found indicating that terrestrial ecosystem segments also yield data which can be directly applied to the field (RÖMBKE et al., 1994).

9.2 MODEL ECOSYSTEM SEGMENTS (= MICROCOSMS)

Model ecosystem segments or microcosms are understood to mean systems which can be controlled and reproduced (in the laboratory, greenhouse or even in climate chambers) to simulate the processes and interactions of the chosen segment of the environment (GIESY, 1980; MORGAN & KNACKER, 1994). In recent years, interest has shifted from the fate to the effect aspect (e.g. HAMILTON & DINDAL, 1989; MORGAN & KNACKER, 1994). Analysis of two biological levels of organization as well as the presence of ecosystemic functions are regarded as a criterion for such (open or closed) model ecosystems. Approaches in which several effect assessment parameters (if possible, structural and functional parameters) can be investigated in parallel, e.g. microbiological as well as zoological, are especially interesting in this respect (HENSCHKE et al., 1989; VAN WENSEM et al., 1991). Unfortunately, the results of most of the methods employed thus far are difficult to compare due to differences in design, mode of application and the question under investigation. Guidelines or largely accepted proposals for procedures exist for both the aquatic (e.g. SETAC, 1991) and the terrestrial medium (e.g. ASTM, 1987).

Specialized Techniques

An attempt has been made since the 1970s to develop ecosystem models which can be used to study dispersion processes under standardized conditions, especially degradation, metabolization and (in the terrestrial medium) leaching (e.g. DRAGGAN, 1976; AUSMUS et al., 1978; GILLETT & WITT, 1980; RODGERS et al., 1980; LEFFLER, 1981; SCHUPHAN, 1985). Since the early 1980s, a large number of such fate-oriented techniques have also been described for the marine compartment of the environment (CLARK & CRIPE, 1993); some of them contained biological elements (e.g. regarding the influence of marine worms of the species *Clymenella torquata* on oil degradation; KOETERING-WALKER & BUCK, 1989). However, most of these small test systems ("ecocores") are more reminiscent of laboratory degradation or biodegradation tests than of microcosms.

One marine test system in the form of a sand-filled aquarium is oriented strictly to effect parameters. In this system, the sensitivity of various animal species to several pesticides was generally at least as high as in acute single-species tests (HANSEN & TAGATZ, 1980). Sensitive species in laboratory tests and sensitive species in the microcosm test often come from the same taxonomic groups (HANSEN, 1989). When the system was compared with the development of benthic coenoses in the field, fewer species but a higher number of individuals were found in the microcosms (TAGATZ & DEANS, 1983).

More recent approaches, such as the closed-system design, take the volatility of pesticides into special consideration when investigating the terrestrial medium (NASH, 1983; BÖHNKE et al., 1989). This technique makes it possible to produce the mass balance of a substance through residue analysis (FIGGE et al., 1985; SCHUPHAN, 1985; GILLETT, 1989), especially if radiolabeled substances are employed (FÜHR et al., 1991). Ecosystem segments are particularly suitable if, like agricultural land, they present a comparatively simple structure (NASH, 1983). Data from such "microagroecosystems" show a particularly good correlation with degradation data from field tests.

Integrated Techniques (Aquatic Medium)

A large number of microcosm tests have been developed in the aquatic medium. Three of them — the SAM, the MFM and the PFU — are worth mentioning in particular. The SAM (Standardized Aquatic Microcosm) system developed back in the 1970s is frequently employed, and a guideline has been formulated for its use in the meantime (TAUB, 1976; EPA, 1987). In spite of the small size (3 liters), good correlations were found to exist between the parameters studied (e.g. abundance and dominance structure of algae and invertebrate populations; nutritional dynamics and fate of the chemical) and pond and field studies (CAIRNS & CHERRY, 1993). The same performance was achieved with the even smaller (1 liter) and simpler MFM (Mixed Flask Microcosm) system, which some authors prefer to the SAM system due to its simplicity and easier interpretation (SHANNON et al., 1986). A similar degree of standardization was achieved with the somewhat larger-scale (7 liters) PFU or "artificial substrate community" test (CAIRNS et al., 1985). This test makes it possible to measure a large number of biological parameters including highly integrative, functional ecosystem parameters in addition to the fate of the chemical. Field effects were forecast with this test procedure (e.g. the effect of effluents on stream organisms; PONTASCH et al., 1989) in some studies, but not in others (PRATT et al., 1989). Based on the experience to date, the number of protozoan species has proved to be the most sensitive parameter in relation to chemical stress, which may also be one explanation for the test's low dissemination: specialists for protozoan taxonomy are few and far between. Very small, simple 2-species systems which primarily examine interactions between predators, prey or competitors under stress have not met with any success (e.g. HANSEN, 1981).

Another approach to the study of pesticides in microcosms was developed by KERSTING (1984): the "microecosystem" (a total of 7.5 liters) consists of three containers that are linked to each other and filled with algae (autotrophic system), *Daphnia* (herbivorous subsystem) and a sand/bacteria mixture (decomposer subsystem), respectively. The measurement parameters were algae growth, *Daphnia* abundance, the concentration of chemicals as well as pH-value and oxygen content. In addition to effects that could be predicted with acute and chronic laboratory experiments, this approach was able to observe ecosystemic changes (e.g. in oxygen content) which no laboratory test would have been able to predict (KERSTING & VAN WIJNGAARDEN, 1992). Generally speaking, however, it must be noted that, in spite of many years of experience, no standardized microcosm method has been established for registration purposes to date.

In the marine compartment, there are also many testing recommendations, most of which simulate neritic zones near the coast and thus primarily make use of the development of benthic fauna as test parameters (CLARK & CRIPE, 1993). Marine microcosms are currently not required anywhere, other than in the United States, to assess the efficiency and ecotoxicological hazard of bioremediation products after oil spills (NETAC, 1991).

One interesting, very complex system makes use of the seaweed *Thalassia* including its epiphytes in 7-liter containers, in which the development of fauna as well as the degradation of organic matter can be investigated in addition to diverse physiological parameters (e.g. the chlorophyll content in the plant). In other words, structural and functional parameters are employed on a parallel basis (MORTON et al., 1986). Comparisons of various parameters with measurements in the field revealed differences in the macro-invertebrate coenosis, but no adverse effects on the plants, at least over a period of six weeks (CLARK & MACAULEY, 1990). In the same system, the fate of radiolabeled test substances, such as Tributyltin, was studied in addition to biological parameters (LEVINE et al., 1990). Another example is related to the fate of Methylparathion and three environmental chemicals and their effects on populations of microbes or microbial enzymes in small (2-3 liter) containers (PORTIER, 1985). An integrative approach of this type is much rarer in marine microcosms than it is in limnic test systems.

Although larger units (approx. 0.1-13 m^3) are referred to as mesocosms in a summary compiled by CLARK & CRIPE (1993), they are included here because they are systems reconstructed on land with a relatively high degree of control over the environmental conditions (e.g. artificial lighting). In most of these studies, the MERL system being the one most frequently employed (GEARING, 1989), the fate and effect of environmental chemicals (especially petroleum products) on benthic and plankton organisms were examined in parallel (CLARK & CRIPE, 1993). The validation of this MERL system was especially successful in a comparison of the fate and effect on phyto- and zooplankton after exposure to oil residues in the microcosm and in the field (Baltic Sea, ELMGREN & FRITHSEN, 1982). Another test, in which the fate and effects of Atrazine (on aquatic plants, for example) were studied in an hierarchically arranged series of micro- and mesocosms (0.07-500 m^3) (KEMP et al., 1980, 1982), is also worth mentioning.

Integrated Techniques (Terrestrial Medium)

Integrative approaches in which fate and effect parameters are studied in parallel are rare in the terrestrial medium (e.g. BENGTSSON et al., 1988; VAN WENSEM, 1989). Effect parameters can also be examined in a few closed systems which were originally designed for fate tests including bioaccumulation studies (e.g. HEISE et al., 1988). An ASTM and EPA guideline focusing on the environmental fate of chemicals in intact soil columns (VAN VORIS et al., 1984) has existed ever since 1987. Nonetheless, it has hardly been applied despite considerable advances related to the effect parameters and their validation through field tests (KNACKER et al., 1990; KNACKER et al., 1991). There are plans to formulate this testing recommendation (TME method; Figures 9.1 and 9.2) in Germany as a guideline for the assessment of environmental chemicals (MORGAN & KNACKER, 1993).

In a much smaller system which was used to assess the impact of the fungicide Triphenyltinhydroxide, VAN WENSEM et al. (1991) showed that, when isopods were added to this artificial sand/tree litter system, effects of the chemical on litter decomposition and nitrogen cycling became recognizable which had *not* been noticeable in single-species tests with isopods alone or in straight degradation tests.

9.3 ECOSYSTEM SEGMENTS IN THE FIELD

In the interest of improving the interpretation of laboratory test results, an attempt to study the fate and effects of chemicals under field-like conditions was already made at an early stage of ecotoxicology. The focus of this interest, especially in the registration process, was aimed at tests on the degradation and, in terrestrial systems, on the mobility of pesticides, since the overriding influ-

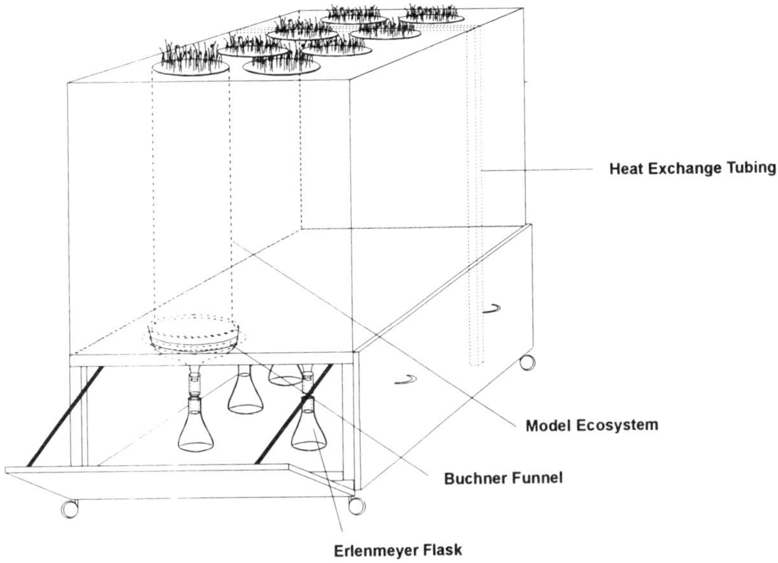

Figure 9.1 Schematic diagram of the TME method.

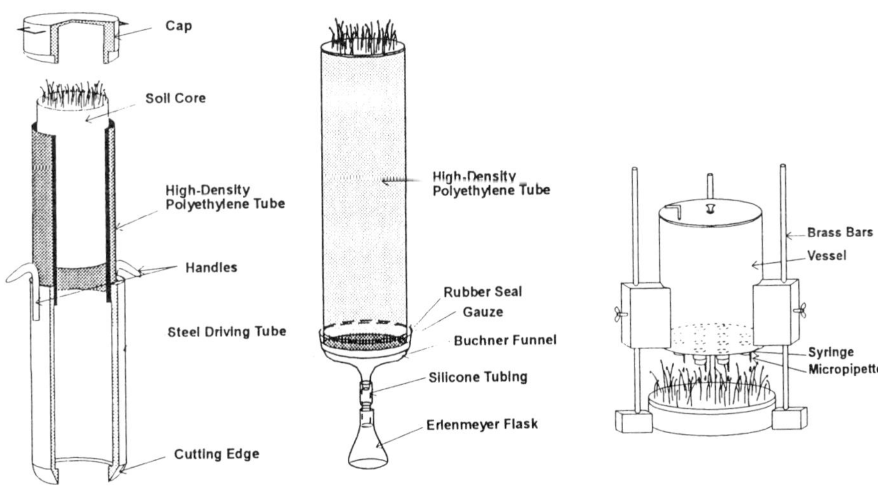

Figure 9.2 Detail diagram of the TME method.

ence of environmental and substrate conditions is obvious here. A major advantage of this method is that potentially dangerous substances can be tested without any risk of environmental contamination under real climatic conditions, a fact which is particularly relevant for the testing of genetically engineered organisms (GEMs; FREDERICKSON et al., 1990).

Specialized Techniques

There are so many possible testing procedures in the aquatic medium that it is difficult to cover them in an overview (SOLOMON & LIEBER, 1988; CAIRNS & CHERRY, 1993). In the simplest case, for example, unstressed animals or algae from laboratory cultures can be exposed to the water examined in the field, either in small cages or porous bags in rivers or lakes (EIDE et al., 1979; SOLBE, 1993). Similar tests were conducted in the marine compartment in a framework polluted with a pesticide and extrapolated for field conditions targeted for colonization by the widest variety of invertebrates (TAGATZ et al., 1987).

In the terrestrial medium, special techniques for the investigation of mobility are often referred to as lysimeters (LEAKE, 1991). Guidelines (EPA, 1982b; BBA, 1990a) have been available for these tests since the early 1980s if not before, as well as more precise technical specifications (JANCZYK, 1983). Virtually all of the known mobility tests which make use of soil columns (even on the very small laboratory scale) to assess chemicals can be referred to as lysimeters. In this context, however, lysimeters are defined more precisely as large (usually about 1 m^3), natural soil cores which are set up in the field and can be used to examine the leaching of a substance and its metabolites under realistic conditions, even over a period of several years (Figure 9.3; HERZEL & SCHMIDT, 1979; FÜHR et al., 1991; LEAKE, 1991). The fact that some of these soil cores are weatherproofed or transported to special facilities to alleviate the problem of leaching is irrelevant in this context.

Lysimeter tests play an important role in the validation of models on leaching in particular (BERGSTROM, 1990; KLEIN, 1991; KÖRDEL et al., 1991). Conversely, complex models were also successfully consulted for the interpretation of lysimeter results (ECETOC, 1992). It is important to bear in mind that, compared to the real situation in the field, such studies can only be considered representative to a limited extent, given the restricted scope (high costs!) for testing. In England, for example, the soils used in lysimeter studies *without* impermeable layers (i.e. excluding a lateral flow) cover only about 30-40% of the land used for agricultural purposes (HOLLIS & CARTER, 1990).

Although definitely comparable to the lysimeters in terms of their surface area (not volume!), the approximately 10 different semi-field tests performed with beneficial organisms are much more heterogeneous (HASSAN, 1985, 1992). The tent tests with honey bees are probably the most well-known, in which approximately 3 × 4 × 2 m gauze nets are stretched across plants in bloom that have been treated with the test substance. The measurement parameters are mortality, the analysis of flight activity, the bees' behavior as well as brood development (BBA, 1991c; FELTON et al., 1986).

In all of these tests, the parameters vary significantly in terms of duration and proximity to the field, depending on the biology of the species to be tested (JEPSON & MEAD-BRIGGS, 1992a). The carabid test (Figure 9.4), which is

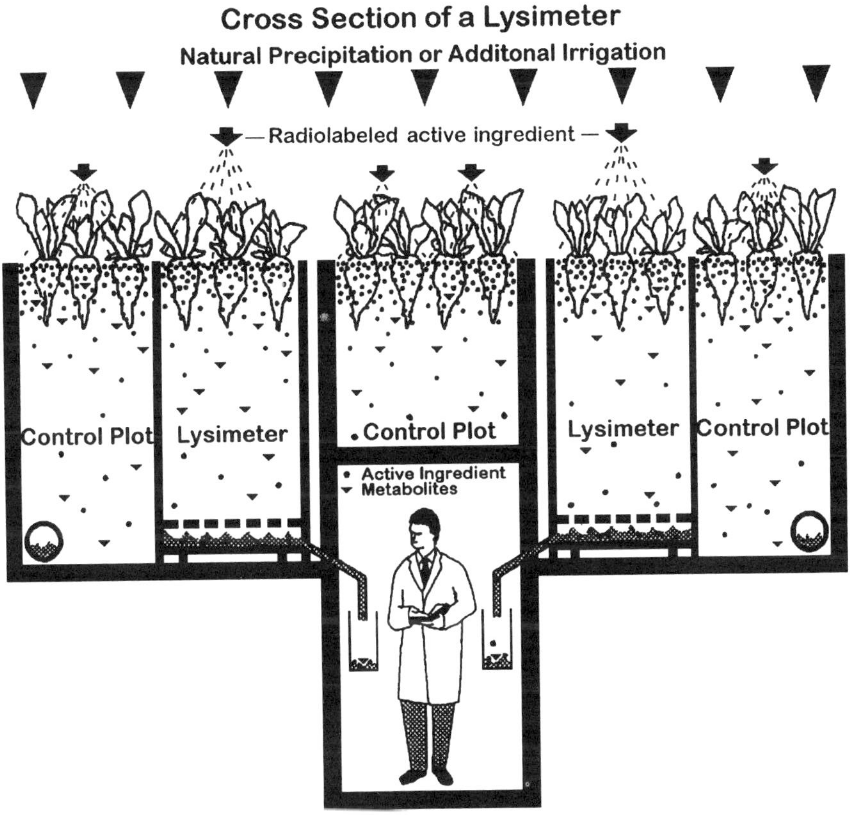

Figure 9.3 Schematic diagram of a field lysimeter.

to be defined as a BBA guideline in the near future, is presented here as a good example of short-term tests with organisms exposed to realistic conditions (ABEL & HEIMBACH, 1992). In this test, 6-10 ground beetles in a 0.25 m^2 steel frame driven 15 cm deep into the ground in a wheatfield, for example, are exposed to normal environmental conditions for a period of two weeks. The chemical to be tested is applied in accordance with agricultural practice. EVERTS et al. (1989) made use of various spider species in semi-field cages as part of a similar approach.

Although the author referred to the procedure as a field test, restricted field areas of about 100 m^2 in which as many species as possible are to be analyzed with various methods for the effects of pesticides on beneficial arthropods would also come under the category of "special mesocosms" as we have defined it here (JEPSON, 1993).

Another possibility is offered by earthworm tests for the study of contaminated sites in the framework of the American SUPERFUND program, in which

Figure 9.4 Carabid semi-field cages.

individuals of the species *Lumbricus terrestris* are exposed to field conditions in relatively small, restricted containers (CALLAHAN et al., 1991). The worms in these tests are used as combined reaction and accumulation indicators.

Integrated Techniques

As a forerunner of today's techniques, the fate and effect of heavy metals and oil products were studied in the 1970s with water columns (some of them including sediment) enclosed by plastic bags (BEERS et al., 1977). Generally speaking, the well-known approaches can be divided into three groups:

- "artificial streams" (e.g. EATON et al., 1985)
- "artificial ponds" (e.g. CROSSLAND & WOLF, 1985)
- "enclosures of natural systems", e.g. sectioned off in natural lakes or in stationary wadden seas ("Bremerhaven"-Caissons, FARKE et al., 1984; SOLOMON et al., 1986; BRAZNER et al., 1989)

In studies on the influence of realistic herbicide concentrations on plankton coenoses in a Canadian lake, plastic bags 5 × 5 × 5 m in size were used to show that the substance can have both direct and indirect effects on the zooplankton (THOMPSON et al., 1993). Guideline proposals as well as a large number of test results (usually on pesticides) exist for artificial ponds in particular (TOUART, 1988). Due to the size of these test systems (approx. 10,000 to 900,000 liters), all of them were carried out with relatively low numbers of replicates (2 to 4), and the testing period usually ranged between four and five months. The focus of most studies was the analysis of invertebrate coenoses and, to a somewhat lesser extent, algae coenoses.

In the aquatic medium, and not only in the actual mesocosms, various parameters are generally studied, i.e. effect and fate studies are integrated. A large number of mesocosm studies were conducted in conjunction with pesticide registration in the United States during the 1980s, particularly at the EPA's request (often in very large ponds with more than 400 m^2 surface area). However, current discussion is aimed at smaller systems with multiple replicates in order to improve the cost-benefit ratio. In other countries, studies of this type are only required in exceptional cases.

There are no known examples of this category of semi-field tests (integrated mesocosms) for the terrestrial medium.

10 FIELD LEVEL

10.1 GENERAL OBSERVATIONS

The ecosystem in its natural context is the most important level for ecotoxicological issues (KIMBALL & LEVIN, 1985), although tests at the population level are obviously easier to conduct (MORIARTY, 1988). The problem is that every ecosystem is extremely complex, with multifarious reciprocal relationships which cannot be fully understood, even in the absence of pollution. Nor is it possible to predict exactly how an ecosystem will behave (MORIARTY, 1988; CAIRNS & PRATT, 1989). This also makes it impossible to establish any absolute threshold values for pollution in ecosystems, since an ecosystem's sensitivity to chemicals can change dramatically with time. For the same reasons, the number of routinely required test procedures for ecosystems, especially according to guidelines, has generally remained extremely low to the present day. In contrast, the number of field studies devoted to examining the fate or effect of different chemicals at various sites with respectively individual methods and priorities is so great as to be almost unmanageable.

Furthermore, the lack of ecological base data poses another major constraint to field-testing. It is not known what many ecosystems look like in their normal state (standard), which makes it difficult to determine the range of natural variation (TORSTENSSON, 1988) or quality criteria (FÖRSTNER et al., 1987). Since it is not possible to analyze ecosystems in their entirety, certain physiological, chemical and biotic structures and related interactions must be selected from the wide variety of all those which exist and must be defined (= operationalized) in order to describe the normal state of an ecosystem and any potential deviations from it (FRÄNZLE, 1983). The segment of an ecosystem which is selected and quantifiable in this way is then referred to as the "reference ecosystem" (MATHES et al., 1991), which can serve as the standard of comparison for pollution situations at other sites with the same type of ecosystem. Only chemical effects which exhibit distinct variations in terms of their intensity or the duration of natural variation (e.g. in relation to an unpolluted control surface or a median value over several years) can also be recognized as such and, in turn, evaluated (SCHAEFFER et al., 1990). This is particularly true with regard to microorganisms, whose density may fluctuate by as much as 90% due to natural stress factors alone (DOMSCH et al., 1983).

The difficulties just mentioned are related not least of all to the fact that there are only a few field projects in which entire ecosystems were studied in quantitative terms and over longer periods of time (e.g. ELLENBERG et al., 1986; BECK et al., 1988). In addition, most ecosystems have been strongly influenced anthropogenically in the course of recent decades or have simply not yet been examined. This is especially true of tropical habitats, where approximately 1 million species have been described to date, but as many as 20 million species are presumed to exist (ADIS, 1989; GROOMBRIDGE, 1992). The general lack of basic taxonomic research is particularly deplorable (HERMAN & HEIP, 1988). It is also difficult to manage the interaction of many different kinds of anthropogenic influences. The depletion of species in central European farming regions is a case in point, where the effects of pesticides are compounded by many other factors (air pollution, land-zoning, less crop rotation; KÖNIG, 1985).

In all probability, the standardizability of field tests is subject to such severe constraints that, for all practical purposes, studies can only be conducted on a case-by-case basis; in other words, the scope of the examination in each case has to be agreed to by the respective producer and the environmental authority (KLEIN et al., 1988). Nonetheless, in spite of all the difficulties involved, ecosystem studies will continue to play a role in validating laboratory and semi-field test results, since the complexity of an ecosystem cannot be adequately simulated at other levels (PERRY & TROELSTRUP, 1988).

10.2 EXAMPLES OF FIELD STUDIES

Only a few years ago, field tests with pesticides were conducted by applying a substance and counting the obvious signs of damage, such as dead fish floating on top of the water in streams or dead birds in fruit orchards (HARDY, 1987). Using different studies of pesticide effects on soil fauna as an example, BOSTROM & LOFS-HOLMIN (1987) were able to show how little the earlier studies adhered to fundamental rules of research, such as reproducibility, for instance. As already pointed out at the beginning of Chapter 9, there are a large number of field-testing methods. However, field studies are not always the same as "ecosystem studies".

Even today, the incidence of drastic damages to larger animals is still a reason for the precise planning of field studies (e.g. for the localization of an emission source, SOLBE, 1993). In England, for example, cadavers of vertebrates that are found by chance undergo pathological analysis or residue analysis. It is also possible to estimate the size of populations of wild vertebrates in polluted and unpolluted regions (e.g. BUNYAN et al., 1981). This method has been modified in the meantime such that wild catches (e.g. field mice) are brought into the laboratory from polluted areas to compare their reactions with test organisms raised in the laboratory (including biochemical methods; e.g. WALKER, 1988).

Generally speaking, however, it is difficult to analyze the population dynamics of wild vertebrates, especially small mammals (COX & SMITH, 1990). This is why MINEAU & PEAKALL (1987) suggested that enzymatic procedures with relatively few animals are not worthwhile for the analysis of effects on birds after the application of insecticides in forests. Instead, costly, long-term procedures employing mark-and-recapture techniques or the measurement of reproduction parameters should be carried out, which could be used to assess the effect of the substances on population level. Both approaches have much in common with monitoring programs (see Chapter 10.3). Residue analysis and statistical methods are also very suitable within the framework of these programs due to the known applied amount and the long testing period.

A few of the procedures with vertebrates have been codified in the framework of EPA guidelines (EPA, 1982c). However, due to the large number of sites and vast areas which are necessary for the statistical accuracy of such tests, not to mention the difficulties of interpretation, the EPA itself has recently questioned the need for these tests. As an alternative, SOMERVILLE & WALKER (1990) have proposed more intensive probing of fewer sites, taking as many species as possible into consideration. GREIG-SMITH (1990b) provides a detailed description of both approaches, including practical experience with the "intensive" methods.

Otherwise, due to the complexity of the subject of investigation and the resulting expense (time and costs), only very few field tests are regularly required in the framework of registration procedures, and even fewer have been standardized (CAIRNS, 1989). In Germany, England and the United States, tests on the environmental fate (degradation, metabolism, also mobility in the U.S.) of pesticides (not environmental chemicals!) are required in the field, usually at several different sites, if a potential risk is indicated by laboratory results (BBA, 1986a; SMITH et al., 1990).

Biological effect tests in the field have only been required to date in the event of a presumed risk to honey bees (BBA, 1991c). A comparable test for earthworms has been developed in the meantime (BBA, 1994a, 1994b). In the bee tests that have been mandatory since the end of the 1970s, mortality, observations of the behavior (e.g. foraging in the crop) and reproduction (estimates of brood status in frames) are assessed over a period of four weeks after the application of a pesticide (FELTON et al., 1986). There are various recommended tests in the area of beneficial arthropod tests, most of which are still in the validation phase (GREAVES et al., 1988; SMART et al., 1989). In the meantime, a field method devised in the framework of the IOBC has been formulated as a guideline in England for the study of side-effects of pesticides on important predators, such as spiders, carabids, ground beetles and ladybugs, in grain crops (especially wheat) (WD7/7G, Ministry of Agriculture, Forestry and Fishery; JEPSON & MEAD-BRIGGS, 1992b). Common ecological field techniques, such as pitfall traps, are employed in this test. Special attention is devoted to the size of the test surfaces (more than 1 ha recommended), since effects determined for mobile organisms may soon be offset by recolonization if the test area is too small.

In most cases, the taxonomic classification of the animal content does not go beyond the family level, especially in view of the high numbers of catches and the difficulty of identifying them. The macrofauna are an exception. It is usually relatively easy to identify the species of ground beetles, for example.

In more recent, so-called integrated approaches, the effects of pesticides on beneficial organisms are studied parallel to laboratory tests by means of monitoring and field studies (LUFF et al., 1990), especially to compare conventional and ecological agromanagement techniques (STEVENSON & WALTERS, 1983; CROFT, 1990). The English Boxworth project is a noteworthy example. In this project, the effects of several pesticides on land areas under conventional and integrated management were studied with a variety of biological parameters over a period of five years (HARDY, 1986; GREIG-SMITH, 1990a).

For the aquatic sediments compartment, the need to integrate biological (laboratory/field) and also physico-chemical data was emphasized in particular by CHAPMAN (1989, 1990a,b). Parallel to the tests in mesocosms (artificial ponds specifically) which were called for to an increasing degree during the 1980s, a few long-term (2- to 8-year) field studies were conducted in which entire lakes were manipulated, for example, with regard to their pH value (e.g. BREZONIK et al., 1987; SCHINDLER et al., 1985). One of these experiments suggested that highly integrative functional parameters, such as decomposition or nutrient cycles, are less suitable as indicators of stress on the system than the structure of biocoenoses of small organisms with a rapid rate of reproduction (SCHINDLER et al., 1985).

In similarly elaborate field tests, an attempt has been made in Germany since the early 1980s to find indicators (= parameters) for the prospective assessment of ecosystem tolerance to stress (MATHES et al., 1991). Various terrestrial ecosystems (e.g. a beechwood soil or an urban ruderal site) were treated with chemicals and the reaction of the system was analyzed as comprehensively as possible (zoologically, botanically, microbiologically, chemically) in comparison to control sites (BECK et al., 1988; MATHES et al., 1991; Figures 10.1 and 10.2). Especially litter decomposition (e.g. with litter bags; Figure 10.3) or the feeding activity of the whole soil biocoenosis (e.g. with bait lamina tests (VON TÖRNE, 1990)) can be recommended as functioning methods for the ecosystem level. For many of the ecosystems studied in the process, a normal state (standard) first had to be defined (= reference ecosystem). As was to be expected, given the great heterogeneity of the ecosystems studied (incl. pre-existing damage) and the stress situations involved (chemical, frequency of application), it did not prove possible to identify any "universal" systemic indicators of chemical stress independent of the respective site (SCHLOSSER, 1988). On the other hand, the research did show that field

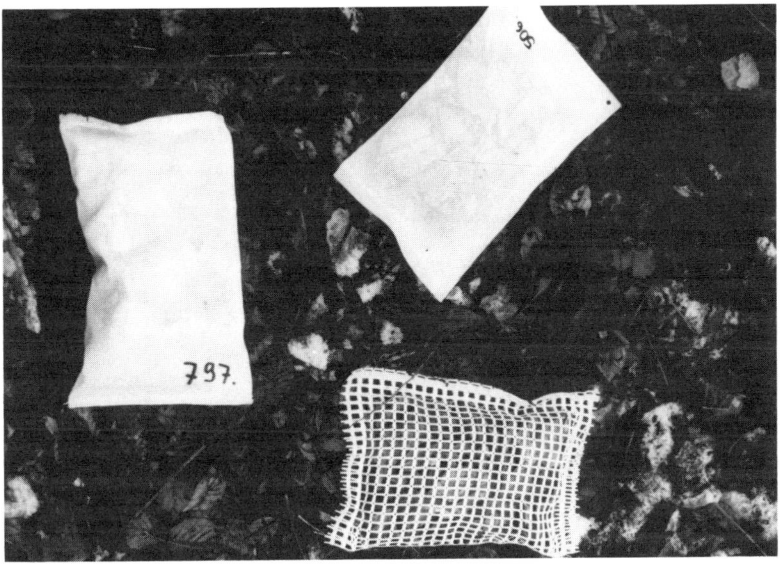

Figure 10.1 Test application of a pesticide.

Figure 10.2 Investigation site in a beechwood forest.

Figure 10.3 Litter bag method for the study of decomposition of organic matter.

studies are generally subject to the following conditions (MATHES et al., 1991):

- The spatial and temporal variability of ecosystem reactions is very high (= poor reproducibility); thus, chemical effects are extremely dependent on the respective climatic conditions, e.g. due to the different exposure situation.
- Due to this variability and effects which are often not manifested until after a long period of time, ecosystem studies must be long-term in design.
- Results from one type of ecosystem are difficult to transfer to other ecosystems, even if the ecosystems would appear to have certain aspects in common.
- A joint analysis of pollutant effects on species in terms of structural changes in communities of organisms (e.g. dominance shift; GRAEFE, 1993) *and* functional parameters (e.g. changes in litter decomposition) is possible. However, transitions between these two dimensions are flexible, especially if microorganisms are being studied.
- There is a considerable need for research with regard to analyzing the normal state of ecosystems as well as identifying additional measurement parameters ("ecotoxicological research is also ecosystems research"; MATHES et al., 1991).

With a similar goal in mind, HECK et al. (1992) and RÖMBKE et al. (1994) studied the suitability of soil fauna as indicators and the functional parameters in forest and grassland biotopes under anthropogenic stress. These studies assign an especially high priority to the simultaneous implementation of monitoring programs in the field, causal-analytical tests in semi-field conditions (e.g. microcosms), and in the laboratory. For obvious reasons, studies on this scale are not feasible if only for ethical reasons and certainly not in the framework of routine chemical experiments.

10.3 MONITORING STUDIES

10.3.1 Definition and Problems

Aside from the testing (analysis, assessment and resulting prognoses) of individual foreign substances, the task of ecotoxicology is to monitor the fate and effect of such substances directly in the environment. In addition to determining the "normal state", the purpose of this observation or monitoring activity is to verify the accuracy of previous forecasts and provide the basis for precautionary measures to safeguard against catastrophes. There is practically no place on earth where it could be said that the environment is not influenced by what are often large-scale human activities. This makes it difficult to distinguish the effect of additional sources of stress on the environment from other changes in the environment.

There are three different ways of monitoring the fate and effect of pollutants (MORIARTY, 1988):

- surveys, i.e. a measurement that is taken to record data only on a current state at a given point in time;
- surveillance, i.e. repeated measurements taken to record changes over periods of time;
- controls (monitoring per se), i.e. ongoing measurements for comparisons between a current state and a previously determined standard.

Whereas in the 1970s attention focused on the detection of chemicals in organisms, the issue has shifted in recent years, along with the discussion revolving around laboratory tests. Today, one tends to ask whether and in what way chemicals have an impact in the environment and how this impact should be recorded and analyzed. Put another way, the original emphasis on identifying accumulation indicators shifted to reaction indicators in the 1980s. An increasing effort is being made to merge the two aspects in the framework of a holistic assessment of ecotoxicological hazard potential.

One current question of interest is focused on whether the kind of information yielded by monitoring programs, e.g. on the concentration of a pesticide in an organism, can be obtained by different means — for example, by assessing accumulation values based on analytical measurements taken in the respective environmental compartment (HOPKIN, 1993a). In another relatively new approach, an effort is being made to determine whether monitoring programs can be used to identify populations which have developed a resistance to certain chemicals for genetic reasons — an extremely important aspect for pesticides (e.g. on the varying sensitivity of mosquitoes to organophosphate insecticides, DEVONSHIRE & FIELD, 1991).

Given sufficient resources, monitoring programs cannot only identify levels of pollutants which enter the environment but also determine their fate and changes over time as well as their biological effects. If only because of the vast number of potential pollutants and possible effects, not to mention the costly and time-consuming process of analyzing them, it is rarely possible to achieve or maintain an ideal state of knowledge on any one of them. In comparison to other environmental chemicals (with the exception of mass pollutants in the air such as SO_2) or pesticides, heavy metals have been the principal object of investigation.

In the planning, implementation and evaluation of monitoring programs, it is extremely important to take other environmental factors such as humidity, pH, salinity, etc. into consideration since they are often much more important for the occurrence, behavior or uptake of chemicals than the properties of the foreign substance itself (therefore, toxicity is a function of bioavailability). The selection of organisms employed as bioindicators has a major influence on the results of monitoring studies, as one can see from the following examples:

- closely related taxonomic species can be extremely different in terms of their reaction or bioaccumulation (e.g. various isopod species towards heavy metals; HOPKIN, 1990);
- an individual's physiological state, age or experience can definitely influence the uptake of substances (DEPLEDGE, 1990);
- in the case of invertebrates in particular, it is often not known what they feed on, which makes accumulation studies difficult to interpret. For example, evidence of a chemical in the body tissue of an animal is not, in and of itself, sufficient to pinpoint the source of emission unless at least the food preferences and mobility of this organism are also known.

Aside from these factors, the effects of which are predictable, at least to some extent, experience has shown that individuals of the same species may differ in the field with regard to their indicator characteristics, with no plausible explanation for it. As a consequence, it is hardly feasible to estimate concentrations of chemicals in organisms based on abiotic compartmental measurements (HOPKIN, 1993a). On the other hand, this means that, in addition to an average degree of stress, the spectrum of reactions or accumulations must also be known in order to assess the hazard potential of the relevant chemical.

Methods which directly record the chronic effect of foreign substances in the framework of monitoring programs are becoming increasingly prevalent (such as the "Scope for Growth" method with mussels, cf. Chapter 8.2.2), whereby the transition between semi-field and field tests is flexible. The most widespread of these methods is the exposure of animals in small cages in the field, although the animals are taken from the laboratory or from sites where the level of exposure is known. One example is the collection of data on the growth rate in small crustaceans (MALTBY et al., 1990); another is the change in the rate of respiration in trout (SEAGER & MALTBY, 1989). The problem involved is how to predict the long-term impact of the measured changes for the population as a whole.

In the meantime, fish (and water fleas) in particular are being used in the continuous monitoring of effluents in largely automated treatment plants, where, for example, the animals' swimming activity serves as a measurement parameter (PUZICHA & KNIE, 1994). Methods in which physiological parameters of fish can be measured with microelectronics without impairing the animals in any way (remote-sensing; DEPLEDGE & ANDERSEN, 1990) are even more future-oriented. Such methods enable sublethal parameters to be used directly for environmental monitoring.

10.3.2 Examples of Monitoring Studies

Monitoring studies cannot be systematized in the same way as tests. From the standpoint of design and implementation, tests are subject to much tighter, hands-on controls than monitoring studies. Therefore, a variety of monitoring studies are described in the following subsections according to the environmental medium in which they are conducted.

Aquatic Medium

So many different groups of animals have been employed in monitoring programs in the aquatic medium that they can hardly be counted. In the limnic compartment, plankton organisms (e.g. protozoans and small crustaceans), insect larvae (e.g. diptera), annelids (e.g. tubificids or leeches) and fish in particular are most frequently employed. The accumulation of chemicals is difficult to predict since it is often not known whether exposure takes place via the food supply or directly through the water (via respiratory organs or the body surface, VAN HATTUM et al., 1991).

The effects of foreign substances, especially in connection with waste-dumping and oil-drilling, were also examined in the marine compartment, focusing on sublitoral and benthic coenoses (e.g. CLARK, 1989). The most well-known example is no doubt the investigation of the effects of the extremely toxic anti-fouling agent Tributyltin (HOPKIN, 1993a). In addition to the macrobenthos, small sediment inhabitants, such as nematodes or annelids, were also used as monitoring organisms (BILYARD, 1987; GRAY et al., 1988; BELLAN et al., 1988). Molluscs became the most frequently employed indicator organisms — a testkit ("Mussel Monitor") was even developed for them and not only because they are consumed by human beings. In the meantime, so much data has been compiled on the mussel *Mytilus edulis* (e.g. BERGEN et al., 1993), which can be found around the world (but is said to involve a complex of three species; LOBEL et al., 1990), that "global mussel watch schemes" have been set up for periods of many years (BAYNE, 1989). Marine vertebrates have also been used as bioaccumulation indicators for a long time. Many foreign substances build up in the large fat deposits of penguins or dolphins, for example, (VETTER & LUCKAS, 1991), and stay there for many years, even after the source of contamination no longer exists, due to the absence of degradation enzymes (HOPKIN, 1993a).

As far as the use of meiofauna in marine monitoring studies is concerned, the question of whether a specification up to species level is necessary is still being intensively discussed (CRANSTON, 1990). The following example shows how difficult it is to answer this question: Various studies which took the same of group of organisms at species level (nematodes), and then aggregated them into higher taxonomic units (familiy level), resulted either in a considerable loss of information in doing so (WARWICK et al., 1990) or no change whatsoever from the original species-level findings (HERMAN & HEIP, 1988). Since, as already mentioned, even closely related species may exhibit sharply varying accumulation behavior, for example, a case is being made for retaining the species level in monitoring studies.

The use of whole groups of species as reaction indicators was formalized for the limnic and marine compartments (RAMADE, 1987a). Examples of such reaction indicator systems are the saprobe index developed nearly 100 years ago (KOLKWITZ & MARSSON, 1908, 1909) and related indices to determine the quality grade of bodies of water in which the occurrence of the widest variety of plants (especially one-cell algae) and animals (especially invertebrates such as protozoans, tubificids or insect larvae) are put in relation to a particular level of pollution (BAUR, 1987; METCALFE, 1989). This approach undoubtedly has a practical value, but the validity of the individual indices is always limited to the region examined. Moreover, it is rarely possible to correlate biological effects and results of chemical analyses with one another directly.

Today, the diversity indices, such as that of Shannon-Wiener, which are also frequently employed in the aquatic medium in the framework of monitoring programs, are not considered to be of much use for the reasons already mentioned (see Chapter 6.6.2). Nor can any of the biotic indices, which make various uses of the sensitivity differences in various species as measured or presumed to exist in the laboratory, be recommended (e.g. the relationship between *Assellus* sp. and *Gammarus sp.;* WHITEHURST, 1991) or the composition of the annelid coenosis (BELLAN et al., 1988). It cannot be assumed that the indicator species are equally sensitive to all foreign substances and their interactions.

Terrestrial Medium

The use of plants, e.g. grasses, moss, fungi (especially as indicators of radioactive substances) or lichens (ARNDT et al., 1987; BURTON, 1990) as active and passive pollutant indicators is widespread. The lichen indication will be discussed in more detail here, since it is a perfect example of the benefits and problems involved in the use of bioindicators. The evidence of damage to lichens from chemicals (e.g. flue gases), a fact which has been known since the mid-19th century (NYLANDER, 1866), has been utilized as an indicator for a vast number of studies, especially in urban scenarios, in which mappings of the effects on lichens were correlated with the content of air pollutants (SCHÖLLER, 1993). As of about the 1970s, moreover, a reverse approach was used to gauge the stress situation at the target site based on the occurrence of lichens. In one modification of this approach, (apparently) healthy lichens were transplanted from regions with pure air to sites for which levels of stress were to be assessed based on the reactions of the exposed lichens (BRODO, 1961). At the same time, laboratory tests determined that lichens manifested a high, but species-specific sensitivity to air pollutants, especially sulfur dioxide (TÜRK et al., 1974). Consequently, the observed effects on lichens were quantified using indices, the most well-known probably being the IAP (= Index of Atmospheric Purity), which has been modified many times (LE BLANC & DE SLOOVER, 1970).

Growth anomalies in lichens as an early indicator of forest damages are only one example of the many successful uses of lichens in the United States (SCOTT & HUTCHINSON, 1990). One problematic aspect, however, is that their occurrence was often correlated monocausally to the concentration or composition of air pollutants, without taking other ecological factors (e.g. microclimate) into account (SCHÖLLER, 1993). Another complication is posed by the fact that lichens are very complex organisms (being a symbiosis between fungi and algae, which are very different in terms of their nutrient physiology), whose biology and taxonomy entail many problems (JAHNS, 1989). Although the determined correlation of certain lichen species with the atmospheric stress in various regions is highly contradictory in some cases, lichen mapping has continued to serve as a standard procedure up to the present day, because it is simple to implement. In view of the problems mentioned, it is recommended that the monocausal model of lichen distribution (exclusively as a function of the air quality, especially the SO_2 concentration) and thus also the quantification be discarded. Instead, the diversity of influencing factors should be monitored parallel to the ongoing, especially long-term, lichen mapping, and the cause-effect relationships between the occurrence of these organisms and those factors should be researched (SCHÖLLER, 1993).

The conclusions drawn from the example of lichen utilization are also relevant for many other indicator systems. However, this does not mean that field-monitoring should be questioned as a whole. The point is to ensure that its implementation and interpretation are practiced with circumspection. The critical importance of monitoring programs to back the results of other tests is underscored by the fact that such programs are being required to an increasing degree for pesticides (and, it should be added, for environmental chemicals) even *after* they have been registered (or notified, as the case may be). Such programs are thus being implemented on a large scale (MATHES et al., 1991).

Very different species of animals can be used for monitoring, depending on the region, foreign substance and issue involved. In the terrestrial medium, for example, ciliates, earthworms, snails, isopods, gamasids, collembola, carabids, butterflies and small mammals, but also parts of larger animals (roe deer organs, bird feathers), are suitable (HOPKIN, 1993a). In the meantime, it has been possible in many cases (especially with heavy metals) to correlate effects on soil fauna (also small mammals) with a certain source of emission. However, a large database is available from long-term studies on the side-effects of pesticides in agricultural systems, especially with beneficial organisms (KÖNIG, 1985; JEPSON, 1993).

The use of earthworms as combined reaction and accumulation indicators to monitor wastes represents another interesting approach. Studies in this area of monitoring activity have relied on either the indigenous populations which (still) occur or animals bred in the laboratory and exposed in containers (CALLAHAN et al., 1991). Due to the extreme complexity of soil, it is difficult to compare the composition of species at a given site with that of a given standard. The only way to alleviate this problem is through long-term monitoring of organisms concurrently with specific abiotic factors, especially the climate.

11 LEGAL FOUNDATIONS

11.1 HISTORICAL OVERVIEW

Chapters 6 to 10 described the ways in which ecotoxicological properties of pesticides can be recorded and monitored, as well as the limitations of the various approaches employed. It is implicitly assumed that legislators are aware of the problems involved in applying the results of tests from lower levels to higher levels of biological systems. Moreover, they have to recognize that ecotoxicology harbors an element of uncertainty when it comes to assessing the ecosystem level, and therefore must rely on principles of the natural sciences. This has put lawmakers in an extremely difficult position since the structure and function of ecosystems are not easy to define in legal terms. As already emphasized at various points in this discussion, ecotoxicological methodology evolved from questions which lawmakers addressed to the field of ecology in search of an answer. In the meantime, ecotoxicology has developed a dynamics of its own and produced a store of information and knowledge that is useful for regulatory agencies. This not only means that ecological issues are generally given more consideration in all areas of society and in legislation. It has also led to the formulation of specific criteria which have to be met by new chemicals as well as by the so-called existing chemicals before they are put on the market or are allowed to remain there.

Pesticides have been subject to legal regulations for a relatively long time, as one can see from the historical overview of policy in the United States in Table 11.1 (COSTA, 1987; HARWELL, 1989; CROSBY & MINYARD, 1991; KOHN, 1991, JELLINEK & GRAY, 1992):

This overview shows that standards were defined by U.S. lawmakers during the 1970s — not least of all in response to pressure from the public — which later formed the basis for legislation in Western Europe (EU, Scandinavia, Switzerland, see SCHÜEPP et al., 1990, for example), and most recently, for some of the legislation in Eastern Europe as well (DAWSON, 1990). The United States, especially the EPA and to some extent also the FDA, has continued to function as a trendsetter in the elaboration of guidelines. In view of this fact, our discussion of legislation related to ecotoxicology will be limited to a peripheral outline of the situation in the United States, with the

Table 11.1 Historical Overview of Pesticide Policy in the United States

1910	Passage of the Federal Insecticide Act to prevent risks to human health.
1938	Passage of an addendum to the Pure Food Act of 1906 for the prevention of food contamination from pesticide residues.
1947	Revision of the 1910 law as the Federal Insecticide, Fungicide and Rodenticide Act (FIFRA); the U.S. Department of Agriculture (USDA) is responsible for enforcing it; the problem of residues in food is dealt with in another law (Federal Food, Drug and Cosmetic Act [FFDCA]).
1959	Expansion of the FIFRA's purview to other pesticides, such as nematicides.
1962	Publication of Rachel Carson's *Silent Spring*, which helped to trigger public discussion of environmental problems.
1969	First moratorium on the use of a pesticide (DDT).
1970	Establishment of the Environmental Protection Agency (EPA), henceforth responsible for pesticide registrations inter alia, instead of the USDA; and for stipulating tolerance levels in food (instead of the FDA); shortly thereafter, ban on alkyl mercury compounds in seed treatment.
1973	Ban on DDT; thereafter, increasingly frequent restrictions or bans on certain pesticides.
1975	Ban on Aldrin and Dieldrin; publication of EPA guidelines for the registration of pesticides in the United States; RPAR (Rebuttable Presumption Against Reregistration) is set up, a procedure requiring submission of evidence to dispel reasonable doubts before a pesticide can be reregistered.
1977	Passage of the Toxic Substance Control Act (TSCA) for environmental chemicals; no application of seed treatments containing mercury; the concept of Integrated Pest Control is established.
1978	Passage of further FIFRA revisions for pesticide registration and publication of a list of pesticides for limited application only.
1982	Publication of EPA guidelines supplementing the FIFRA (e.g. Pesticide Assessment Guidelines Subdivision E: Hazard Evaluation: Wildlife and Aquatic Organisms).
1985–88	EPA publishes TSCA implementation rules.
1988	FIFRA supplemented by agenda for reregistration of pesticides, which is to be completed by 1997 — it affects a total of 1,300 active ingredients (LENG, 1991).
1992	EPA adopts new criteria for data acceptability to expedite the reregistration process, which is way behind schedule (new final deadline: 2002).
1988–93	Growing trend towards adoption of additional requirements at state level (e.g. California's Green Plan).

main focus on specific sections of German legislation characterized by an especially high density of regulations (Figure 11.1). It should be noted, however, that the American fixation on the aquatic medium is proving to be

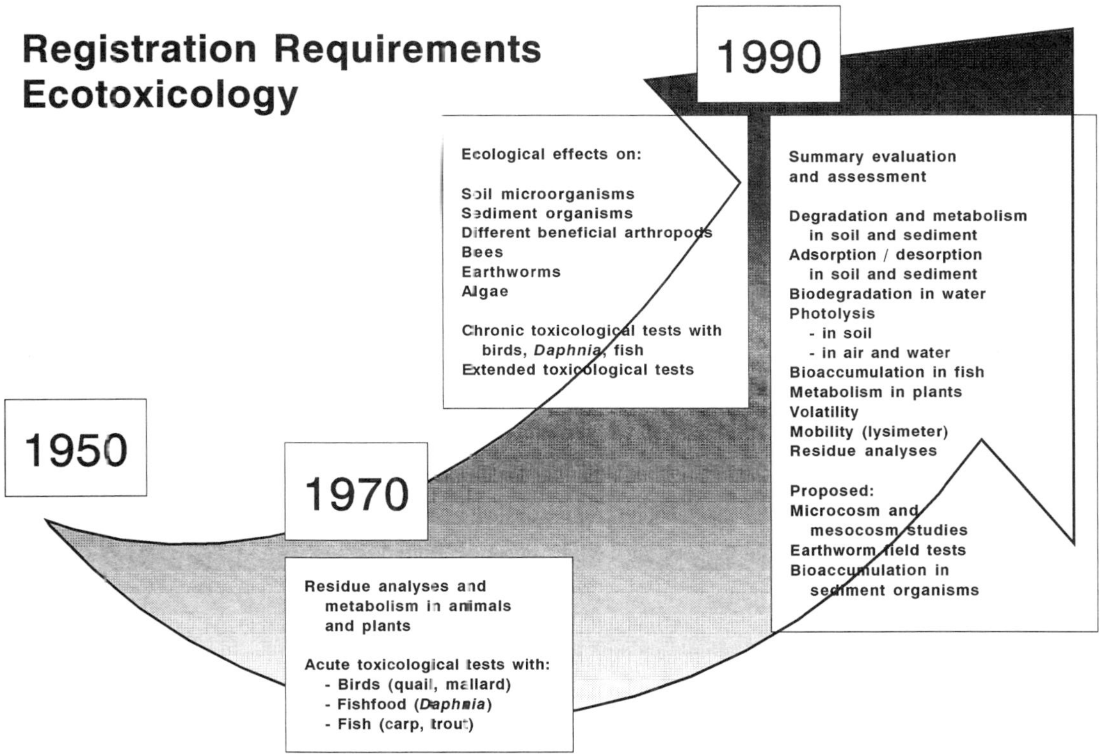

Figure 11.1 The expanding scope of ecotoxicological registration requirements in Germany from 1950 to 1990.

increasingly problematic. Compared with the European regulations, moreover, the U.S. registration process is very costly in terms of bureaucratic red-tape and paperwork (WALKER, 1990). At present, American activities are focused not so much on the drafting of guidelines but on the interpretation of data up to the risk assessment (NORTON et al., 1992).

Non-Governmental Organizations

A number of non-governmental organizations are working in close cooperation with international governmental authorities such as WHO or OECD to bring ecotoxicological issues to a broader public, also in the field of research, and to improve international cooperation (KAPPELER, 1979). The Scientific Committee on the Problems of the Environment (SCOPE) is one example. As part of the International Councils of Scientific Unions (ICSU), it organizes conferences and has published a number of basic works on the scientific foundations of ecotoxicological assessments (e.g. BUTLER, 1978, 1984; SHEEHAN et al., 1985; TARDIFF, 1992). Additionally, the Society of Environmental Technology and Chemistry (SETAC) has been working to promote ecotoxicological awareness ever since the early 1980s. However, links to the research sector, such as the pesticide chemistry section of the American Chemistry Society (ACS), have had a decisive influence on the design of test procedures and criteria due to their personal and scientific contacts with the EPA (KEARNEY, 1991).

Similar forms of cooperation also exist between international organizations and industry. One example is the European Chemical Industry Ecology and Toxicology Center (ECETOC) (founded in 1978), which speaks on behalf of European industry as a whole on questions concerning the environmental hazard from chemicals. ECETOC has published ecotoxicological methods, including the results of validation ring tests, in a series of Technical Reports. The Groupement International des Associations Nationales de Fabricants de Produits Agrochimiques (GIFAP) was founded as early as 1960 to represent the interests of 90% of pesticide producers worldwide (European section: ECPA = European Crop Protection Association).

These two institutions close to industry have been working actively in the area of methodology development, test standardization, training, etc. Specialized working groups deal with questions of labeling or resistance formation. GIFAP has also issued a large number of guidelines tailored specifically to conditions in the tropics and subtropics. They cover practically all aspects of pesticide use (e.g. transport, quality control, user protection), but ecotoxicological aspects are only mentioned in passing ("Guidelines for the safe and effective use of pesticides" — GIFAP, 1989).

One finds that the laws and guidelines on ecotoxicology which have been passed in the various countries or by international organizations have many things in common. All of them start with the problem of how to recognize, evaluate and forecast the possible damaging impact of pesticides or other chemicals in the environment. As shown in the preceding chapter for ecosystems, this objective can only be achieved at present — if at all — at a very high cost, a cost too high for routine practice. This set of circumstances led to the development of a tiered test program of ecotoxicological studies, initially in the individual industrialized countries and later in coordination with international institutions. These programs are very similar in terms of their basic foundations. The overall program approach to hazard and risk assessment (FAVA et al., 1987) is described in Chapter 12.

11.2 CHEMICALS LEGISLATION

At one time, every country had its own national chemicals legislation. In the meantime, however, these regulations have become increasingly harmonized in the framework of international committees in recent years. In the United States, for example (Toxic Substances Control Act — TSCA), an attempt was made to develop a special test scheme for each individual substance. Industry was opposed to this approach, since the costs that would be incurred for the registration of a chemical were almost impossible to estimate. In Europe, on the other hand, efforts on the part of individual countries as well as at EU level were already aimed at developing a tiered test program which would be the same for all substances, at least at the lower levels, with no need for special test standards unless there was a specific cause for doubt. A completely different approach was adopted in Japan, however: the persistence of a substance in the soil as well as the bioaccumulation of a substance in fish were regarded as key parameters for the classification of a chemical (KITANO, 1990)

Europe

EU Directive 92/32/EEC for the alignment of the laws, regulations and administrative provisions relating to the classification, packaging and labeling of dangerous substances is the leading guideline for chemicals legislation in Europe. It is the seventh revision, on 30 April 1992, of Directive 67/548/EEC; the sixth revision 84/449/EEC, on 18 September 1979, also provided the framework for the German Chemicals Act (ChemG — Law on Protection from Hazardous Substances, 14 March 1990). The essential components of the 92/32 directive are:

- mandatory notification, requiring an assessment of the substance's toxicity and ecotoxicity, correlated to the tonnage produced;
- labeling regulations;
- authorizations to impose bans and restrictions;
- standardized test methods and rules for the implementation of these tests in accordance with the principles of "Good Laboratory Practice" (GLP).

The actual evaluation approaches used to assess the hazard potential based on the information that has to be submitted for registration are not explicitly contained in the guidelines or legislation to date. Although they are still the subject of scientific discussion (e.g. PARLAR & ANGERHÖFER, 1991), general principles for the "assessment of risk to man and the environment of substances in accordance with Council Directive 67/548/EEC" were not enacted into law until 1993 (EC, 1993).

The laws applying to chemicals have emerged over the last 20 years in a process of close interaction with guidelines of international institutions. The most important of these are the OECD "Guidelines for Testing of Chemicals", which include the rules of "Good Laboratory Practice" (GLP). This set of rules and guidelines has been continuously expanded and refined since 1984, and they are recognized throughout most of the world today. Directive 88/320/EEC regulates the mutual recognition of GLP certifications.

The ecotoxicological methods printed in Appendix V, Part C of EU Directive 84/449/EEC are more or less equal in status to the OECD guidelines for the regions of the European Community. This so-called "Base Set" largely corresponds to tier 0 of the German Chemicals Act. After running into a few coordination problems, the OECD-EU rules have now been harmonized for the most part.

EU Directive 92/32/EEC (seventh amendment to Directive 67/548/EEC) extends the catalogue of methods under 84/449/EEC, with considerable improvements in usefulness. Other changes have been made with regard to the treatment of chemicals which are only produced in small quantities, attempts to avoid test duplications (e.g. for animal welfare reasons) and the introduction of a symbol to identify environmental hazard (Figure 11.2). As mentioned before, refinements and expansions of this set of rules during the 1990s will be aimed at risk assessment, i.e. the evaluation of test data.

In addition, standardized test guidelines as well as suggestions for generally binding definitions of terminology are presented in the International Standards Organization (ISO) (e.g. ISO, 1989; CAIRNS, 1989). Such standardized systems exist in Germany (Deutsche Industrienorm; DIN) and also in the U.S. (American Society for Testing and Materials, ASTM).

Germany

Under the German Chemicals Act (ChemG, 1989, redrafted on 14 March 1990) and in harmony with EU legislation, the number of ecotoxicological tests required for notification depends on the volume of substance produced, in addition to various chemical and human-toxicological information. At the base level, i.e. for virtually all substances, data must be supplied on biodegradability as well as on acute aquatic toxicity (e.g. *Daphnia* test). At the next higher level, i.e. as of a production volume of 100 metric tons per year (or 500 metric tons in total), tests on bioaccumulation, chronic aquatic toxicity as well as on the effect on algae, plants and soil organisms are required. Follow-ups — in other

Figure 11.2 Hazard symbols. The "N" for "dangerous for the environment" was not introduced until EU Directive 92/32 was enacted.

words, more complex tests of longer duration — are necessary, especially if the substance can enter the environment in larger quantities, is difficult to degrade or has a potentially toxic effect on aquatic organisms (LANGE, 1991).

From the scientific point of view, these standard laboratory tests are obviously not adequate for an assessment of environmental risk. Level 2 studies are not required until the annual volume reaches or exceeds 1,000 tonnes (or 5,000 metric tons in total) or if there are specific indications of hazard. Ecological considerations, such as proximity to the field, can be taken into consideration only in level 2 studies. However, level 2 has never been required to date, if only because many substances were registered shortly before the law went into effect. Up to 1982, for example, approximately 400 chemicals were registered per year whereas, from 1983 to 1991, only 600 substances in all were registered in the EU (BROWN, 1992). The tests in question are generally oriented to the corresponding OECD or EU guidelines, but have been slightly adapted or refined by the German Federal Environment Agency (Umweltbundesamt, UBA; see RUDOLPH & BOJE, 1986). The criteria for assessing the data submitted are updated in Germany at irregular intervals, in line with the state of ongoing discussion in research and practice (UBA, 1986a, 1990).

The chemicals laws of all European countries are accompanied by a large number of other laws and ordinances which essentially amount to a conversion of the various EU directives into national law. In Germany, for example, these are the symbols for hazardous substances (Figure 11.2 as well as R[isk] and S[afety] phrases), the emission protection ordinances, technical rules for storage and fire prevention, for transports of hazardous goods, maximum concentrations at the workplace, emergency ordinances, etc. In the broadest sense, the following domains are also affected by chemicals: environmental impact assessments, environmental liability, product liability, waste and water legislation, animal welfare and conservation of nature as well as laws governing export trading (e.g. the ban on exports of technology which could be used to produce chemical weapons).

United States of America

Comparable U.S. laws which include ecotoxicological aspects are as follows: the Toxic Substances Control Act (TSCA), which regulates the marketing of chemicals; the Clean Air Act (CAA), Resource Conservation and Recovery Act (RCRA), Safe Drinking Water Act (SDWA) and the Comprehensive Environmental Response, Compensation and Liability Act (CERCLA). Based on the CERCLA, the EPA's Office of Solid Waste and Emergency Response (OSWER) also forecasts the ecological impact of wastes, especially at disposal sites.

The Clean Water Act (CWA) is especially important. The EPA (Office of Water) is also responsible for these ecotoxicological studies. In addition to chemical-specific data, the toxicity of whole effluents is the principal focus. Of course, it is extremely difficult to examine these sources, which are quite variable in terms of time and space, or even to predict their ecosystem impacts. For this reason, the EPA's efforts are currently focused on defining criteria for the potential colonization of a body of water, including sediments. More specifically, "biocriteria" are to be found which define the water quality in quantitative terms via the structure and function of the respective biocoenosis, thus making it possible to evaluate even complex stressors (such as effluents with many individual components, BASCIETTO et al., 1990).

Independently of the respective law, the EPA has been developing ecotoxicological activities to an increasing degree in recent years, especially in the area of hazard and risk assessment and evaluation. As late as 1987, one of the EPA's technical advisory councils pointed out that the agency should be concentrating more strongly on general environmental aspects and less on the immediate protection of human health (quoted from ICU 5/90).

The Ecotoxicity Subcommittee was founded the same year. It is responsible for the development of criteria and guidelines in all areas under the EPA's jurisdiction. This subcommittee has, for example, created a master plan for ecotoxicological hazard/risk forecasts which is oriented to the level of biological systems. It can be used for "top-down" (i.e. starting with field studies) as well as "bottom-up" (i.e. starting with laboratory tests) approaches. In any case, however, it strives for an overall view which includes the ecosystem level (BASCIETTO et al., 1990). It must be supplemented by field monitoring of the substance's use, the results of which are included in the evaluation which forms part of the regular review for registration.

Existing Chemicals

The chemicals legislation of many countries — including the German Chemicals Law (Law on the Protection against Hazardous Substances, since 1990) — distinguishes between new chemicals which are coming out on the market for the first time and so-called existing chemical substances. In the EU area, there are about 100,000 such chemicals (defined in EU Directive 79/831/EEC as those which were in circulation before 18 September 1981), which are compiled in a separate register (EINECS = European Inventory of Existing Chemical Substances). The problems involved with these substances have to do with the fact that there are so many of them and because of serious gaps in information about them (OECD, 1986). In the past, follow-up studies on individual existing substances were conducted on the basis of random observation. Current EU-coordinated efforts based on international cooperation are now under way as an initial step aimed at systematic compilation of information on those substances which are produced or introduced in quantities of more than 1,000 tonnes per year.

Other than in Germany (Advisory Committee on Existing Chemicals of Environmental Relevance = Beratergremium für umweltrelevante Altstoffe, BUA), special committees have been created, at least in the United States (Interagency Testing Committee, ITC), Canada (the Canadian EPA's Priority Substances Program) and at EU level, which are designed to select those of the many substances already on the market which should be subjected to intensive testing due to reasonable doubt. Chemicals governed by separate regulations (e.g. pesticides), natural substances (e.g. amino acids) or inorganic compounds (e.g. heavy metals) would be excluded from consideration, as extensive collections of data often already exist for them.

The German BUA had initially selected 60 substances from the total of 4,554 considered environmentally relevant for priority examination. The goal was to obtain at least a set of base data on these substances as the foundation for an assessment. In the meantime, the German Federal Environment Agency has conducted a preliminary assessment for approximately 400 substances (UBA, 1992); detailed reports are available for about 100 substances. Data on exposure (occurrence in water, soil and air as well as degradation in water and air) and on effect (bioaccumulation potential, acute aquatic toxicity (*Daphnia*, fish) and mammal toxicity as well as indications of mutagenic or cancerogenic effects) were taken as selection criteria. In the meantime, the assessment principles for this area have been established by the UBA (1992). Up to 1990, the ITC in the United States has selected 151 out of 386 possible candidates for further tests (in Canada, the number is 44). There has been growing resistance from U.S. industry to bear the costs of these additional tests on its own, since the chemicals in question are often in widespread use on a global scale. For this reason, there have been efforts to coordinate the EPA's and CEPA's activities with the corresponding activities in Europe (KEITH, 1990).

11.3 INTERNATIONAL PESTICIDE REGULATIONS

As a matter of principle, pesticides are also governed by chemicals legislation, especially with regard to production, transport and storage. Moreover, they are dealt with in a number of special laws and thus belong to those chemicals, in addition to pharmaceuticals, whose registration, distribution and application are most intensively regulated by law. Since plant protection agents are sold on an international scale and the residues on or in foodstuffs are relevant for consumers around the world, international organizations and agreements are playing an increasingly important role. However, this will not render national legislation superfluous, since the protection of health and the environment comes under the jurisdiction of the sovereign states, even in confederations such as the European Union. An overview of the regulatory agencies in the various countries is provided by KIDD et al. (1989).

11.3.1 European Union

As we have already seen in the case of chemicals legislation, regulations in the European Union are based on a number of basic directives which have been supplemented by additional Council or Commission directives. The main directives applying to pesticides date back to the year 1976, when the EC Commission submitted three drafts (PETZOLD, 1990, 1991, provides an overview):

- Directive on the approximation of the laws of the Member States relating to the classification, packaging and labeling of dangerous preparations (78/631/EEC)
- Directive for the restriction or prohibition of pesticide active ingredients (79/117/EEC)
- Directive concerning the placing of plant protection products on the market (registration directive; 91/414/EEC)

The biocide directive 78/631/EEC is closely related to the directive for the labeling of chemicals (67/548/EEC) and has been amended several times in conjunction with this guideline. The provisions were converted into German law with the ordinance on hazardous substances.

Over the years, the number of active ingredients affected by Directive 79/117/EEC (prohibiting the sale and use of plant protection products containing certain active substances) has grown considerably. Traces of ß-HCH found in Lindan (γ-HCH), for example, were prohibited in 84/491/EEC and the application of DDT was banned entirely by 86/280/EEC. The last revision of Directive 79/117 is contained in Directive 91/188/EEC. Directive 79/117/EEC has been enacted into German law through the pesticide application ordinance.

At about the same time as these two directives were first drafted in 1976, the Commission submitted an official proposal for a directive that would apply to sales of pesticides with the aim of establishing the same criteria for the assessment and evaluation in all EU countries. This was also intended to ensure that the same tests would be recognized in all of the member countries. This not only results in considerable savings for producers; it also helps to reduce the number of animal experiments significantly. Another aim was to dismantle obstacles to trading in commodities within Europe.

Despite the obvious advantages, it took 15 years before the proposal was finally passed in the form of a Directive (91/414/EEC). Within this period, the registration requirements and criteria stated in the original plan were expanded and made stricter (KAUS, 1993). This development went hand in hand with a higher degree of awareness on the part of authorities, producers and the public, especially with regard to ecotoxicology issues. However, only the directive went into effect in 1991, without the six appendices, which contain important regulations for its implementation. These appendices will probably be reworked in the form of separate directives. Specifically, they are as follows:

Annex I: Active substances authorized for incorporation in plant protection products (so-called positive list of active ingredients).
Annex II: Requirements for the dossiers to be submitted for the inclusion of an active substance in Annex I.
Annex III: Requirements for the dossiers to be submitted for the authorization of a plant protection product (preparations).
Annex IV: Standard risk phrases.
Annex V: Standard safety phrases.
Annex VI: Uniform principles for the evaluation of plant protection products.

The wording of Annex I has been a recurring point of contention because it is believed that active ingredients or products which are banned in only one country (e.g. Atrazine in Germany) would be forced on these countries if they were registered at EU level. In fact, however, the national authorities, which ultimately have the last word on a registration, still have the possibility of excluding an active ingredient for special climatic, pedographic or application-related reasons.

There is basically a consensus among all parties involved, to the extent that they accept the use of pesticides at all, about the need for the following standards (KAUS, 1993):

- The residues of an active ingredient in edible fruits, in edible products of animal origin or in the environment must not have any damaging impact on the health of humans or animals or on the environment. It must be possible for this factor to be measured using generally common methods if they present a possible hazard.

- Based on the scientific and technical data, it must be possible to assume that the pesticides produced from these active ingredients will not have any damaging impact on the health of humans and animals, nor any unacceptable adverse influence on the environment.

Differences do exist, however, with regard to the interpretation of precisely what is a "damaging impact" or an "unacceptable adverse influence". This can be made clear in particular from the discussion of threshold values for pesticides in groundwater, which form an integral part of the "Uniform Principles" (Annex VI of Directive 91/414/EEC).

Groundwater — the most important resource for drinking water in many cases — enjoys special protection under pesticide legislation. An entry of pesticides by leaching, for example, must therefore be seen in connection with other directives, which are currently in the process of being harmonized (e.g. the drinking water directive 80/778/EEC, the surface water directive 75/440/EEC and a proposed soil protection directive). Toxicologically founded threshold values for each individual active ingredient are at odds with the call for a general ban on application. It is not yet clear whether the leaching behavior will be used as the cutoff criterion for pesticides.

Annexes II and III of 91/414/EEC contain a large number of testing requirements which can be used to arrive at a differentiated assessment of ecotoxicological behavior and effect. Therefore, it would only be logical for the Uniform Principles to permit a case-by-case decision in which especially favorable properties in one area requiring a decision (e.g. protection of beneficial arthropods) are weighed against a certain disadvantage in others (e.g. the leaching behavior). This can then be tied to restricted applications in water reservoir areas, for example.

The standard of ecotoxicological data required in the framework of EU Directive 91/414 for pesticides is exemplary throughout the world. Table 11.2 contains a list of these criteria (Annex II for active ingredients). It should be borne in mind that no standardized methods are available as yet for some of these criteria. This is particularly true with regard to the studies required for microorganisms, as well as for field studies in general. It should also be noted that the relative importance of aquatic procedures is diminished in this directive.

As far as biological products such as viruses and microorganisms are concerned, the data requirements differ from those for chemicals (Table 11.2). Pesticides containing or consisting of organisms altered by bioengineering are also covered by Directive 91/414/EEC. However, in the case of a decision on the registration of an agent, a risk assessment comparable to that in the Directive on the deliberate release of genetically modified organisms into the environment (90/220/EEC) is an absolute prerequisite.

Nevertheless, there is need for more research, especially in the areas given in Table 11.3. Aside from these specific research objectives, it is

Table 11.2 Ecotoxicological Requirements from the EU Pesticide Registration Directive 91/414, Mainly Annex II Part A (7 and 8) (Slightly Abridged)

Studies on the active ingredient:

- Acute oral toxicity on birds and effects on reproduction
- Acute toxicity in fish and *Daphnia magna*
- Chronic toxicity in fish
- Effects on fish reproduction and growth rate, as well as for *Daphnia magna*
- Bioaccumulation in fish
- Effects on algal growth
- Acute toxicity in honeybees and other beneficial arthropods (e.g. predators)
- Toxicity in earthworms and in other non-target micro or macro soil organisms
- Effects on other non-target organisms (flora and fauna) believed to be at risk
- Effects on biological methods for sewage treatment
- Field trials with the formulation (e.g. studies on acceptance of bait, granules, or treated seeds by birds or toxicity in foraging bees)

Fate and behavior in the environment:

- Rate and route of degradation (to 90% degradation) including identification of the processes involved and identification of metabolites and breakdown products in at least three soil types under appropriate conditions
- Adsorption and desorption, as well as mobility, in at least three soil types and, where relevant, adsorption, desorption and mobility of metabolites and breakdown products
- Extent and nature of bound residues
- Rate and route of degradation in aquatic systems — biodegradation, hydrolysis, photolysis, including identification of metabolites and breakdown products
- Adsorption and desorption in water (sedimentation) and, where relevant, adsorption and desorption of metabolites and breakdown products
- Rate and route of degradation in air (for fumigants and other volatile substances)
- Distribution and dissipation of the preparation in water, soil and air

Criteria for microorganisms and viruses:

- Acute oral toxicity and/or pathogenicity and infectivity for birds, fish, *Daphnia magna* and aquatic micro-organisms (if appropriate), honeybees, earthworms, soil micro fauna and flora in general, as well as other non-target organisms believed to be at risk, such as important parasites and predators of target species
- Extent of indirect contamination to adjacent non-target crops, wild plants, soil and water
- Effects on other flora and fauna, as well as effects on algal growth
- Spread, mobility, multiplication and persistence in air, water, soil
- Information concerning possible fate in food chains
- In cases where toxins are produced, data as outlined above for chemicals, where relevant

Note: This part does not apply to genetically modified organisms covered under Directive 90/220/EEC.

Table 11.3 Need for Further Ecotoxicological Research within the Pesticide Registration Process

Environmental fate and exposure:
Bioavailability and accumulation of chemicals in sediments
Bioavailability as a function of soil properties
Role of volatile substances in aquatic systems (both biotic and abiotic)
Transport processes in special ecosystems such as wadden sea, deep sea, etc.
Transport and behavior of metabolites
In general, influence of environmental factors on exposure
Effect:
Dependence of the effect on the time pattern of exposure
Identification and measurement of stress at the ecosystem level
Development of sublethal test systems
Toxicity of mixed compounds
Influence of various external factors such as climate on the probability of effects
Modeling:
Modeling of accumulation processes
Field validation of exposure models
Guidelines for micro- and mesocosms (definition of these terms), especially the simultaneous analysis of effect and environmental fate
Ecological bases:
Ecology of microorganisms in water and soil
Identification of further sentinel or key species
Similarities and differences in various ecosystem types

necessary to compile and critically evaluate the existing literature, e.g. on ecosystem studies.

11.3.2 Council of Europe/EPPO

The Council of Europe has published criteria for the registration of pesticides in regular intervals, e.g. 1984, and, most recently, in 1992. According to German authorities, they provide the basis for own decision-making schemes in developing countries, especially in the Near East and North Africa (COUNCIL OF EUROPE, 1984). In Chapter V "Guidance on Environmental Phenomena and Wildlife Data" of the 1984 publication, ecotoxicological aspects are taken into detailed consideration, especially field tests. The methods are derived from the laws of the member countries or correspond to the EPA or OECD guidelines. On the whole, the paper presents a collection which roughly amounts to a summary of the OECD and the German ChemG (base level and level 1) taking BBA methods (bees, beneficial arthropods) into account.

The cooperation which began in the 1980s between the Council of Europe and the European and Mediterranean Plant Protection Organization (EPPO, one of the Regional Plant Protection Organizations (RPPO) of the FAO) led to the publication in 1993 of a "Decision-Making Scheme for the Environmental Risk Assessment of Plant Protection Products". The objective was to harmonize criteria and test schemes with regard to the environmental hazard from pesticides on an international scale. These efforts focused not so much on the stipulation of specific test guidelines which, as mentioned, were derived from standardized guidelines such as those of the OECD. Rather, the assessment of specific problems with a substance were the focus of interest here (SMITH, 1990).

11.3.3 OECD

Since 1971, the Chemical Group of the OECD (Organization for Economic Cooperation and Development) has been working with a "Chemicals Test Program" which serves above all to develop and standardize methodology (UBA, 1984). One of the first recommendations made by this group applied to the "Minimum Pre-Marketing Set of Data (MPD)" necessary to market a substance. In addition to the physico-chemical properties of a substance, ecotoxicological effects (biological effects as well as the fate) were initially the focus of interest. At the same time, the OECD organized efforts to improve the exchange of data and introduce an internationally binding GLP system (KAPPELER, 1979). The latter was finally passed in 1982 and has formed the basis for mutual test recognition since that time. Tests for the analysis of long-term trends in the fate of environmental chemicals (especially heavy metals and organochlorine compounds) via a monitoring program (fish, birds) are even older (1972).

The activities of the OECD are primarily geared to chemicals in general. Pesticides did not meet with particular interest until the early 1990s. The Pesticide Forum confers with the Chemicals Group. In addition to the national delegations (representatives of the responsible ministries and companies), other institutions such as the EU Commission, the World Wildlife Fund, GIFAP/ECPA and others are represented in the Pesticide Forum.

The main items on the OECD's program agenda with respect to pesticides are, in descending order of priority:

- test guidelines
- data requirements
- hazard/risk assessment
- re-registration
- risk reduction

Ecotoxicological procedures are described in sections 2 (Effects on Biotic Systems) and 3 (Degradation and Accumulation) of the OECD Guidelines for

Testing of Chemicals. An updating commission of the OECD is concerned with the ongoing development and refinement of this set of guidelines, which also forms the foundation for most EU directives.

11.3.4 UN and FAO

Initial, intermittent contributions on ecotoxicology have been forthcoming since the mid-1950s from the UN Scientific Committee on the Effects of Atomic Radiation (UNSCEAR), which has published a number of basic works on anthropogenic impacts on the environment (KAPPELER, 1979). In 1976, UNEP (UN Environment Program) created the basis for an International Register of Potentially Toxic Chemicals (IRPTC) headquartered in Geneva. Its function is to serve as a kind of central clearing house for various national collections of data (HUISMANS, 1980). In addition to substance-related information (physico-chemistry, toxicology, also ecotoxicology, albeit not under this name), production data, guidelines, designations, etc. are collected. All UN activities related to this data pool are coordinated by UNEP in the framework of the Global Environment Monitoring System (GEMS).

UNESCO has taken up ecotoxicological issues in at least two parts of its "Man and Biosphere" (MAB) projects:

Project 9: Ecological Assessment of Pest Management and Fertilizer Use on Terrestrial and Aquatic Ecosystems.

Project 14: Research on Environmental Pollution and its Effects on the Biosphere.

Ecological base data on many developing countries are only available as a result of the promotion of such field projects. The MAB program has also served to promote activities such as the conference held in Alexandria, Egypt in 1983 on "Environmental Hazards of Agrochemicals in Developing Countries".

Since the early 1950s, various committees of the Food and Agriculture Organization of the UN (FAO) have been working on the side effects of pesticides, often in collaboration with experts from the World Health Organization (WHO). Up to the early 1970s, however, such research activities focused almost exclusively on the direct risk to human health from residues in food (PIETERS, 1977).

In 1985, approximately 100 member states of the FAO passed a resolution (Resolution 10/85) establishing an "International Code of Conduct on the Distribution and Use of Pesticides" as an orientation for governments with insufficient pesticide legislation, insufficient government consultancy and control as well as a low level of training for users. This — voluntary — codex contains a catalogue of detailed recommendations (from application to international trade) which are addressed to exporters and importers alike, pesticide producers and users. They thus define areas of responsibility (LEBER, 1987). The basic idea is that pesticides will be necessary for a foreseeable period of time (PAN, 1989; LEISINGER, 1989). The code was officially recognized by 25 non-governmental organizations, including the international association of the pesticide-producing industry (GIFAP) as well as by national representatives of industry (VON LOEPER, 1986).

Section 4 stipulates that producers must conduct all studies on the efficacy and persistence of a substance according to "established and proven scientific procedures and according to GLP". The results must be supplied to the authorities of the importing countries. Moreover, the latter are to be supported by industry in conducting further registration studies, e.g. field investigations. The governments of the import countries were, in turn, required to build up a monitoring system of their own. This includes the creation of legal foundations for the registration and development of executing organizations.

In 1987, two years after the resolution was passed, the FAO Code's implementation was reviewed, among others by the Pesticide Action Network (PAN, 1987), an environmental non-governmental group, taking four Latin American, three African and six Asian countries as examples. The result was that, in many developing countries, the terms of the code were not yet known very well. Even where national laws or guidelines had been created, there was often no possibility for enforcement or controls. In many cases, pesticides highly toxic for humans (Ia and Ib according to WHO classification) could still be sold on the open market. Moreover, it was not unusual for branches of international firms to conduct inaccurate or misleading advertising, often against the parent company's wishes. In the meantime, it can be assumed that international companies are for the most part complying with the FAO Code. The companies regard this as a part of their product stewardship.

In 1990, the exchange of information between supplier and recipient countries in the framework of the PIC procedures was established as a new arrangement. PIC — "Prior Informed Consent" — means that, if a chemical of any kind is subject to severe restrictions or is banned for health-related or environmental protection reasons in one of the participating countries (in the case of older substances, in at least three countries), the receiving country must be notified of this fact. The fact that a chemical is on the PIC list does not automatically mean that trading in this chemical is ruled out. This is determined on a case-by-case basis and will largely depend on the evaluation arrived at by the importing country as a result of the risk-benefit evaluation (HCN, 1992). DDT, for example, is still on the market in some developing countries, although it has been banned in the industrialized countries for a long time. Even in the developing countries, however, it is hardly used in agriculture. Instead, it is utilized in WHO-recommended applications in which persistence is considered desirable, namely, for household hygiene.

In March, 1985, the FAO also issued its own set of guidelines — "Guidelines on Environmental Criteria for the Registration of Pesticides" — which are to be supplemented by the respective regional requirements. They focus on two aspects:

I. Ecological criteria for the registration of pesticides
II. Guidelines for suitable test methods

Part I enumerates the usual physico-chemical parameters as the base data for a given pesticide. The section on "Environmental Fate" mentions degradation, metabolism and leaching including the substrate-specific analytical methods as the decisive factors. It is noteworthy that, in addition to soil and water, plants are also required as a substrate in degradation and metabolism tests. Acute *Daphnia* and fish tests as well as acute tests with birds and honeybees are mentioned for biological tests similar to those of the EU. In contrast to the very detailed physico-chemical studies, the biological aspect is clearly underrepresented. The methods listed in Part II essentially correspond to the commonly employed EPA and OECD methods, partly including German or Dutch methods. Despite the lack of standardization, field studies are considered necessary and feasible.

These data, together with information on the conditions of application, are intended to permit a prognosis of the environmental fate and any potential effects. If there is an indicated potential for accumulation, additional tests may be required which have a higher ecological relevance, such as chronic tests with *Daphnia* in the aquatic medium or microflora and earthworm tests in soils. Unlike other guideline recommendations (except the new EU regulations), a function test (degradation rate of organic material in litter bags) is explicitly mentioned here as a possible expansion on the test spectrum. The set-up basically comes very close to the most recent EU requirements (cf. Table 11.1).

In the FAO's design, the results of these tests would provide the basis for an evaluation, i.e. a forecast of whether observed negative impacts can be avoided by changing the application scheme. The goal is to strike a balance between the environmental aspects and other aspects involved in the utilization of this substance. The principle of "cost-benefit analysis" is explicitly mentioned. In cases of doubt, especially in the event of possible long-term effects, it is recommended that the substance be admitted for a limited period of time, pending the results of experience in the field. Data of this type could be obtained in the form of surveys, among other things, the results of which could in turn be used as the basis for monitoring programs. In many cases, it will probably be necessary to study the ecology of the respective system beforehand in order to have a comparative standard of measurement for any potential biological effects.

The FAO was one of the first to realize that, in addition to chemical pesticides, biological methods (and predators, pathogens and parasites in particular) have ecotoxicological impacts which can go beyond the target organism. A guideline for the registration of biological pest control methods was published by the FAO in 1988, and the FAO is currently working on a corresponding code of conduct for the importation and release of pest-control organisms (FAO, 1992).

11.3.5 Germany

Comprehensive ecotoxicological studies are required in Germany for pesticides in particular (PflSchG of 1968, revised 15 September 1986). MILLES (1991) and MICKLITZ (1991) give an overview of the historical development in crop protection as well as the legal aspects. Environmental law is surveyed by BURHENNE (1989). Since EU registration Directive 91/414 went into effect in the summer of 1991, the German government has failed to revise its pesticide legislation and has already been reprimanded for this by the EU Commission several times. However, although a draft of this law has been presented, a revision cannot be expected to be forthcoming before 1996 (IVA, personal communication).

The pesticide legislation in Germany is the oldest body of legislation in which ecotoxicological issues were taken into account. A law which governed the ban on DDT, at least in the first version, dates from 1972 (UBA, 1984). Until 1986, the Biological Federal Agency for Agriculture and Forestry (BBA), an arm of the Ministry of Agriculture, was solely responsible for enforcing the pesticide laws. Since 1986, precisely because of the increased emphasis on ecotoxicological aspects, the Federal Environment Agency (UBA), which belongs to the Ministry of the Environment, has had the status of an advisory registration agency. The Federal Institute for Health Protection of Consumer and Veterinary Medicine (BgVV), an agency of the Ministry of Health, is otherwise responsible for health protection.

In the 1986 version of the pesticide law, an attempt was made to give the principle of hazard prevention a crucial priority. In other words, during the registration process, it must be demonstrated that there is no reason to fear any damaging impacts on the health of humans or animals or on the groundwater. In addition, other impacts on the balance of nature which are not acceptable based on the current state of scientific knowledge must not occur. The "balance of nature" in this context is not a biological but a legal term. If anything, it can be described as the "soil, water, air, animal and plant species as well as the interactions between them".

By this definition, ecosystems are "highly complex, highly dynamic and ultimately unique local manifestations of the balance of nature" (WEIDEMANN & MATHES, 1991). The term "impact", also defined in legal terms, would then refer to any effect on even an individual in a population. The decision as to whether an impact of this type can be scientifically substantiated or not primarily depends on the likelihood that these impacts will occur (cf. the German High Court's "Paraquat decision" of 10 November 1988, quoted in ROTHERT et al., 1990). This in turn depends on the toxicity and the availability (exposure) of the substance for the population under consideration.

According to the legal interpretation of "current scientific state of knowledge", an impact recognizable as damaging is given if it is offset prior to a renewed active effect by the agent. One example would be the recolonization of a treated plot of land from neighboring plots. As far as the functions of an ecosystem are concerned, BBA (ROTHERT et al., 1990) assumes that the role of a species within a complex of species involved in a specific function is interchangeable — which definitely cannot be regarded as a safe assumption in any general sense.

In spite of all efforts to make scientific criteria the basis for registration, aspects of a more or less economic nature are ultimately decisive in the framework of a risk assessment. Although these economic interests are only supposed to be valid in cases where no other alternatives are available for the substance to be studied, it is fairly obvious that damaging impacts on individual species or on the "balance of nature" will only lead to a denial of registration in the rarest of cases. No consideration is given to the fact that, while all registration studies presume that the prerequisite of "proper" application will be met, the only way legislators can control this is by maintaining the maximum levels of pesticide residues on the harvested products.

It is the first time that a goal as comprehensive as "protection of the balance of nature" has been firmly established in the law (HEIDLER, 1987). Since this level of complexity cannot nearly be approximated with the methods currently available, the only alternative is to examine partial aspects, i.e. primarily in acute laboratory tests. These few data must be compared with a large number of assumptions, e.g. on spray-drift losses, in order to arrive at a registration decision. However, the law does leave an option open for more far-reaching studies, which has led to stronger BBA involvement in the development of methodology.

The package of physico-chemical and biological tests, or more precisely aquatic tests (algae, *Daphnia*, fish; both acute and prolonged), which is for the most part common on an international scale, is required for laboratory tests in Germany. In addition, earthworm and beneficial arthropod tests (2 or 3 out of approx. 20 different methods, depending on the application culture, HASSAN, 1988, 1992) have been required since late 1989, although the selection depends on the application time recommended for the substance and also on the crop. Prior to that time, beneficial arthropods, usually insects, were only covered by tests with honeybees (laboratory, semi-field and field).

If standards under the pesticide law are not met during the registration process, the authorities in charge (BBA, UBA and BgVV) can issue restrictions on use (e.g. minimum distances from bodies of water) or bans on application. The latter does not include production and sale. Restrictions must be clearly indicated as such on the label with so-called safety phrases (HEIDLER, 1987). In 1986, 28 substances (including groups of substances such as cadmium compounds) were prohibited from use based on this law, and 10 others were made subject to restricted application. Another 31 substances are subject to restricted application in the vicinity of water reservoirs, for resource conservation reasons.

Exports, especially to non-EU countries, are governed by section 23 of the pesticide law. It expressly specifies mandatory observance of the FAO "Code of Conduct". For example, suspected negative impacts on the balance of nature must also be identified. In special cases (albeit only by legal ordinance with the approval of the German Bundesrat), exports can be prohibited entirely.

11.3.6 United States of America

In the United States, pesticide use is governed by the Federal Insecticide, Fungicide and Rodenticide Act (FIFRA); pesticide sales are subject to the Toxic Substances Control Act (TSCA). Both laws require the provision of ecotoxicological data, with a great deal of bureaucratic red tape (MARCO et al., 1991). In addition to a large number of laboratory tests, semi-field and field studies are required more frequently in the United States than in other countries. As far as exports of pesticides are concerned, both houses of Congress passed legislation prohibiting exports of substances not registered in the United States (AGROW, 17 August 1990).

The emphasis on aquatic procedures (cf. FIFRA: Pesticide Assessment Guidelines: Subdiv. E: Wildlife and Aquatic Organisms) is noticeable. For a short time now, the Environmental Protection Agency (EPA), which is responsible for its enforcement, even has a guideline for the implementation of tests in the mesocosm zone, i.e. tests under field-like conditions in natural or artificial lakes (CAIRNS & MOUNT, 1990). The role of sediments is also moving into the focus of further EPA activities in the area of test standardization to an increasing degree.

Until very recently, only birds and mammals were taken into account in the terrestrial medium (HOFFMAN et al., 1990). Mammals include rodents from long term laboratory broods in particular, but the suitability of a mink for standard tests is being considered. One change came about due to the problem of wastes, i.e. the effects of pollutant compounds in landfills. Since no EPA guidelines were available to cover this, OECD methods were used, and own methods were developed to some extent (e.g. plant tests; e.g. MILLER et al., 1985). Tests with other non-targets, e.g. beneficial arthropods, are only required for tests of biological agents. There are ecotoxicological test requirements specifically for the latter; with few exceptions, however, they are no different than those for chemicals (cf. Pesticide Assessment Guidelines: Subdiv. N: Microbials).

Since pesticide registrations are normally granted for an indefinite period of time, a large number of data (listed in document 40CFR Part 158) must be submitted to the EPA (Office of Pesticide Program/OPP), involving a considerable amount of red tape and paperwork. The individual guidelines (naturally with GLP) generally resemble corresponding European or OECD rules, but are often not compatible with them. Since 1981, the EPA has published the rules in the form of "Pesticide Assessment Guidelines", but the law only specifies quality standards and the external format of the report to be submitted.

Much of the public concern over pesticides stems from the fact that people became aware that some of the pesticides currently in circulation were first registered a long time ago and have never been subjected to a modern evaluation. This is what led to the FIFRA amendments, better known in the meantime as the "re-registration process". This procedure is based on the following basic ideas (GRAY, 1991):

- Existing information about a registered pesticide should be brought up to date by defining the current requirements and ensuring that corresponding data are available to fulfill these criteria.
- These data must be reevaluated with respect to the current registration criteria in order to decide whether a change in registration status is necessary.
- Changes may have to be made in the instructions for use or the precautionary advices which appear on the label, or even in the registration status itself.

The possibility for re-evaluation in five-year intervals was provided for in the FIFRA from the very beginning, in other words since 1947. However, it was never recognized by the USDA as an instrument for the consideration of new evaluation criteria. Finally, in 1972, the EPA was ordered to review all 30,000 to 40,000 pesticides by 1976. However, it was not until 1988 that the EPA submitted the final version of the re-registration guidelines. Congress then amended them by setting time limits for the completion of the revision process. Producers subsequently shelved tens of thousands of pesticides which were not likely to qualify for re-registration in order to save fees and the costs for the necessary studies. The range of active ingredients thus became narrower and the number of indication gaps grew — a trend that was accelerated by short patent terms and high test costs. For many smaller producers and second registration applicants, this meant ceasing production (LENG, 1991).

The re-registration procedure for all pesticides registered prior to November 1984 has become extremely complicated, and there have been many complaints about needless losses of time and the exclusion of effective products. On the other hand, a great deal has been learned about efficacy, side-effects, environmental fate, long-term impacts on the ecosystem and health risks (GRAY, 1991). While the re-registration program has been in the process of elaboration, producers have invariably been faced by the problem of constantly changing standards (moving targets), with no end in sight. As a case in point, the GLP guidelines were not published in their final version until August 17, 1989 (LENG, 1991). Consequently, the whole re-registration process scheduled for completion in 1997 will last at least until the year 2002.

12 ASSESSMENT AND EVALUATION

12.1 VARIOUS APPROACHES

In addition to identifying and monitoring the fate and effects of foreign substances, ecotoxicology must assess the potential hazard of these substances to the environment based on scientific criteria. Evaluation goes one step further than assessment in that it takes additional — e.g. economic — criteria into consideration (WEIDEMANN, 1990). If assessment and evaluation, moreover, form part of a registration process, legal expertise will also be required (RUDOLPH & BOJE, 1986). It is difficult for ecotoxicology to define criteria which are sufficiently unequivocal to serve as the basis for legal decisions, if only for the following reasons (MORIARTY, 1988):

- The behavior of foreign substances can differ depending on the prevailing environmental conditions and their inherent physico-chemical properties; in other words, the bioavailability is subject to wide variations, depending on which compartment of the environment is involved.
- Depending on the degree of exposure and the genetic make-up, individuals and populations react in different ways to foreign substances.
- The complexity of the reciprocal effects and synergisms with other stress factors in what are usually hundreds of species and their environment is extremely great.

Consequently, when assessing the ecotoxicological hazard potential, an attempt must be made to weigh a large number of different data to yield an overall result that will permit the respective chemical to be classified in an understandable way. The core of the problem here is that, for reasons inherent in scientific theory, there is no way to *prove* that a substance is *not* hazardous. Nor will it be possible to eliminate this contradiction through the increasing use of models, which, despite or even because of their complexity and the complexity of the subject investigated, cannot be validated in the classical sense (BARNTHOUSE, 1992). At best, they can be used in an iterative process of approximating real conditions; however, the assumptions and prerequisites on which they are based are often not disclosed.

For the sake of completeness, classification procedures which have not met with any widespread use will be presented before going on to the currently most accepted methodology (Hazard and Risk Assessment).

(1) Extremely simple methods which are usually employed for the "screening" of environmental chemicals in the EU notification procedure (assessment of individual, isolated chemical properties such as the volume produced or the absolute concentration data; e.g. more than 100 mg/l or 1000 mg/kg) will not be considered in any detail here.

(2) In Germany, the "Environmental Hazard Profile" approach was developed for a relative classification of the risks posed by organic chemicals (KORTE et al., 1978). Results from the following tests are incorporated in this substance-specific "profile":

- metabolism in a mammal organism (rat)
- bioaccumulation in fish or algae
- biodegradation in sludge
- abiotic degradation in the atmosphere

Effect studies do not receive sufficient consideration when this approach is selected. In spite of the explicit reference to the *ecotoxicological* hazard potential, the investigation of environmental fate and, indirectly via the bioaccumulation studies, the human being as a target of protection are overly predominant. The subsequent addition of a *Daphnia* test, as well as the introduction of a three-tier classification scheme (low, unclear or high hazard potential), did little to eliminate this deficiency (KLEIN et al., 1984). However, due to the large number of environmental chemicals studied in this way, correlation factors, e.g. between the log P_{ow} and the degradation rate, were defined which proved helpful for the development of QSAR models (FREITAG et al., 1985). A recent overview on this and related classification schemes is given by PARLAR & ANGERHÖFER (1991).

(3) Another approach which is not employed very often is the use of ecotoxicological acute toxicity tests as well as toxicological data in the automation of assessments (PAASIVIRTA, 1988). The problem with such approaches, however, is not the preparation of corresponding computer programs but the generally poor condition of the data (with regard to quantity and quality) as well as the classification of existing data in an assessment scheme. ZITKO (1990) compared various possibilities for the relative assessment of environmental hazard potential and concluded from the differences ascertained

that such classification lists cannot take the place of expert decisions in individual cases.

(4) The concept of "ecological health" is a relatively new, preventive approach to environmental protection against chemical impacts. It reveals a number of analogies to the diagnosis of human health and not only in the terminology used (KARR, 1993). Instead of testing single species or the fate of a chemical in specific concentrations in the laboratory, the idea is to rely on monitoring studies, since they are the only means of collecting data on the ecosystem as a whole. In concrete terms, this would require the collection of data on the following five aspects (example: aquatic medium): water quality, habitat structure, flow regime, energy flow and biotic interactions. The information obtained in this way is then formalized into an index (e.g. the index of biotic integrity = IBI; KARR, 1991), which can be used to assess the state of a specific site. In the meantime, the method has been applied on several occasions in the United States, especially in connection with waste-water problems (e.g. KARR et al., 1985), although one cannot speak in terms of any general acceptance for it as yet.

The approach to the definition of "ecological health" is criticized harshly by SUTER (1993) for the following, so stated reasons:

- "Ecosystems are not organisms; so they do not behave like organisms and do not have properties of organisms such as health."
- "Attempts to operationally define ecosystem health result in the creation of indexes of heterogeneous variables. Such indexes have no meaning; they cannot be predicted, so they are not applicable to most regulatory problems; they have no diagnostic power; effects on one component are eclipsed by responses of other components; and the reason for a high or low index value is unknown."

The same criticisms have already been discussed in connection with the use of indexes in general (see Chapter 6.6.2). It should simply be pointed out here that this kind of criticism does *not* discredit the use of monitoring data. Their use is essential for every assessment of ecotoxicological hazard potential.

12.2 SIMPLE EFFECT CLASSIFICATION SYSTEMS

A classification of the ecotoxicological hazard potential is mandatory for the registration or notification of pesticides and environmental chemicals, respectively. This chapter provides examples of some older classification systems, which are limited to effect comparisons. However, all of these procedures may be entirely adequate to assess individual aspects of ecotoxicological hazard potential (overview in VAN STRAALEN et al., 1994). None of them, however, can be used to arrive at an overall assessment of the chemical.

Table 12.1 Assessment Classes of Results from Tests with Beneficial Organisms (Data in Comparison to an Untreated Control)

Laboratory tests (percentage mortality or diminished beneficial capacity):	
Not harmful	less than 50%
Slightly harmful	50–79%
Moderately harmful	80–99%
Seriously harmful	more than 99%
Semi-field and field tests (percentage mortality or diminished beneficial capacity):	
Not harmful	less than 25%
Slightly harmful	25–49%
Moderately harmful	50–75%
Seriously harmful	more than 75%

Water Quality Numbers

One example of a strict classification of effects is contained in the German Drinking Water Ordinance (TrinkwV of 22 May 1986). Simple ecotoxicological tests such as biodegradation or acute bacteria and fish toxicity are used in conjunction with toxicological data to divide chemicals up into one of four classes of water hazard. Ratings are given on the basis of the test results and are then averaged; the resulting mean value, the water quality number (Wassergefährdungszahl, WGZ), identifies the water hazard class. In unclear cases, further tests — on the chronic toxicity or on bioaccumulation, for example — may be required. A similar procedure is also employed in the United States (Water Quality Criteria; cf. RIPPEN, 1987). In one such procedure recently proposed for the State of Florida, the leaching potential of a pesticide is offset against its chronic toxicity values (e.g. carcinogenity). The results are then incorporated into a system of relative values that is used to arrive at a decision (BRITT et al., 1992).

Beneficial Arthropods

Classification systems based on IOBC methods (two, three or four stages, depending on the country) for the assessment of the side effects of pesticides on beneficial arthropods are a good example of ecotoxicological effect data classification (BOLLER, 1989; HASSAN, 1992). Hazard assessment is relatively simple with these approaches, since the exposure is assumed to be equivalent to the highest recommended application quantity. Based on the observed effect (usually mortality) on the test organisms in standardized laboratory tests, the first decision is whether further tests must be conducted in the framework of a tiered test program (laboratory → semi-field → field).

Ultimately, the substance is divided into one of four classes of environmental hazard (Table 12.1). Various criteria are used to define the four classes: if the chemical was tested in semi-field or field conditions, the values from these tests will be more reliable than laboratory data, since it can usually be assumed that they are more relevant for real application situations. The use of tested pesticides is more or less restrictive depending on the observed side effects in these tests. In any case, registration cannot be denied solely on the basis of tests with beneficial arthropods, at least not in Germany; in other words, they do not serve as a "cutoff" criterion. If data are available on the exposure and behavior of beneficial organisms as well as laboratory tests with the same species, it should be possible to generate so-called "hazard indexes", i.e. simple estimates of the hazard potential (JEPSON et al., 1990).

Modifications of this classification scheme are urgently required:

- The number of classes should be reduced since experience has shown that the variability of results obtained with carabid tests renders the use of so many classes impracticable (RÖMBKE & HEIMBACH, 1994).
- The definition of the individual classes must be changed since seriously harmful effects may indeed exist below an effect of 100%. This value has been established against the background of pesticide use which is limited in terms of both time and space, i.e. a rapid recolonization is assumed in the case of these mobile organisms. This prerequisite, however, does not apply for environmental chemicals.
- In addition, it should be pointed out that this classification offers no indication of the dose-response relationship since, as already mentioned, only one concentration (the highest recommended application quantity) is tested.

Microorganisms

A scheme developed by DOMSCH et al. (1983) to classify the results of laboratory tests with soil microorganisms was adopted by the BBA in Germany as a basis for assessing the side effects of pesticides. This scheme is based on a model in which the relationship between the pesticide and induced effects are described in relation to the natural changes of soil microflora (e.g. for the parameters respiration and dehydrogenase activity, Figure 12.1). In this graphic representation, a recovery time of 30 days is postulated, assuming a normal doubling time of 10 days for soil microflora and an 18-day median period of recovery from stress. In other words, a delayed recovery period of up to 60 days would be tolerable, whereas delays of up to 90 days would be considered

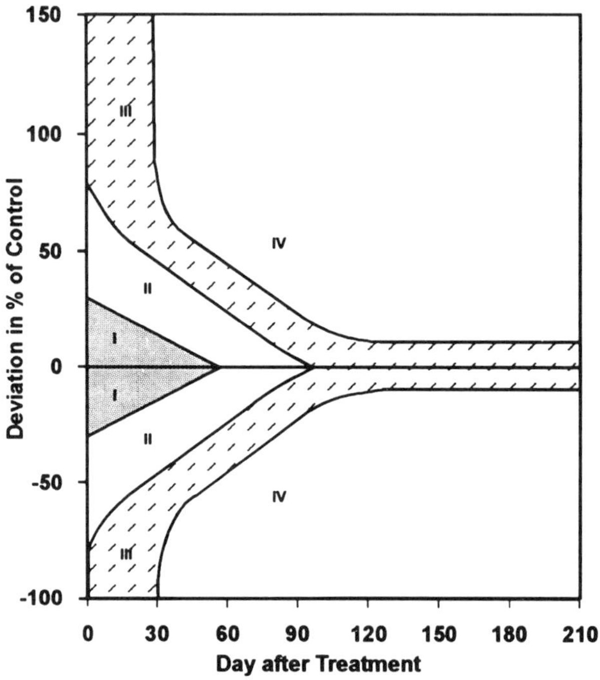

Figure 12.1 Model for the classification of ecotoxicological effects of pesticides on soil microorganisms (DOMSCH et al., 1983; HEITEFUSS, 1987). Zone I = negligible, Zone II = tolerable, Zone III = critical, Zone IV = not tolerable.

critical. Only if effects last longer than 90 days would this have any consequences on registration of the substances. While there is no reason to question the basic premises of this simple model, it does not give sufficient consideration to the occurrence of combined effects or to modifications resulting from site conditions (especially the existing quantity of microbes or the composition of microfloracoenoses; MALKOMES, 1985; SCHUSTER & SCHRÖDER, 1990). The question of scale (What deviation from the control value is still tolerable?) is a matter of controversy, since there is little if any published evidence of "non-tolerable" effects. For this reason, an attempt should be made to integrate the results of the discussion on this issue into the registration process in practice.

12.3 "HAZARD AND RISK ASSESSMENT" — OVERVIEW

In addition to the procedures outlined above, the quantified comparison of exposure and effect concentrations is the most suitable approach for the assessment of an ecotoxicological hazard potential.

The procedure widely accepted today for assessing the environmental hazard potential of chemicals (incl. pesticides) is the "Hazard and Risk Assessment", which was developed in the late 1970s (FAVA et al., 1987). This procedure is "simply a systematic means of developing a scientific basis for regulatory decision-making" (BARNTHOUSE, 1992). The process takes in aspects of all three "classical" partial tasks of ecotoxicology: the identification, prediction and monitoring of undesirable effects from chemicals.

Generally speaking, this approach is suitable for pesticides and environmental chemicals alike, although there are distinct differences between the two groups when it comes to ascertaining environmental concentrations (UBA, 1992). In a more general version, namely, the "Ecological Risk Assessment" also developed by the EPA, chemicals represent merely one of any number of other potential hazards (EPA, 1993). In other words, basically all human activities which can have ecological effects (from the mortality of individuals to changes in ecosystem functions) are assessed. This is not restricted to chemicals. Road construction or grazing, for example, would also be regarded as stressors (NORTON et al., 1992). By the same token, this approach can also be used to assess the hazard to human beings (e.g. ECETOC 1990).

The theoretical foundations of the "Hazard and Risk Assessment" have, for the most part, been corroborated in actual practice, but there is still a considerable need for research (HEIDLER, 1987; BUSCH et al., 1989; OECD, 1989a):

- development of models for the exposure and effect side
- identification of "ecosystemic" parameters
- assessment of safety factors for risk assessment
- validation of existing risk assessment procedures

The individual steps in the process of hazard and risk assessment are classified and defined in different ways, depending on the author. The model which has been selected for presentation here (Figure 12.2) is based on EPA and EU policies (EPA, 1992; EC 1993). It comprises four steps, in which the first three are devoted to the assessment of a chemical's hazard potential and the fourth to the evaluation.

Step 1: Hazard Identification

The first step of the assessment process is devoted to formulating the problem, in other words to "hazard identification". Identification means ascertaining which environmental media or compartments a chemical has entered or may enter and which organisms potentially exhibit effects (e.g. based on targeted application).

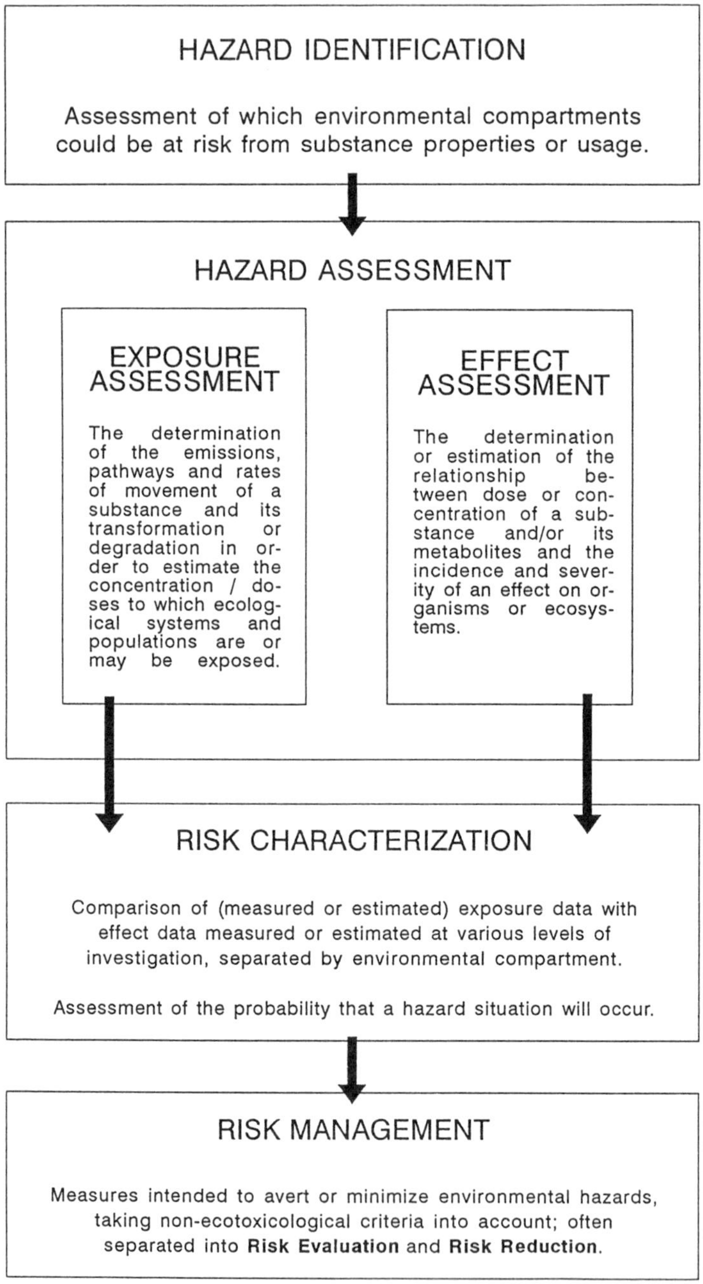

Figure 12.2 Overview of the Hazard and Risk Assessment process (modified from EPA, 1992).

Step 2: Hazard (Exposure and Effect) Assessment

In the second step, two factors are examined separately, but together are referred to as the "hazard assessment" (UBA, 1984; WEBER & BARBEN, 1984; VAN LEEUWEN, 1990):

- the measurement or assessment of the concentration of a substance in the various environmental media or compartments to which organisms are potentially exposed (exposure assessment);
- the measurement or assessment of the concentration in which a substance, either alone or together with others, may have an effect on organisms or ecosystems (effect assessment).

This step should be repeated consecutively several times, depending on the quality of the data obtained (e.g. acute, chronic or field data), taking the best information yielded in each case for both sub-areas. This iterative process of approximation to obtain the best possible information must be based on close coordination between both parts of the hazard assessment (NORTON et al., 1992). The individual steps, with higher-quality data at each successive step, have practical counterparts in the tiered test program described in Chapter 7; they are referred to by various authors as the screening, intermediate and comprehensive phase (OECD, 1989a). Not only individual values, but also entire sets of data can be compared with each other with the help of models. As already mentioned in the presentation of the various test systems, far more examples are available for limnic systems than for sediments, marine or terrestrial systems.

Step 3: Risk Characterization

In the risk characterization, the exposure data (e.g. concentration of a chemical in the water) are compared with — species-specific — effect data (e.g. LC_{50} or $NOEC_{Reproduction}$) for each compartment of the environment (Figure 12.2). EPA (1992) defines this step as follows: "A phase of ecological risk assessment that integrates the results of the exposure and ecological effects analyses to evaluate the likelihood of adverse ecological effects associated with exposure to a stressor. The ecological significance of the adverse effects is discussed, including consideration of the types and magnitudes of the effects, their spatial and temporal patterns, and the likelihood of recovery."

Therefore, the third step is devoted to assessing the probability that the possible effects will occur under real conditions. However, since direct field studies are only available in the rarest of cases, the extrapolations necessary for this assessment are covered by safety factors or model assumptions. Put more specifically, a measured or estimated environmental concentration (PEC = Predicted Environmental Concentration) is compared with a measured or estimated concentration in which there is no effect on organisms in ecosystems

(PNEC = Predicted No Effect Concentration; less often also referred to in the literature as ERC = Ecotoxicologically Relevant Concentration (KLEIN et al., 1992)). If the comparison of PEC/PNEC values < 1, it can be assumed that the affected substance presents a low ecotoxicological hazard potential; if > 1, further studies or measures are necessary.

In this comparison, it is important to bear in mind that the toxicity of a chemical — given the same test conditions, with test organisms which are as much alike as possible (e.g. in physiological state or age) — should not fluctuate very much from one test to another. On the other hand, the exposure of an organism in the field can vary sharply, depending on the environmental conditions, the characteristics of the substance or the biological state of the coenoses which are composed in different ways in each case (RAND & PETROCELLI, 1985). Put another way, the bioavailability of a substance under real conditions is the single most important factor in determining the ecotoxicological hazard potential of a chemical in the field. Of course, every species has its own range of control or safety mechanisms in relation to chemicals (partly at the individual level, partly at the population level) (KETTRUP et al., 1991).

Although limitations on extrapolation must be borne in mind, this step pays too little attention to ecological parameters since only the specific exposure and effect data are compared. Some ecological parameters which might be suitable for a more in-depth assessment are:

- the diversity of ecosystem structures
- the degree of "proximity to nature"
- the preservation of ecosystem functions
- the ecosystem's capacity for regeneration

In particular, some aspects — such as joint analysis of several media, the period of exposure and the persistence of the effect on the ecosystem — are not given sufficient attention (FAVA et al., 1987). Discussion has only just begun to address the aspect of the different time patterns at the various levels of investigation (ranging from several days for laboratory tests all the way to the years or centuries, for that matter, which it takes for an ecosystem to regenerate, BURGER & GOCHFELD, 1992).

Step 4: Risk Management

In the last step, namely, risk management (which is often subdivided into risk evaluation, decision-making or risk reduction), appropriate measures indicated by the results from the preceding steps must be defined and carried out in order to prevent or at least minimize the hazard deemed probable by the risk assessment. Non-scientific criteria such as cost-benefit considerations are also included in this evaluation step. This evaluation supersedes the assessment because it takes political, legal and economic criteria into consideration as well as the ecotoxicological parameters.

Each of the four steps is discussed in the following chapters. However, the focus is on the discussion of the various methods used for effect assessment (Step 2).

12.4 HAZARD IDENTIFICATION

In the Hazard Identification step, very basic data are compiled in order to determine whether there is any chance of the chemical ever entering a specific compartment of the environment and, if so, in which compartments. In particular, this step relies on data regarding the purpose for which the substance is to be employed (e.g. as an intermediate product in the chemical synthesis or as a pesticide) as well as the production techniques. Both aspects should be delineated and examined as precisely as possible by location and region. An identification of the ecosystems or ecosystem types at risk would offer the best foundation for further assessment, since the test programs could then be optimally tailored to them.

This is also the step in which a decision must be made as to which practical tests should be conducted to yield the desired information. Before this can be done, however, the Assessment End Points should be clear; in other words, it is essential to know what kinds of changes may occur in the ecosystem or, as the case may be, which of them can be recorded and monitored. This makes it possible to select laboratory effect tests for the environmental compartment to be observed. QSAR studies or initial considerations with regard to suitable model approaches often form a part of this step in the Hazard and Risk Assessment.

Generally speaking, the step is used to identify and make a rational selection of tests which are necessary and worthwhile but also manageable and within reason. The options can thus be narrowed down, because it would be impossible to conduct every conceivable test for the large number of chemicals to be classified in all compartments.

12.5 HAZARD ASSESSMENT — EXPOSURE

12.5.1 Basics of Determination

According to MATHIES (1991), the overall goal of determining the environmental exposure can be summarized as follows:

- determination of the concentration in abiotic and biotic environmental compartments;
- improvement of the understanding of processes which determine exposure;
- identification of the stress level of ecosystems.

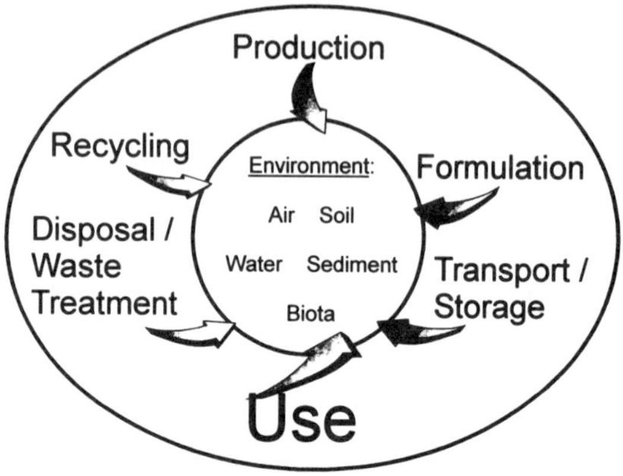

Figure 12.3 Emissions of a substance into the environment (modified from ECETOC 1993).

The exposure to environmental chemicals can be determined in various ways. Monitoring data from the field (e.g. concentrations in surface waters) are the most suitable for this purpose, but they are not available for chemicals which are to be newly registered nor for the vast majority of already existing substances (as many as 100,000 substances in the EC alone!). The only other alternative is to estimate the level of concentration in the various environmental compartments (= Predicted Environmental Concentration; PEC) using models. These models especially make use of data from laboratory tests on environmental fate (SCHEUNERT & KLEIN, 1985; SAMIULLAH, 1990). In the past, they have focused on persistence in the aquatic medium (biodegradation and bioaccumulation; WEBER & BARBEN, 1984; KLÖPFFER, 1989).

In order to estimate the concentration of a chemical in various compartments of the environment (PEC), the first question is how much of the substance is reaching the field (e.g. production volume per year). This figure is used to differentiate whether the entry of a chemical into the environment at specific locations (e.g. production sites, sewage plants) is diffuse or punctual (i.e. strictly localized). Pesticides belong in the second group because the quantities and conditions of their application are prescribed for the most part and can thus be taken over just as they are to serve as the PEC. A certain element of assumption is inevitable nonetheless (e.g. about the depth of a pesticide's penetration into the soil). In theory, with the exception of accidents and improper applications, the concentration of pesticides in the field is relatively easy to estimate due to the manufacturers' recommendations. In practice, however, the assessment is rendered all the more difficult by the fact that emissions of various kinds may occur in the environment at different points during the life-cycle of a chemical (including waste treatment and disposal as well as recycling) (Figure 12.3).

Based on the same considerations, in Germany the concentration of pesticides in soil is estimated as the PIEC (= Predicted Initial Environmental Concentration) on the following assumptions:

- even distribution in the upper 2.5 cm of the soil;
- in crops with low cover: whole applied volume reaches the soil surface; cuttings are necessary if the cover is high;
- in canopy cultures, it is assumed as a round figure that 50% of the applied quantity reaches the soil.

12.5.2 Use of Models

Generally speaking, a distinction is made between local, regional and global models of exposure. Local models describe the behavior or fate of chemicals in the area immediately surrounding the point of entry in the environment, which is assumed to be homogenous for the most part. The intermediate (regional) zone is currently the most difficult to model since environmental heterogeneity as well as temporal variability have the greatest influence here (GDCh, 1993). Global models deal primarily with the — usually airborne — transport of persistent substances on a continental scale. Only very general observations can be made at this level.

In the meantime, examples of successful validations with lysimeter tests are now available for local models on the environmental fate of a substance (e.g. SESOIL or PRZM for leaching). Data which are as precise as possible should be collected for the chemical properties and environmental variables; otherwise, even small differences can add up and ultimately lead to misinterpretations (KLEIN, 1991). In the verification of prognoses made with models such as PRZM or GLEAMS on the leaching behavior of various herbicides, for example, which were based on data of very good quality, the values for the first 1 to 3 weeks after application were accurate by conditions in the field but diverged increasingly from reality at later points in time (MUELLER et al., 1992; cf. Chapter 8.1).

On a larger scale, the level of exposure is assessed on the basis of the quantities which enter the environment, the type of source as well as the physico-chemical properties (e.g. the log P_{ow} or the Henry constant) of a substance (CALAMARI & BACCI, 1987; DVWK, 1989; BLUME, 1990). In ECETOC (1993), a distinction is made between screening, confirmatory and investigative phases with varying degrees of complexity, depending on the quality of the data (emission and physico-chemical data or fate data).

The "fugacity approach" is probably the most widespread of the commonly employed exposure models (Figure 12.4). It has been differentiated according to various levels of complexity (Level I, II, III) as well as in terms of special scenarios (MacKAY, 1991). The only result initially yielded by the lowest levels of complexity is the relative distribution between the media or compartments water, soil, sediment, air and biota after an assumed thermodynamic state of equilibrium has been reached (e.g. Level I, MacKAY, 1991). For this assessment of exposure, the data generally available on the following parameters are required as input data:

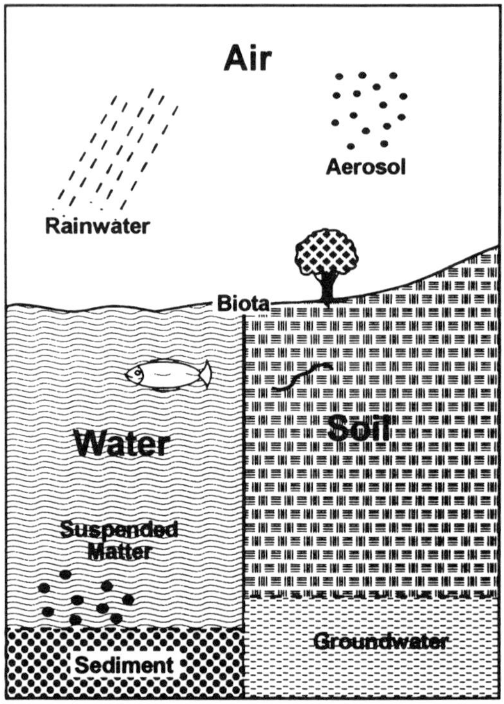

Figure 12.4 "Unit world" (adapted from MacKAY, 1991).

- molecular mass
- water solubility
- vapor pressure
- n-octanol/water partition coefficient

The relative distribution also provides an indication of the environmental compartment in which effect tests should primarily be conducted. In Germany, it is assumed that an environmental compartment should be examined for exposure if the calculated estimate is that over 20% (in air) or 10% (in water or soil) of the chemical quantities come together in that compartment (UBA, 1992). If a compartment meets this description, in-depth assessments should be conducted there with local models. If there is no indication that an excessive proportion of the produced quantity is persistent, a safety factor (1 to 10) can already be used at this stage.

For the differentiated calculation of absolute concentrations of foreign substances in various compartments, additional information is required which is often extremely difficult to obtain, e.g. on the entry of a substance in the environment by time unit (WEBER & BARBEN, 1984) or on the rates of degradation and volatility. Moreover, site-specific data on the respective regional or local soil and environmental conditions are also necessary in these models, since they may have a critical influence on the persistence or mobility of a substance (JURY et al., 1987). Moreover, the fate of a substance can depend to a very strong degree on which environmental medium it entered first.

It should be borne in mind that even these models of exposure must be validated, for example, using monitoring data (overview in OECD, 1989b; ECETOC, 1992). One problem with these models is that the environmental media (air, water, soil) differ in terms of the time period during which they are to be observed (ranging from seconds to years) as well as in degree of heterogeneity (e.g. with respect to the chemical, physical or biological processes influencing the chemical behavior). Generally speaking, there are positive correlations between the estimates derived through the use of exposure models with values measured in the aquatic medium, but sharp deviations in terrestrial systems (MATHIES, 1991).

If at all possible, estimated values (PECs) should be corroborated by monitoring data from the field, assuming that there is a positive correlation, if the difference between the assessed and measured concentration is no greater than a factor of 10 (UBA, 1992). In the case of many chemicals, however, it is not surprising to find that this discrepancy is greater since it is extremely difficult to correctly estimate time-dependent processes such as degradation or metabolism from what is normally a very narrow database.

12.6 HAZARD ASSESSMENT — EFFECTS

12.6.1 Basic Considerations

The assessment of a chemical's effect is based on the determination of a dose-response relationship. A concrete measurement is preferable to an estimate in any case. The determination is generally based on standardized laboratory tests which take substance-related physico-chemical characteristics into consideration. Although results of such tests can hardly be applied to field conditions (even the question of which criteria should be applied for a validation is a subject of controversy, LIVINGSTON & MEETER, 1985), relative assessments of the effects of foreign substances can be made if there are a sufficiently large number of test results. They can also serve as a basis for extrapolations to other species, parameters or the next level of investigation. Conversely, this means that the ecotoxicological soundness of a substance cannot be judged on the basis of such tests alone (RUDOLPH & BOJE, 1986; HALTRICH, 1991).

It has become historically established practice to employ aquatic toxicity tests to assess the environmental hazard of a substance. The approach has obvious limitations, bearing in mind that all of the species in aquatic ecosystems are to be protected on the basis of only three tests (with algae, *Daphnia* and fish). The situation is similar with regard to the relevance of test procedures for sediments (e.g. VAN DER KOOIJ et al., 1991), whose importance for the whole aquatic medium (especially with regard to the bioavailability of chemicals) has often been underestimated in the past (MAKI et al., 1987).

Compared with the medium of water, the soil medium poses a number of special problems. In 1990, VAN LEEUWEN concluded that an ecotoxicological assessment of effects was not feasible for soil due to the lack of suitable testing methods. As shown in Chapters 8 to 10, ideas for tests in the soil medium already exist, but have hardly been required on a regular basis to date and have thus also not been validated as yet. For this reason, there is a need for a rapid improvement of the data available for this medium. The use of aquatic data to estimate active concentrations in the soil, as discussed by LÖKKE (1994), is not considered worthwhile due to the many different assumptions which are necessary for such an approach.

Despite all of these difficulties, there is for the most part a consensus in the international discussion with regard to the fact that, to assess a chemical's effects, it is imperative to ascertain the precise exposure concentration at which no effects on individual organisms occur (NOEC = No Observed Effect Concentration). In theory, this value is valid for all species in the compartment; in practice, however, it is based on 3 to 5 test species per compartment, at best. Theoretically, the NOEC value represents those concentrations up to which the control mechanisms against the chemical affect functions on various levels of organization. In practical terms, the NOEC specification is subject to methodologically determined limitations (cf. Chapter 6.5.2). The lowest NOEC value for each species, parameter and exposure pathway observed is selected, which then serves as the basis for assessing the substance (MORIARTY, 1988). These NOECs can only be directly measured in relatively elaborate, chronic tests with sublethal parameters (e.g. growth, reproduction). Although these tests are theoretically possible on all three levels of investigation (laboratory, semi-field, field), nearly all of the available data are taken from laboratory tests. If data from studies at higher levels are available, these will of course be more relevant than the results of laboratory tests. In this assessment, one should be mindful of the fact that effects may occur in different but more or less related ecosystems (e.g. spray drift or run-off of pesticides in aquatic systems, ecological risk assessment on a regional scale, cf. GRAHAM et al., 1991).

Studies in which only one concentration (derived from the highest allowable volume of application as in the case of pesticides) is investigated are not suitable for the determination of a NOEC value. Instead, the dose-response relationship must be ascertained for at least five concentrations. It has been suggested that these tests be replaced by studies at molecular level since the latter are said to be especially simple and sensitive (e.g. KETTRUP et al., 1991). Due to the difficulties involved in the interpretation of changes which occur at molecular level, this is not considered a meaningful alternative. It should also be pointed out that there are certain, generally acknowledged compartment-specific ceilings for effect concentrations (OECD, 1989a). No

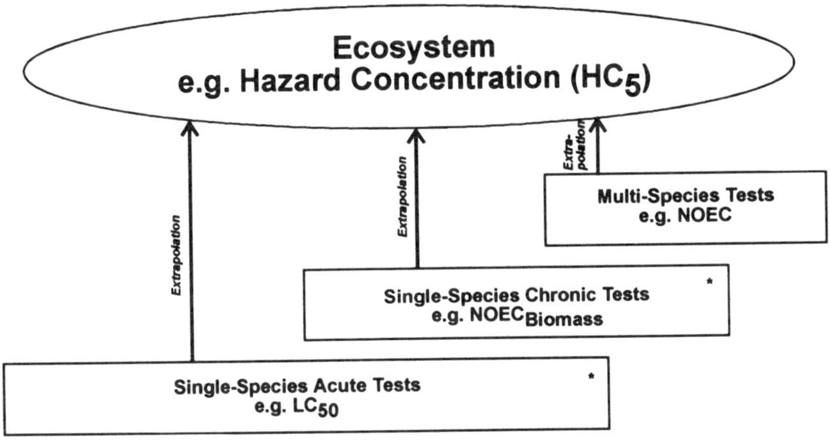

Figure 12.5 Extrapolation steps in the Hazard and Risk Assessment (* = including QSAR estimates).

further evaluations are considered necessary for concentrations in excess of 100 mg/l water or 1000 mg/kg soil.

Since no data from chronic laboratory tests, much less field data, are available for most chemicals, various methods of gauging the No Effect Concentration are suggested in the literature (cf. VAN STRAALEN et al., 1994). Given the rapid pace of developments, however, it is not possible to provide a complete overview

12.6.2 Problems of Predicting No Effect Concentrations

Precisely in view of the large number of chemicals to be investigated, there has been a growing attempt in spite of the obvious difficulties to predict effect concentrations (analogously to NOEC values), which are then called PNECs (= Predicted No Effect Concentration). A distinction is made between three different sets of issues when making this forecast (Figure 12.5):

- the extrapolation from acute to chronic laboratory tests;
- the extrapolation from one species to other species;
- the extrapolation from laboratory tests to field conditions.

If a field study is available, it can be assumed that a fourth extrapolation step (from one ecosystem type to another) would be superfluous, since it would be impossible to cover all of the possible field situations, even under optimal test conditions.

The three levels of extrapolation will be discussed in detail on the following pages.

Extrapolation of the Results of Acute Tests (e.g. LC_{50}) to Chronic Effects (e.g. $NOEC_{Reproduction}$)

Theoretically, there are several ways of extrapolating from acute to chronic effects (GIESY & GRANEY, 1989). In any case, it is necessary to work with safety margins or statistical methods. Despite a few examples in which such approaches were successfully employed, caution is advised since there is no scientific basis which "conclusively permits an extrapolation from acute to chronic effects in the same or other species" (BIAS, 1993).

Based on practical experience (almost exclusively from aquatic tests), it can be assumed that the two values ought not to diverge very often by more than one order of magnitude (SLOOFF et al., 1986). VAN LEEUWEN et al. (1992), for example, only found factors ranging between 2 and 11 for seven aquatic species in a comparison of the NOEC to the LC_{50}. A slightly different picture is given by VAN GESTEL et al. (1992), who found ratios between 5 and >100 in earthworm tests with heavy metals and pesticides. RUDOLPH & BOJE (1986) compared the results of fish, *Daphnia* and algae tests for 20 environmental chemicals and found that acute and chronic effect thresholds may even differ by as much as a factor of 100,000! BARNTHOUSE et al. (1987) also considered the extrapolation from acute values to chronic data (the LC_{25} in their case) to be the step with the highest degree of uncertainty in predicting the toxic effect of chemicals on fish.

Extrapolation of Test Results with One Species to Other Species (e.g. From Daphnia *to All Invertebrates in Limnic Bodies of Water)*

Species-specific differences can (but need not!) diverge widely. It was concluded from the results of older studies in the aquatic medium with relatively toxic substances that an acute test with one species was sufficient to predict this effect in other species (KENAGA & MOOLENAR, 1979). Plant growth tests also showed that the toxic effects of 16 herbicides on 151 dicotyledonous species differed by no more than a factor of 10 (FLETCHER et al., 1990). In a comparison of the acute effects of 410 chemicals on 66 animal species, however, MAYER & ELLERSIEK (1986) found that there was no species, family or class which has the most sensitive reaction in every case (also cf. CAIRNS, 1986a). The differences between species amounted to a factor of 100 on average and in individual cases were as high as a factor of 100,000. RUDOLPH & BOJE (1986) came up with similar results as did SLOOF et al. (1986) — though not quite as extreme — when they interpreted aquatic acute tests with 25 and 15 environmental chemicals, respectively.

However, it is not clear how many different possible species should be tested per environmental compartment in order to cover a sufficiently large spectrum of sensitivity. It should be borne in mind that laboratory tests of course represent worst-case scenarios in many respects: the exposure is constant during the test, and the organisms have no means of retreating. Based on these considerations, some authors consider the choice of three very different species (algae, water fleas, fish) adequate for the limnic compartment of the environment (FISCHER & HEIMBACH, 1991), whereas others (sometimes including the sediment compartment) call for tests with up to eight species (STEPHAN et al., 1985; cf. Chapter 8.2.1). For environmental chemicals in the terrestrial medium, a tiered scheme using five species including plants, micro-organisms, earthworms, and two beetle species has been proposed (RÖMBKE et al., 1993). For other compartments of the environment, however, it is not yet clear how many or what types of species should be tested.

Extrapolation from Laboratory to Field Conditions

FLETCHER et al. (1990), who evaluated studies with 13 plant species and 17 chemicals, also found that the determined effect concentrations varied by no more than a factor of 2 between the laboratory and field level. SLOOFF et al. (1986) came to a similar conclusion after reviewing the literature yielded on the aquatic milieu for the effects (given as NOEC) of 38 environmental chemicals and pesticides in acute tests and tests at ecosystem level (e.g. in microcosms). There are fewer examples available for the terrestrial medium, but, in the case of pesticides in the field, for example, earthworms did not show any effects which had not already been detected in the laboratory test at the same concentration (HEIMBACH, 1992). The remarks on the worst-case situation in the laboratory test are basically also valid for this extrapolation step. This is demonstrated by the fact, for example, that the toxicity of pyrethroids for aquatic organisms is very high in the laboratory but very low in the field — due to high adsorption to sediment particles (HILL, 1989).

Although the differences revealed in the examples mentioned were small, despite all of the open questions (e.g. with regard to the parameters compared, the prediction of NOEC values in the ecosystem tests or the difference from real field situations), there are other examples (primarily from the aquatic medium) in which it was not possible to predict which test level (laboratory or field) would react more sensitively (MALTBY & CALOW, 1989).

Conclusions

On the whole, past experience has shown that there is no adequate scientific basis for the individual extrapolation steps (OECD, 1989a), although extrapolations of NOECs (e.g. on the basis of acute tests) are possible for many, but not all, chemicals. In any case, the results of such predictions must be taken with a grain of salt since the predicted value and real effects may deviate significantly in an appreciable number of cases. The most important reason for these deviations is probably indirect effects of the chemical, which are in turn often manifested in various kinds of exposure situations. False prognoses in extrapolations, given the same exposure situation, are in all likelihood primarily a function of the chemical's active mechanism (the more specific they are, the greater the anticipated differences, for example, in the comparison of various species).

When discussing the uncertainties, one should not lose sight of the fact that, even if the testing system and tested chemical are the same, tests conducted in different laboratories can result in fluctuation bands averaging < 10, and in very extreme cases up to 250 (RUDOLPH & BOJE, 1986). For this reason, however, at least 10 should be figured into all extrapolation calculations as a method-specific factor of uncertainty. On the other hand, it is much more difficult to quantify the combined influence of the other three extrapolations, as already pointed out. From that standpoint, the positive assessment given to the usefulness of LC_{50} data from acute tests by SLOOFF et al. (1986) must also be discounted, as they compared the uncertainties in the three extrapolation steps based on different sets of data from the aquatic medium. In their view, the number of acute tests with taxonomically and physiologically different species should be increased instead of increasing the number of chronic or, for that matter, ecosystem tests.

For all of the difficulties involved in the assessment of effect concentrations, prognostic methods are the only feasible and practical option available to authorities for assessing the ecotoxicological hazard potential of chemicals, especially in view of the large number of chemicals that need to be evaluated and the complexity of the ecosystems to be protected.

12.6.3 Prediction of Effect Concentrations Based on Chemical Properties

Another approach to the prediction of NOECs is to correlate well-known ecotoxicological effects (e.g. from bacteria tests) with certain physico-chemical properties (especially the n-octanol-water partition coefficient) of the respective foreign substance (e.g. HAWKER & CONNELL, 1985; DEVILLIERS et al., 1987). Put more generally, the QSAR (quantitative structure-activity relationships, Figure 12.6) are regarded as a method which is used to determine a certain chemical property, either by means of a parameter which is easy to measure or calculate or directly from the molecular structure (HERMENS, 1989).

Positive correlations were found to exist for environmental chemicals such as phenols in water (e.g. correlation between increasing toxicity and increasing lipophilia, NENDZA & SEIDEL, 1988). Chemical-specific QSAR were derived on the basis of these relationships which made it possible to predict their bioaccumulation, for instance. QSAR proved to be especially reliable for

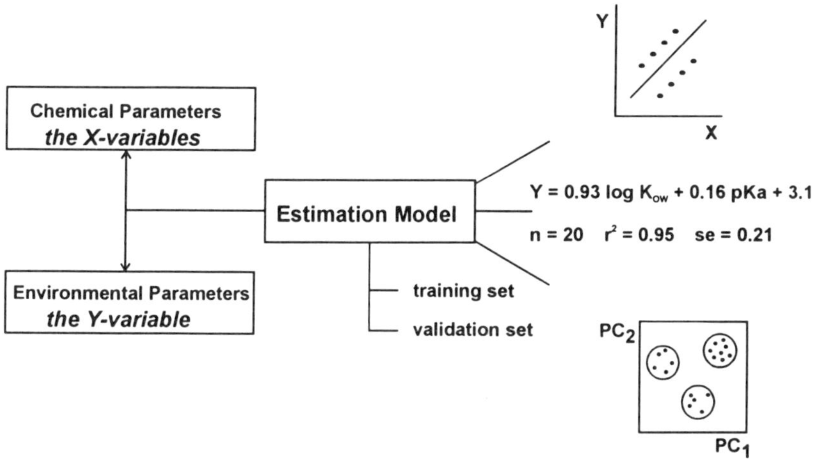

Figure 12.6 Schematic explanation of QSAR (adapted from HERMENS, 1989).

environmental chemicals with a narcotic mode of action (VAN LEEUWEN et al., 1992). The toxicity for *Daphnia* was also predicted on the basis of a few test results for many (not all!) substances in the same class of chemicals (TOSATO et al., 1991). It is much more difficult to do this in the case of organophosphates, however. These substances are metabolized in the organism, and the enzyme systems involved and the sites of effect are not always known (DE BRUIJN & HERMENS, 1991).

These QSAR generally cannot take the place of pesticide tests, although there have been many attempts to use them in that way (KOCH & NAGEL, 1988; CALAMARI & VIGHI, 1990). BIAS (1993) attributes these difficulties not least of all to the sharp divergences in the sensitivity distribution of the organisms, due to stochastic events. Although the derivation of effect concentrations from molecular structures is not (yet) suitable for routine application, it could be used to set priorities for further investigations (e.g. among the large number of existing chemicals, KLEIN et al., 1988).

12.6.4 Two Concepts for the Extrapolation of Effect Concentrations

In view of the problems involved in predicting effect concentrations, two different extrapolation strategies were developed for the notification or registration process for chemicals and pesticides, both of which (or in combination, as in Holland) are under discussion by various authorities. The so-called "Safety Factor Concept" as well as the "Distribution Model Method" are used to determine the concentration of a chemical in which no effects or only negligible effects are observed under field conditions. Few if any direct comparisons of the two approaches based on the same set of data have been conducted to date. In a test with more or less randomly selected environmental chemicals and pesticides in the aquatic medium, OKKERMAN et al. (1993) concluded that the experience gained thus far was not adequate to warrant giving general preference to one or the other method. For theoretical reasons, however, these authors tend to prefer the distribution model to the use of largely fixed safety factors. On the other hand, FORBES & FORBES (1993) recommend the "Safety Factor Concept" due to its simplicity.

The Safety Factor Concept (Assessment or Application Factors)

In the Safety Factor Concept, the range of concentration which must in no case be reached in the environment is determined on the basis of existing test data, employing " (un)safety" factors (ranging from 1 to 10,000). The scale of these factors is derived from a separate prognosis of various uncertainties (extrapolation from one species to another, from acute to chronic, from laboratory to field) as well as their subsequent multiplication (RUDOLPH & BOJE, 1986). The ranges of concentration to be avoided per environmental compartment are then computed by dividing the existing test data by the respective safety factor.

Generally speaking, the worse the quality of the data (e.g. the lower the ecological relevance of a test result), the higher the safety factors must be. This means that data from acute tests must be assigned higher factors than those from microcosm studies. In hyperbolic terms, it could be said that safety factors amount to "educated guesses", i.e. they are on a scale which experts deem acceptable in good conscience. Additionally, at least three tests with organisms from two different taxonomic groups should be conducted for each level, whereas one field study per level should be sufficient (ECETOC, 1993). The details of these specifications cannot be decided on any scientific basis; instead, like the scale of safety factors, they are the result of empirical considerations such as the relationship between acute and chronic tests for as many species as possible (BIAS, 1993). In other words, it is a matter of decisions made by committees of experts.

Various gradations of this concept are currently being discussed in the various countries (Table 12.2). At the OECD level, for example, a factor of 1000 is suggested for environmental chemicals if only a few LC_{50} data are available, and, if chronic NOEC values exist, a factor of 10 is recommended. An individual (case-by-case) evaluation must be carried out if field data are available (OECD, 1989a). This approach can be refined further (e.g. by the EPA or the OECD) for the aquatic medium due to the relatively good quality of the data. These organizations suggest a factor of 1 for data from field studies.

ECETOC (1993) modified this approach by quantifying the respective safety factors on the basis of data compiled in a research study of the literature available (Aquatic Hazard Assessment Database (AHA)) on environmental

Table 12.2 Overview of Various Safety Factors Depending on the Origin or Quality of the Data (Based on Various Sources)[1]

Data source	EPA (1984)	OECD (1989a)	ECETOC[2] (1993)
Fewer than 3 acute test values	1,000	1,000[3]	200
At least 3 acute test values	100	—	—
Fewer than 3 chronic test values	10	100	5
At least 3 chronic test values	10	10	—
Existence of field tests	1	1	1

1 QSAR forecasts could be employed at all levels except the field, in addition to test data.

2 See Table 12.3 for the calculation.

3 If warranted, the factor recommended by OECD (1989a) may be raised to 10,000; however, this should definitely be an exception. Before applying such high factors, the quality of data should be upgraded by additional tests.

chemicals (without inorganica and metals). As of this point, the assessment factors were no longer based entirely on predictions (Table 12.3). Inter-species disparities are no longer taken into consideration as an uncertainty factor since the scale of these deviations are supposedly covered by the spectrum of (aquatic) standard species employed (algae, *Daphnia* and fish). Instead, the relationship between acute and chronic, chronic and ecosystemic as well as the ecosystemic and "real field NEC" (which cannot be determined) is estimated. The recommended factor is oriented to the 90th percentile values of the respective set of data. Based on these values, which seem low in comparison to other recommendations, the PNEC or lowest LC_{50} value would be divided by 200, the lowest NOEC would be divided by 5, and a field-study NOEC would be divided by one (Table 12.3).

For all of the reservations about the use of such factors, they can nonetheless indicate the scale above which chemical effects must be anticipated (RIPPEN, 1987; DÜRRSCHMIDT, 1987). In any case, the results of field

Table 12.3 Safety Factors Based on the AHA Database (ECETOC, 1993)

Transition from	Sources	Range	Factor
Acute to chronic (LC_{50} to NOEC)	5–58	0.1–668	40
Chronic to ecosystemic (NOEC to NOEC)	13 *	0.1–6.3	5
Ecosystemic to field	—	—	1

* Including 10 pesticides

monitoring studies (such as chemical concentrations in various environmental compartments) should be included as far as possible. FORBES & FORBES (1993) also suggest that median or geometric mean of a test dataset be used instead of the lowest NOEC as the calculation basis to which the safety factor is applied, in order to avoid the random results that occur with sharply fluctuating NOEC values.

As already mentioned, this approach can basically be used for environmental chemicals and pesticides alike. However, the question is whether the safety factors for pesticides can be lower than those for environmental chemicals:

- The stressed area is smaller and, in spite of spray-drift, etc., easier to overview.
- A recolonization is more likely to be possible due to the time intervals between applications.
- Because pesticides are normally designed to have a selective effect and the state of knowledge about them is comparatively good, pesticide effects are easier to assess.

In any case, increased attention should be devoted to the duration of stress in the case of both groups of chemicals. Where pesticides are involved, it is assumed that the site and the populations will recover by the time of the next application (ROTHERT et al., 1990). However, this assumption — which is doubtful in monoculture — is clearly not valid for environmental chemicals.

The Distribution Model

The basic idea of this approach goes back to KOOIJMAN (1987), who, however, made use of LC_{50} values. It seeks to protect a previously defined part of the biocoenosis by means of a model for the sensitivity distribution of all species in an ecosystem based on several NOEC values (VAN STRAALEN & DENNEMAN, 1989; WAGNER & LÖKKE, 1991; ALDENBERG & SLOB, 1991). Put another way, an attempt is made to extrapolate from tests with chronic data to the field situation (Figure 12.7). The objective is to determine the "5 percentile Hazard Concentration" (HC_5), at which the NOEC is exceeded for a maximum of 5% of all species in a biocoenosis. The higher this figure, the lower the presumed ecotoxicological hazard potential of the respective chemical.

This approach is based on the following considerations:

- Ecosystems are exposed to "stress" even without human intervention, i.e. a 100% preservation of the status quo is neither possible nor desirable.
- If the structure (= species inventory) of the ecosystem is protected, the functions are protected as well.

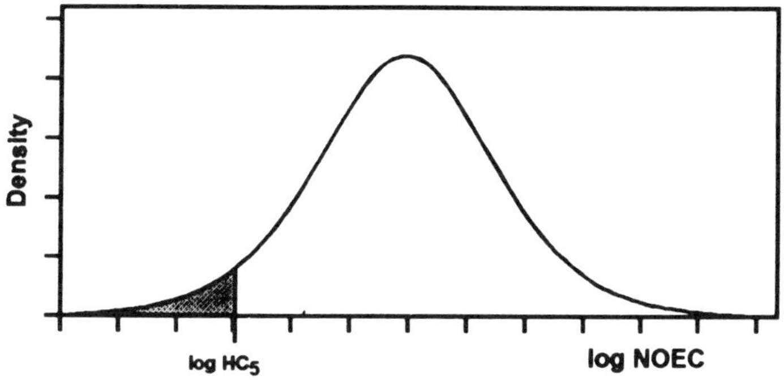

Figure 12.7 Graphic representation of the distribution model with definitions (according to VAN STRAALEN & DENNEMANN, 1989).

More specifically, this approach presumes that the variability of the sensitivity between various species (inter-specific) and within the same species (intra-specific) is symmetrically distributed. The assumption of such a distribution is realistic based on considerations by KOOIJMAN (1987), using data from SLOOFF & CANTON (1983). Analogous to the estimation of EC_{50} values in laboratory tests, the HC_5 is derived directly from these distribution curves (Figure 12.8). Depending on the author, this distribution is described as "log-logistic" (not much different from the normal distribution but easier to calculate; VAN STRAALEN & DENNEMAN, 1989; ALDENBERG & SLOB, 1991), "log-normal" (WAGNER & LÖKKE, 1991) or "log-triangular" (EPA, 1984; STEPHAN et al., 1985). Since the last-mentioned approach implies the indication of threshold concentrations (points intersecting with the x-axis) at which effects do not yet or no longer appear, and the line of argumentation is based on family level rather than species level, it will no longer be considered in the following discussions.

In a comparison of various methods, and using a set of aquatic data (SLOOFF & CANTON, 1983), WAGNER & LÖKKE (1991) found that their own method as well as that of ALDENBERG & SLOB (1991) yielded practically identical hazard concentrations (HC), whereas the VAN STRAALEN & DENNEMAN method consistently yielded somewhat higher values, i.e. the hazard potential was underestimated in this case. In a comparison of QSAR toxicity estimates as well as HC_5 calculations for approximately 100 environmental chemicals with a narcotic mode of action in the aquatic medium, VAN LEEUWEN et al. (1992) were unable to ascertain any appreciable differences between the two calculation methods used (log-logistic versus log-normal). OKKERMAN et al. (1993) obtained the same result using data reported in the literature on the effect of organic environmental chemicals and pesticides in the aquatic system. Generally speaking, the differences between the various methods of calculating the distribution model should be relatively minor, i.e. based

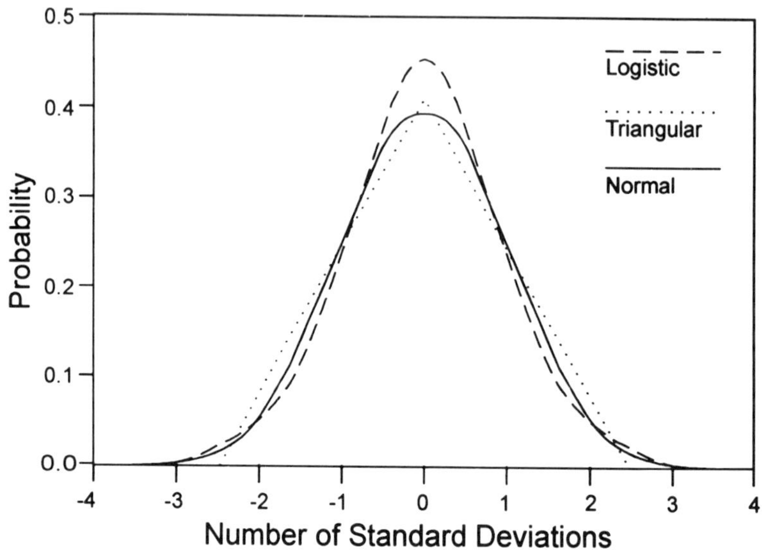

Figure 12.8 Probability density functions with a mean and standard deviation of 0 and 1, respectively (VAN LEEUWEN, 1990; HCN, 1989).

on the current state of knowledge, the calculated HC_5 values lie in the same range (OECD, 1992). The approach taken by VAN STRAALEN & DENNEMAN (1989), which attracted special attention for its use in the terrestrial medium, must be considered slightly less suitable than the other two methods.

The practical problem involved in all of the modifications of this approach is that, unless additional safety factors are to be incorporated, the results are extremely dependent on the number of values available. In order to avoid a high degree of uncertainty, 5-8 NOEC values must be available from tests with various species. This number was calculated both by STEPHAN et al. (1985) and VAN LEEUWEN (1990) on the basis of a set of data from SLOOFF & CANTON (1983), in which as many different taxonomic units, trophic stages and exposure pathways as possible were to be covered. OKKERMAN et al. (1993) point to the fact that, if at all possible, seven sets of data should be available in order to reduce the variability as much as possible, but that five tests would also be acceptable as an absolute minimum for ethical and economic reasons. Some authors, do, however, consider four NOECs as adequate (SLOOFF & VAN DE MEENT, 1990). Neither four nor five seem sufficient because the variability in the calculated concentrations would increase sharply. This is especially problematic if, in addition, these NOECs have already been estimated from EC or LC values from acute tests, which is the case particularly in the terrestrial medium.

The discussion is currently focused on the question of which distribution curve is more critical given a small number of species and whether overprotective threshold concentrations may even tend to discredit the approach (WAGNER & LÖKKE, 1991). Due to the inherent difficulties involved in the use of extrapolation factors, however, data from higher levels of investigation should be consulted for the assessment, as called for in the tiered test program (SMITH & CAIRNS, 1993). Microcosms are suitable for this purpose, as well as data derived from chemicals in the same class. There is another problem, which arises when it comes to assessing chemicals which occur naturally in the environment and, in some cases, are even essential. With this model, the difference between natural background levels and concentrations which are given a "non-hazardous" assessment may be minimal (SLOOFF & VAN DE MEENT, 1990; HOPKIN, 1993).

Aside from the high expectations, the "Distribution Model" method has the following drawbacks (cf. BIAS, 1993; FORBES & FORBES, 1993; HOPKIN, 1993b; SMITH & CAIRNS, 1993):

- there is a need for further evidence to support the assumption of a normal distribution of sensitivity of all organisms in an ecosystem, regardless of their taxonomic differences;
- the selection of test species from the totality of all species of a biocoenosis is not random;
- the various functional roles and the possible exposure pathways are seldom covered by this selection;
- the 5% threshold is artificial (key species might not be taken into consideration);
- interactions of various species are not taken into account;
- laboratory tests serve as the sole database;
- most of the reported uses of distribution models to date (incl. comparative studies with the use of safety factors) are based on the same set of data employed by SLOOFF et al.; in other words, validation studies are urgently required;
- the respective NOEC strongly depends on the measurement parameter employed.

Since other forms of extrapolation are at least equally subject to such shortcomings (including the QSAR estimations), the "Distribution Model" method must currently still be regarded as the best approach. It is also preferable to the Safety Factor Concept (VAN LEEUWEN, 1990; UBA, 1992), although the underlying database is still extremely unsatisfactory. Moreover, it is doubtful whether a damage of no more than 5% is ecologically and ethically acceptable for the most important species (FORBES & FORBES, 1993). Even so, as VAN STRAALEN (1993) rightly points out, this 5% "damaged" does not automatically mean "killed", and the protection level can still be adjusted if key species are affected. Generally speaking, this is why there is still no consensus regarding the suitability of distribution models. In addition to positive views of the approach (e.g. VAN LEEUWEN, 1990), there has been a growing number of calls for further validation studies (e.g. SMITH & CAIRNS, 1993).

In a reversal of the procedure described thus far, the "Distribution Model" method can also be used to establish the degradation rates of pesticides required to be safe (e.g. DT_{50} values). A formula was developed for this purpose, which includes an exponential degradation of the chemical and the log-logistic sensitivity distribution of the species of the ecosystem observed (VAN STRAALEN et al., 1992). In this case, the ecotoxicological hazard potential is at a minimum if the concentration of the substance in the soil has dropped below the respective HC_5 before it is applied again. This would ensure that the NOEC values are not exceeded in more than 95% of the species living in the relevant ecosystem.

12.7 RISK CHARACTERIZATION

In various countries (e.g. USA, Germany, Netherlands) as well as international organizations (OECD, 1989a; EC, 1993), guidelines are currently being prepared for the "Hazard and Risk Assessment" procedure. None of these plans can yet be regarded as complete. For example, there are still no means of covering intermittent exposure conditions. Although all of the guideline proposals are based more or less on the same approach, they differ in terms of the kinds of cases in which they are applied, the purpose and range of safety factors as well as the required base data. ECETOC (1993) provides an up-to-date summary of the most common approaches. It is difficult to compare the various models because the same terms are used with different definitions or vice-versa. For example, usually the PEC/PNEC ratio is used in the risk characterization step (ECETOC, 1993), whereas the German UBA makes use of the Q (= NOEC/environmental concentration; cf. Figure 12.9) for its evaluation (UBA, 1992). It should be borne in mind that the hazard assessment must be carried out in specific compartments. Supra-compartmental assessments or evaluations are conducted at later stages of the overall process.

In the case of pesticides, a very detailed "risk assessment scheme" based on cooperation between EPPO and the European Council (EC, 1993) was recently presented. Although the overall approach used in this procedure is similar to other methods of evaluating the ecotoxicological hazard potential, it differs in terms of the names and definitions of the individual steps (GREIG-SMITH, 1992). The environmental fate of the substance is distinguished according to whether it occurs in the soil, the ground and surface water as well as by its effects on aquatic organisms, soil microflora, earthworms, beneficial arthropods, bees and terrestrial vertebrates. On the other hand, the atmospheric compartment is left out as well as plants, at least until concrete signs of risk

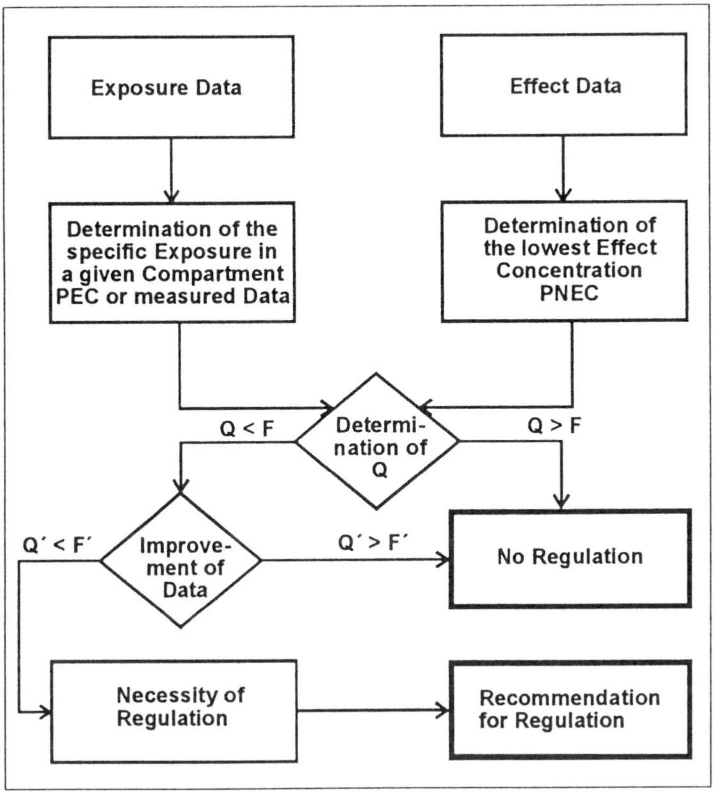

Figure 12.9 Evaluation and assessment of the environmental hazard of wastes by the German Federal Environment Agency (according to UBA, 1992).

appear. Such a detailed comparison is suitable to indicate the relative risk to individual species and can thus be useful for the definition of recommendations pertaining to the application of pesticides in the interest of protecting beneficial organisms (WILES et al., 1991). A transfer of the data or of the results of the risk assessment from one medium to another is not possible, although this has been suggested as part of an effort to simplify the assessment process (e.g. rates of soil degradation from aquatic biodegradation tests, KLEIN, 1991). It is still too soon to say whether the EPPO's procedure will have any chance of becoming standard practice.

In the EU, the concentration at which there are no longer any observable effects on ecosystem level is estimated on the basis of several specific NOEC data (measured or estimated from acute data) as defined by the EU "Risk Assessment Directive" (1993). Depending on the quality of the test result, a factor ranging between 1,000 (based on a LC_{50}) and 10 or 50 (based on NOECs) is used to calculate the "Predicted No Effect Concentration" (= PNEC) from the measured value. If or as more data become available, this factor may change on a case-by-case basis. This PNEC is then compared with the estimated exposure value (Predicted Environmental Concentration = PEC). The PEC/PNEC ratio then indicates whether any hazard potential exists (>1) or not (<1). If a hazard potential is found to exist, and a substance is classified as a "substance of concern", the registration authority must decide whether further tests should be conducted (e.g. in-depth tests, monitoring studies) for it. The ratio of these two values, the PEC and PNEC, determines which steps are subsequently taken in the assessment process. In any event, the exposure side should be analyzed as precisely as possible before proceeding with further effect tests.

In most of the procedures mentioned, decision-making aids for the PEC/PNEC comparison in each of the environmental compartments studied are presented in the form of a decision-making tree ("process maps"), which often distinguishes between existing and new environmental chemicals and pesticides (e.g. OECD, 1989a). These process maps are used, first of all, to determine which tests should be conducted according to which criteria in the framework of a tiered test program. An example of this procedure (EPPO, 1993) is provided for fate in the soil compartment as well as the effect on aquatic organisms (Figures 12.11 and 12.12). Secondly, the maps help to indicate whether the ecotoxicological hazard potential can be regarded as negligible or whether it is warranted to proceed with the risk management step. Such decision-making trees must be designed for each environmental compartment or every group of organisms exposed to a chemical in the same way. The general principle of the process based on a suggestion by the ECETOC (1993) is illustrated in Figure 12.10. The ECETOC publication also contains a few case studies on environmental chemicals in the aquatic medium.

Comparable to the approaches mentioned so far, the EPA's Ecological Risk Assessment is referred to as the "quotient ratio method": a measured exposure value is compared directly with a test result (e.g. a given concentration in water with the LC_{50} in the fish test). The closer this value comes to 1 (or even higher), the greater is the probability that damaging effects will occur (BASCIETTO et al., 1990). In order to even out differences between various species and also between the laboratory and field level, a safety factor is introduced before the final assessment step. This method is used primarily for environmental chemicals. For pesticides, a predicted environmental concentration is compared with toxicity data. The main criticism of this approach, even by the EPA's own "Science Advisory Board", is that it is, on the whole, not sufficiently relevant for the field (BASCIETTO et al., 1990). For this reason, an improvement of the database (e.g. through microcosm studies) as well as the introduction of monitoring studies in the framework of risk management is considered necessary (SMITH & CAIRNS, 1993). In other words, a study of the ecosystem level is advisable when conducting the risk assessment. Another problem with this approach which has been little considered to date concerns

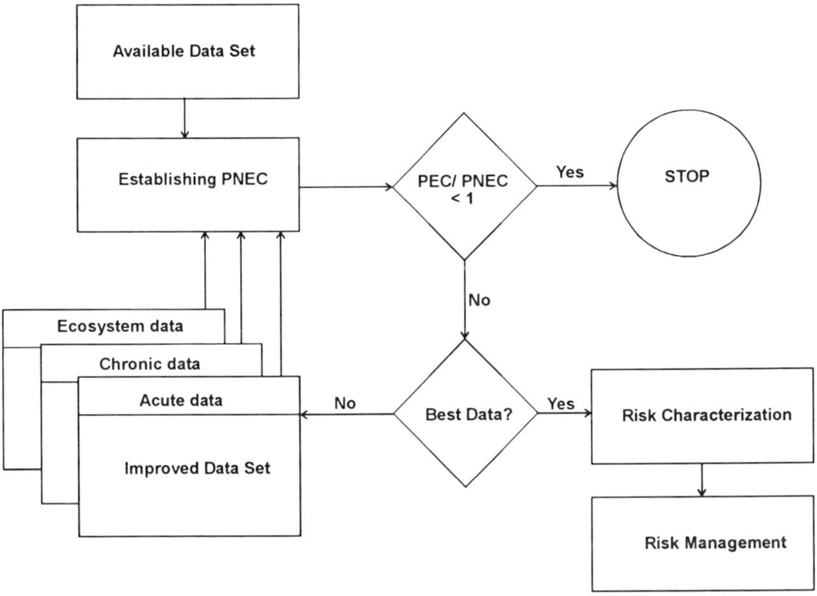

Figure 12.10 Condensed process scheme effects assessment (establishment and refinement of PNEC, ECETOC, 1993).

the transferability of the results, which at best are compartment-specific, from one type of ecosystem to another.

For this reason, the EPA has more recently been considering an "Ecological Risk Assessment Triad" (WARREN-HICKS et al., 1989), in which site- and ecosystem-specific exposure data from the respective compartment are compared with the data collected in monitoring surveys in the field. If the latter yield negative effects, tests (usually in the laboratory) serve to determine the causality between both observations. In other words, the triad approach is more specifically a matter of formalizing the use of monitoring data in the framework of the Hazard and Risk Assessments. Examples of successful applications are available for sediments (Figure 12.13; CHAPMAN et al., 1991) or wetlands (PASCOE, 1993), among others. If data are available for all three sectors (exposure data, laboratory tests, monitoring programs), the resulting assessment of ecotoxicological hazard potential will obviously be much more accurate than it would be on the basis of laboratory tests alone. However, in many cases there is probably not enough information to differentiate the effects of chemicals from natural variability to a statistically significant degree.

In this context, the question has recently been raised in the United States as to which methods will yield stricter quality criteria for assessing the effect of certain chemicals: either the concentration at which no direct effects on organisms can be anticipated are predicted in the framework of "Hazard

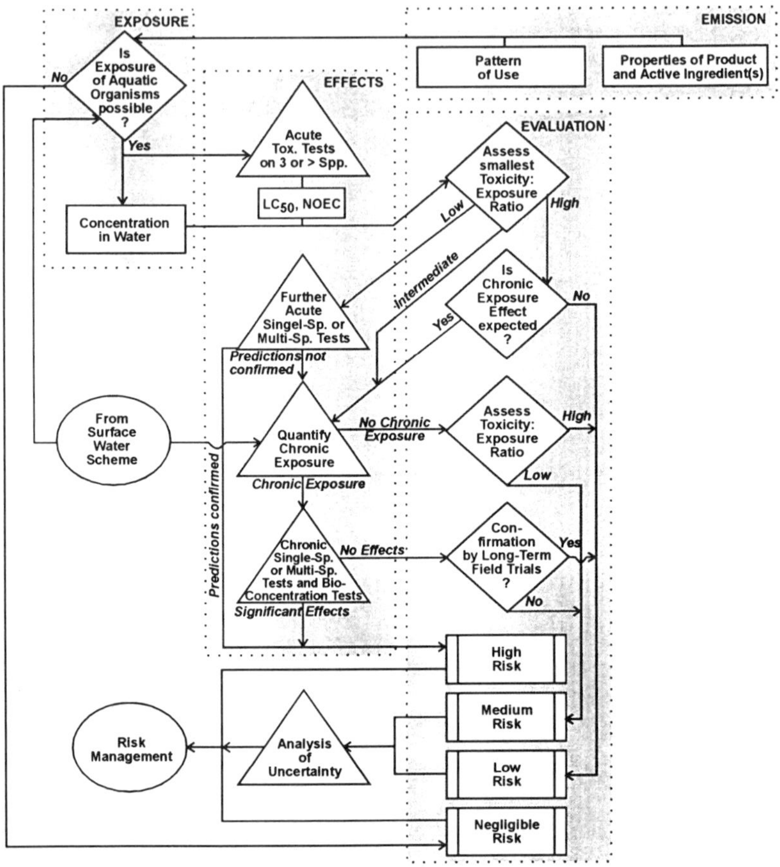

Figure 12.11 Hazard assessment of a pesticide in the soil (EPPO, 1993).

Assessment" or human health is established as the protection target. In this case, the environmental concentration that should not be exceeded is established by assessing the bioaccumulation factors. In the case of non-ionic organic chemicals in limnic sediments, for example, PARKERTON et al. (1993) found that the indirect human toxicological procedure would lead to tighter restrictions. Consequently, the authors recommended that such "indirect" effects be included in the assessment of hazard potential in the future.

12.8 RISK MANAGEMENT (= EVALUATION)

In spite of the often unsatisfactory quality and availability of data, it is ultimately up to the registration authority to evaluate the existing ecotoxicological datasets, including the ratio between exposure and effect concentrations: in short, it must determine whether the substance is "hazardous to the environ-

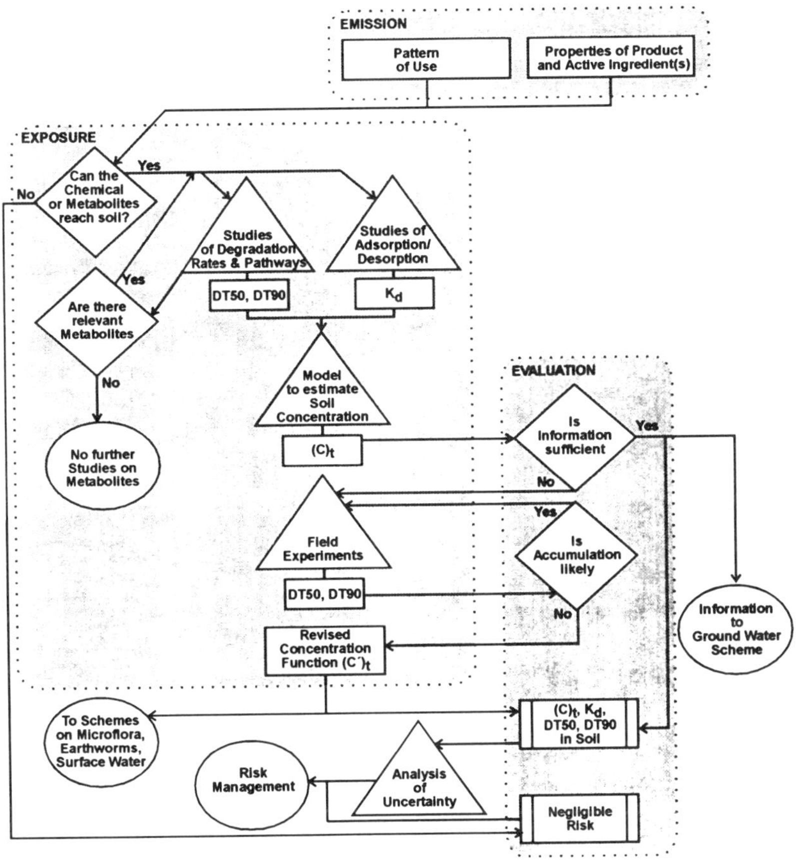

Figure 12.12 Hazard assessment of a pesticide for aquatic organisms (EPPO, 1993).

ment" or not. If at all possible, additional information such as the time pattern of an effect — which often means "expert knowledge" — should be included in the evaluation process (NORTON et al., 1992). If the risk assessment confirms the existence of an ecotoxicological hazard potential, recommendations must be made that will reduce the hazard potential in the framework of Risk Management. This may be done, for example, by rating a substance as "environmentally hazardous", which in turn can lead to restrictions of various kinds (e.g. on the use of a pesticide within a minimum allowable range of the nearest source of water or certain types of waste disposal). In extreme cases, it may mean an outright ban on its application. Restrictions can also take the form of mandatory monitoring for a specified period of time in the framework of monitoring programs with appropriate methods to ascertain any possible effects of routine use (WEIDEMANN & MATHES, 1991).

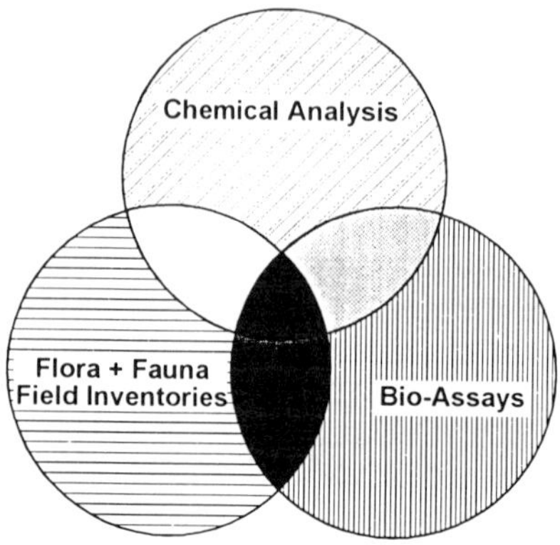

Figure 12.13 Sediment Quality Assay: TRIAD Approach Model (according to CHAPMAN, 1990b).

In industrialized countries, pesticide registrations are generally valid for a maximum of 10 years, but may be revoked at any time if warranted. Regulatory authorities, and producers themselves, monitor the effects of pesticides based on the users' reports (e.g. the occurrence of resistances), and, in several countries, this practice is even legally prescribed. To date, however, it could not be said that there are any monitoring programs which are carried out in a consistent or coordinated manner.

The implications which an evaluation ultimately has for the registration or notification of a chemical may very well depend on other criteria (of a legal, economic or social nature). For example, are there other reasons for permitting the use of a pesticide in spite of the suspected environmental hazard? If so, the benefits of a chemical's use (in the case of a pesticide, its efficacy) should also be considered in a risk-benefit analysis in this step.

The current discussion of the overall drinking water threshold value that would apply for pesticides in the EC demonstrates the degree to which external, political factors intervene in the decision-making process. As opposed to the regulatory principle applied to threshold values for other substances, threshold values for pesticides were not defined on the basis of a differentiated scientific assessment. A uniform threshold value of 0.1 μg/l is applicable for any single active ingredient and 0.5 μg/l for the sum of all active ingredients in the drinking water. These values lie at the detection limit for pesticides. As such, they reflect the underlying conviction that these substances should basically not be tolerated in drinking water, regardless of whether they have any proven toxicological effect or not. The threshold value of 0.1 μg/l for pesticides in drinking water is based on the interest in protecting the health of consumers. There is also a call for the same threshold value to be applied to ground and surface water as sources of drinking water supply. This would mean extending the scope of precautionary consumer protection to the waterworks, which could thus largely dispense with the treatment of raw water to produce drinking water. Although it is unlikely that toxicologically relevant values will be found, given the level and kinds of pesticide application today, a short-term entry in surface water is inevitable during the application phase due to spray-drift or runoff. A strict enforcement of the drinking water threshold value would thus amount to a ban on all pesticides — a conclusion which would not be warranted on the basis of ecotoxicological knowledge or economic considerations.

Environmental chemicals are a somewhat different matter. Substances found to be hazardous under EU legislation (UBA, 1992, ECETOC 1993):

- may be banned until further information for a definitive evaluation is submitted;
- may be distributed only under certain prerequisites or only to certain persons;
- may not be produced or utilized, or only in a certain composition, or only for certain purposes.

Moreover, producers, distributors or users may have to have certain qualifications (e.g. evidence of know-how). In extreme cases, whole production procedures in which hazardous substances occur may be prohibited. The goal of all measures is to exclude the possibility of an environmental hazard by reducing the exposure, in some cases down to the zero emission level, if, as in the case of cancerogenic substances, no effect concentration can be indicated under which a risk can be ruled out (UBA, 1992). However, the simple designation of a substance as "environmentally hazardous" will probably be much more common.

The problem in this connection is the "value" assigned to individual populations, species or system functions beyond the useful value for human beings, which is relatively easy to assess (BALK & KOEMAN, 1984). Due to the complexity of natural ecosystems, it is practically impossible to judge whether, for example, the disappearance of a species of collembola (out of 60 which occur) after chemicals take effect on the forest floor will have a negative effect on the function of the biocoenosis (e.g. by slowing down the litter degradation; BECK et al., 1988). Does this then mean that parts of the ecosystem such as a collembola species are negligible? Put another way, does every species have a value of its own, independently of its "purpose"? These are questions which can no longer be answered exclusively on the basis of scientific research; rather, personal and ethical criteria also play a role (WEIDEMANN et al., 1987; MATHES et al., 1991). In any case, it does not appear permissible to apply purely economic, scientific or aesthetic criteria (MORIARTY, 1988). Yet, it is unrealistic to set absolute targets of protection which are derived from a static picture of nature. Human activities must be evaluated in relation to natural processes. In this sense, the continuing development of ecotoxicology, and the application of this knowledge in the form of laws and guidelines in particular, also depend on the discussion of the ethical standards on which these activities are based (CAIRNS & PRATT, 1989).

13 ECOTOXICOLOGY IN DEVELOPING COUNTRIES

13.1 OVERALL SITUATION

Since 1972 (UN Conference on Problems of the Environment, Stockholm), if not before, efforts have been made to predict the environmental effects of pesticides and environmental chemicals in developing countries (TURTLE, 1977). In the mid-1970s, experts from international (FAO, UNEP, WHO, UNESCO; MAB Program) and national organizations (e.g. USAID, GTZ) as well as GIFAP agreed on an impact monitoring program to study the fate and effects of pesticides under normal application conditions.

In the framework of this project, TURTLE (1977) formulated a few basic conditions for the assessment of the ecotoxicological hazard potential in developing countries. These conditions, which correspond to the tiered procedure presented in Chapter 7, are for all practical purposes still valid today:

- Data on the environmental fate and effects on organisms are to be collected in standardized laboratory tests under tropical conditions.
- Effect tests are to be supplemented by accompanying chemical residue analysis data; if at all possible, these studies should be conducted in laboratories in Third World countries.
- Data from the literature and laboratory experiments are not sufficient for prognoses of environmental hazard potential, which can only be validated with the help of field studies.
- Every ecosystem must be studied separately.

Back in 1977, KOEMAN also called for ecologists (ecotoxicologists in today's terminology) to be involved in all decisions pertaining to pesticide use. The areas of application which were especially problematic at that time are still problematic today (EDWARDS, 1977; MÜLLER et al., 1981; BALK & KOEMAN, 1984; FAO, 1990b; HCN, 1992):

- rice- and cotton-growing;
- combatting locusts (Figure 13.1), mosquitoes and tsetse flies.

Figure 13.1 Combatting locusts in Burkina Faso with a sprayer mounted on an off-road vehicle (courtesy of L. Mosich, GTZ).

The conflict between the purpose of pesticide use and the side effects resulting from its use is quite clear in the case of locust control: until 1988, these insects were combated with ecotoxicologically problematic chlorinated hydrocarbons, especially Dieldrin (BRADER, 1987; PASTRE et al., 1989), although it had long been known that this substance, prohibited in the United States and Germany since the early 1970s, can have long-term effects on non-targets.

Reservations about the problems related to the use of Dieldrin prompted the countries of origin as well as funding organizations such as the FAO to replace it for the most part with other agents, such as pyrethroids (e.g. in Jordan; ALAWI et al., 1990). However, many new pesticides must be applied more frequently or in higher concentrations to achieve the same effects (MÜLLER, 1988). Moreover, in many cases, the substances used as alternatives were also questionable from an ecotoxicological standpoint. Unfortunately, there are no systematic studies on the fate and effect of these substances under tropical conditions (GRANT, 1989a).

Criticism of large-scale pesticide applications (e.g. Fenitrothion and Malathion on 15 million ha in North Africa in 1987/88) is becoming increasingly vociferous in the Third World as well, for two reasons:

- The ratio between the benefits of pest control and the anticipated side effects is becoming exceedingly unfavorable.

- Human- and ecotoxicological side effects are not taken into sufficient consideration (AGROW of 17 August 1990).

The governments of the developing countries, development assistance organizations in the industrialized countries which promote agricultural projects, environmental protection groups and producers of pesticides have recognized that improper use of these chemicals is an infrastructural problem. Various measures providing for storage and user safety, extension services (GIFAP Safe Use Projects) as well as residue controls enable many farmers in developing countries to benefit from the advantages of chemical-based crop protection at medium or high levels of production. However, considerable improvements must still be made in order to cope with the use-related problems in terms of higher safety standards and quality of application, as well as the awareness of possible side effects. There is evidently no economically viable mode of application which is also safe from the standpoint of human health and ecotoxicology in smallholder subsistence agriculture (HCN, 1992).

The use of biological methods might be one solution to this dilemma, but this approach is currently still in the development stages (KOCH et al., 1990). Although it is not yet possible to arrive at any conclusive judgement of their effects and side effects (KRALL & NASSEH, 1990), they will come into increased use in the years ahead. Therefore, at the initiative of FAO and in cooperation with IOBC, guidelines for the import and utilization of biological control agents were developed (FAO, 1992), analogous to the Code of Conduct on the Distribution and Use of Pesticides (FAO, 1990a). This set of rules and regulations is intended to ensure that, just like chemicals, living natural enemies of pests (e.g. viruses, bacteria, fungi, invertebrates and vertebrates) do not induce any negative ecotoxicological effects.

The above-mentioned FAO code, which was enacted in 1986, is an expression of joint responsibility on the part of governments, authorities, producers and distributors as well as users to provide for safe utilization of pesticides in developing countries as well. The large international pesticide producers have given their express commitment to abide by this code. For many developing countries, the code provides the first basic framework for pesticide legislation of their own. In 1990, the FAO code was supplemented by the so-called PIC (Prior Informed Consent) procedure, which gives the developing countries access to the toxicity and ecotoxicity data available on substances they plan to import. Although the PIC procedure applies to all chemicals which are subject to restrictions in the country of origin, it is not sufficient in and of itself to safeguard against hazardous situations (HCN, 1992).

Generally speaking, the probability of exposure is surely higher in the developing countries than it is in the industrialized countries. For one thing, the fate of pesticides (and probably of many environmental chemicals) in the tropics is different from temperate latitudes (e.g. BALK & KOEMAN, 1984;

GUPTA & DURVE, 1987; EL-SEBAE, 1989). For example, the fate of DDT under abiotic conditions in tropical climates is completely different from the fate predicted on the basis of studies in the industrialized countries (MATTHIESSEN, 1985). In particular, the persistence of DDT or Lindan, for example, is much lower due to the more rapid metabolism rate of microorganisms at higher temperatures (EDWARDS, 1977; KAUSHIK, 1991). Due to the higher volatility rates from the body of water, the percentage accumulation in sediments appears to be lower than in more temperate climates (RAMESH et al., 1991). Additionally, photochemical conversion (photolysis) plays a more important role in the tropics due to the stronger light effect than it does in higher latitudes. Although the lower persistence in tropical ecosystems, precisely in the case of organochlorine compounds, is considered positive at first glance, negative effects may occur even at sites which are far removed from the application site, as the accumulation of these substances in organisms of polar regions has shown (TANABE, 1991).

Monitoring has, however, not only revealed a more rapid degradation at higher temperatures but also a higher toxicity: as early as 1958, IYATOMA et al. showed that the effects of Endrin on carp (*Cyprinus carpio*) were several hundred times more toxic at 27°C than at 7°C. The metabolism pathway as well as the ultimate environmental fate are not known for most substances.

Furthermore, the exposure is strongly influenced by the special formulations and application techniques in the tropics (e.g. prolonging the exposure time of non-persistent insecticides in combatting tsetse flies; VOLLNER & GHUDS-ESPHAHANI, 1988). Large-scale applications by helicopter and airplane (Figure 13.2) to combat disease carriers or locusts are particularly problematic, since an entry in natural or quasi-natural ecosystems often leads to contamination of rivers and lakes, with severe effects on aquatic insects and fish. Additionally, sharp decreases in birds were also reported in such cases, compared with small-scale forms of application (KOEMAN et al., 1978). The often much poorer quality of application, due not least of all to the user's inadequate level of training, combined with illiteracy and a low level of safety awareness, is an added factor (MÜLLER, 1988). Under these circumstances in practice, the ecotoxicological effects may be far greater than those which were predicted in the laboratory or under controlled field conditions in temperate climates.

In addition, the substances are often applied much differently than intended in the producers' application scenarios — in much larger quantities, for example (GONZALEZ, 1976), or at the wrong time (HCN, 1992). Since many of the underlying ecotoxicological data on pesticides are only collected for the highest normal application quantity (both in the laboratory and in the field), there is no way to cover eventualities of this type in practice, even with the most stringent guidelines. This situation can only be remedied by building up extension and monitoring institutions with the proper know-how and facilities in the respective regions (IGBEDIOH, 1991). The Swiss-funded cooperation

Figure 13.2 Aerial application of pesticides (courtesy of GTZ).

program which has existed since 1989 between the Groupe d'Ecotoxicologie de l'Ecole Polytechnique Fédérale de Lausanne and the Centro de Estudios y de Control de Contaminantes in Tegucigalpa, Honduras, is one example of what can be done to build up local monitoring institutions of this type.

Moreover, tropical ecosystems are considered to be more susceptible to anthropogenic influences than systems in temperate latitudes (WEISCHET, 1977; ALLEN, 1985). As an example, the number of species — at least in humid ecosystems — is much higher, with a generally lower abundance and reproduction of individual species; consequently, species which have been affected recover more slowly (MACARTHUR, 1969). Many of the potentially affected groups of organisms occur primarily in developing countries (e.g. reptiles), where little if anything is known about their exposure to hazard due to the lack of studies. As another example of the higher risk to species in tropical habitats, there is largely no seasonal synchronization of reproduction. This may have a harmful effect on more sensitive juvenile stages. In Germany, for example, the application of certain insecticides is restricted during the reproduction period (spring/summer) of beneficial arthropods such as bees, whereas, in developing countries, potentially beneficial species may reproduce all year long.

Moreover, factors subject to anthropogenic influences, such as the ratio of treated to untreated surfaces or the temperature at the time of application, also play a role. This complex set of factors can only be assessed in terms of the specific situation, i.e. the interaction of the properties of the substance itself with the specific tropical ecosystem.

Despite all of the differences between the situation in temperate and tropical climates, there are only very few ecotoxicological methods as yet which are especially designed for tropical risk assessments. As a result of various research projects in southern Africa, GRANT (1988) published a collection of methods for terrestrial field studies. The only data available for regulatory purposes, if at all, apply to the aquatic medium. The Handbook of the Use of Pesticides in the Asia-Pacific Region (1987), for example, mentions that there are reservations about the use of organochlorine compounds in countries such as India or Malaysia. With the exception of fish tests (in Korea, Indonesia), monitoring programs are evidently only carried out from time to time. It is characteristic of the situation in the Asian-Pacific region that there is no mention of the term ecotoxicology in this handbook.

Another serious problem arises in comparison to the situation in industrialized countries (EDWARDS, 1986): due to the lack or inadequacy of training and experience in utilization or labeling, serious health damages from pesticides occur much more frequently in developing countries (estimated 3,000 to 40,000 deaths per year; health problems, about 700,000; DÖNHARDT, 1987; HART, 1987; PAN, 1989). In addition to the many substances such as organophosphates, for which the human toxicity has been known for a long time, other substances such as pyrethroids, which have been claimed to be unharmful, have allegedly caused hundreds of cases of poisoning (FORGET, 1991). An overview of ecotoxicological fate and effect data on the tropics can be obtained from the English-language ENVIRON database (available from the Natural Resources Institute, Chatham, Maritime, UK).

The availability and quality of data on environmental chemicals is even less favorable than for pesticides. However, due to the lack of safety regulations as well as the many small production sites which do not have much know-how in dealing with hazardous substances, it must be assumed that a considerable hazard potential exists (HCN, 1992). Although the state of knowledge about the ecotoxicological risks of environmental chemicals is minimal, one exception is the use of water hyacinths as an indicator of stress from heavy metals (GONZALEZ et al., 1989). Most recently, moreover, cases are becoming known in which industrialized countries have attempted to get rid of their environmentally hazardous waste by exporting it to developing countries, which can also be assumed to do serious damage to the environment.

13.2 CASE STUDIES

Few ecotoxicological data have been gathered in the developing countries to date. GONZALEZ (1976), for example, lists no more than three studies for Latin America for the first half of the 1970s. By 1991, no less than 437 publications were already available for the same region (MIGUEL, 1991).

There are almost no standardized field or semi-field tests on the environmental fate or side effects of chemicals for the tropics. Monitoring studies are carried out somewhat more frequently, e.g. on the effect of mercury (PFEIFER et al., 1989) or pesticides (CALLEGARI LOPES et al., 1992) in Brazilian rivers or in the Indian "Ganges Action Plan". The development, validation and use of plants as bioindicators under tropical conditions (e.g. in Brazil) was recently proposed by ARNDT et al. (1991). They stress that they are easy to handle, cheap and less vulnerable to vandalism than measurement equipment.

The effects of pesticides on non-targets are often carried out parallel to effect tests in the framework of monitoring projects (which could be called surveys, if only because of what are often very short observation periods, e.g. in connection with combatting locusts (MÜLLER, 1987, 1988) or tsetse flies (MÜLLER et. al., 1981; EVERTS et al., 1985)). In the last study conducted on the Ivory Coast, the spider species *Aporoptychus lamottei*, which was severely inhibited by various pyrethroids, was identified as a potential indicator species. In compliance with the IOBC's beneficial arthropod test rules, this species has to be tested in the laboratory in order to verify its suitability as a test organism.

Another example is the study of side effects from pesticides in the framework of the onchocerciasis program (LAHR & DE JONGH, 1990; YAMEOGO et al., 1993). For a long time, studies on the effects of pesticides in the developing countries were virtually limited to economically important species such as bees (ADEY et al., 1986). An attempt is now under way to compile data systematically on possible side effects from the use of beneficial organisms (certain insects, such as diptera or hymenoptera species) which consume or thrive as parasites on grasshoppers (FAO, 1990b). Generally speaking, possible side effects have been studied more frequently in combatting disease carriers than, for example, those of herbicides in agriculture (HCN, 1992). Data on the use of pesticides in rice crops are the most likely to be available in this area (e.g. ARUNACHALAM & LAKSHMANAN, 1990).

Moreover, earthworms, which are important for many soil functions, have been studied with relative frequency: results are available from India (e.g. BHARATHI & SUBA RAO, 1984), Egypt (GHABBOUR & IMAM, 1967) and South Africa (e.g. VENTER & REINECKE, 1985), among others. Side effects on other groups of soil organisms such as collembola have already been investigated (e.g. in Brazil: GUARRA et al., 1982). Studies on the reaction of termites, which are important for the soil ecosystem in tropical areas in much the same way as earthworms are in the holarctic, are virtually non-existent. Investigations such as those conducted in Nigeria, where the long-term effects (e.g. a slow process of litter decomposition) of DDT on soil functions were examined (SWIFT et al., 1977), are rare. Cooperation between local organizations and international research institutions are very useful for costly studies of this kind (e.g. the International Institute for Tropical Agriculture, Ibadan, Nigeria, or the Joint FAO/IAEA Division of Nuclear Techniques in Food and Agriculture, Vienna, Austria). One noteworthy positive development is that, most recently, efforts have been made (e.g. by the English National Resources Institute, Overseas Development Administration, Chatham Maritime, Great Britain) to gather at least sample data on key ecosystem functions with parameters such as soil respiration, nitrogen fixation or the degradation of organic substances (GRANT, 1989a). A project funded by the FAO/IAEA (Vienna) is under way in eight African countries simultaneously to study the effect and fate of organochlorine pesticides in farming systems (especially Lindan in maize cultures) with various methods.

As mentioned, the field studies usually involve short-term comparisons between treated areas and untreated controls (MÜLLER, 1988; INGLESFIELD, 1989). These studies have a limited relevance for the following reasons:

- there are no base data on the ecology of the test areas (e.g. on normal population fluctuations);
- the study period is often too short;
- the methods employed are not determined so much by the issue involved but rather by their practicability under special conditions (use of semi-quantitative procedures);
- the scope of early studies in particular was limited to mammals and birds, i.e. there is no ecosystem approach;
- for historical and methodological reasons (just as in the industrialized countries), aquatic studies are more frequent than terrestrial studies.

Although there is at least one example in which the difference in resistance to pesticides led to effects on mammals (dominance shifts in two rat species; WOOD & CHUNG, 1990), frequent use of vertebrates in such short-term projects is not advisable, as can be shown in the case of birds:

- It is rarely possible to establish a connection between exposure (type and duration) and effect due to the mobility of vertebrates.
- A statistical validation is not possible due to the low number of test objects; increasing the number of test animals caught soon raises species and animal welfare problems.
- Issues dating back to the early days of ecotoxicology (accumulation in food chains due to the high persistence of pesticides in use at that time) have become less important in the meantime or are seen in a more differentiated light (more of a food web than a chain).

Attempts to assess ecotoxicological hazard potential do not serve much purpose if they are merely intended as an appendix to classical efficacy tests (e.g. MÜLLER, 1987).

13.3 RECOMMENDATIONS

The final report of MAB Project 9's consultation with experts (UNESCO) in 1975 already provided — albeit under a different name — a list of ecotoxicological research priorities, most of which are still valid today. For example, this list contains most of the research recommendations that were made by a working group

Table 13.1 Recommendations for Ecotoxicological Research in the Tropics

Assess the influence of tropical environmental factors on the fate (e.g. degradation and metabolism) of pesticides.
Develop mathematical models to assess the fate of pesticides in organisms as well as in or between ecosystem compartments.
Collect base data on microorganisms, flora and fauna at sites which are representative of larger areas (reference ecosystems), as far as possible at species level.
Select tropical species based on the criteria that are valid for test organisms and the development of laboratory tests with these species.
Adapt semi-field methods for fate and effect studies.
Study effects of pesticides on non-targets in the field (monitoring in terrestrial and limnic systems); with special consideration for the role of tropical beneficial organisms.
Assess long-term effects of pesticides on important ecosystem structures and functions (e.g. biological productivity), if at all possible at the same sites.
Determine interaction of effects of foreign substances in local and regional contexts, e.g. influence of aquatic systems due to input from terrestrial systems.
Conduct special studies of ecosystem sinks (e.g. limnic sediments).
Prepare ecotoxicological assessment and evaluation criteria adapted especially to tropical conditions.
Evaluate pesticide effects in relation to other forms of anthropogenous environmental influences.

set up under the direction of Prof. Dönhardt in the framework of an ATSAF conference in 1987 (DSE/ATSAF, 1987). This chapter outlines the priorities for ecotoxicological studies, taking these sources as well as own evaluations into account (Table 13.1). They can be classified roughly in terms of the areas of ecological basic research or studies in the framework of a registration process (incl. assessment criteria and monitoring programs). It should be noted that parts of this catalogue may also be used, depending on regional situation, stress pathway, prior knowledge and chemicals employed. GRANT (1989b), for example, presented a detailed plan of procedure based on very similar considerations for ancillary studies in connection with combatting locusts.

In summary, it can be said that there is a deficit of basic knowledge, appropriate ecotoxicological methods and corresponding assessment criteria in the developing countries. As in the industrialized countries, a tiered test program, ranging from pre-registration laboratory tests all the way up to field monitoring of usage, should form the foundation for the definition of ecotoxicological measures. A decision should be made — on a case-by-case basis for the time being, and later on the basis of more experience and criteria

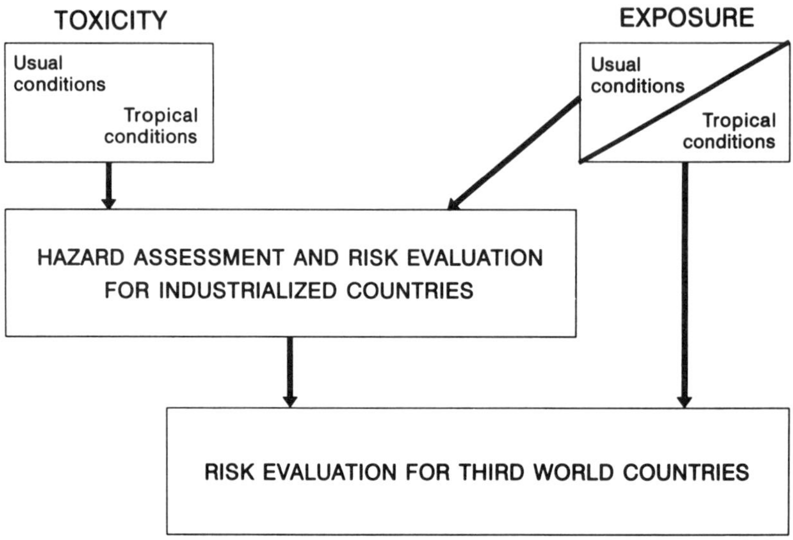

Figure 13.3 Evaluation of risks posed by hazardous substances in the Third World takes specific exposure conditions into consideration (modified after HCN, 1992).

which have yet to be formulated — as to which of the tests required in the producer countries should be repeated for a relative assessment under regional conditions. This could lead, for example, to the introduction of a termite test (still to be developed) for mandatory assessment of the effect on soil organisms in developing countries, instead of the earthworm test that is applicable in industrialized countries (HCN, 1992).

In contrast to the assessment criteria, which differ depending on the test system and parameters (e.g. Which value must not be exceeded by the mortality of a microorganism population if it is to recover between two applications?), there are no specific evaluation criteria for tropical and sub-tropical ecosystems: the claim to protect the structure and function of ecosystems (ecosystem integrity) against damaging effects of chemicals holds true for all systems in the same way.

As presented in a recently published report of the Dutch Health Council (HCN, 1992), the focus has to be placed on assessing the changed exposure situation (Figure 13.3). This does not mean that, as listed in Table 13.1, further study of the effect of chemicals on organisms should be neglected. In other words, the objective must be to adapt the methods and concepts in the industrialized countries to the special conditions in the Third World, such as the lower-quality modes of application due to limited financial and personal resources. The main differences, depending on country, climatic region or ecosystem, must be taken into consideration for each situation (PELFRENE, 1987).

REFERENCES

Abel, C. & Heimbach, U. (1992): Testing effects of pesticides on *Poecilus cupreus* (Coleoptera, Carabidae) in a standardized semi-field test. In: Guidelines for testing the effects of pesticides on beneficial organisms: Description of test methods. IOBC/WPRS Bull. XV/3: 171-175.

Achtziger, R., Nigmann, U. & Zwölfer, H. (1992): Rarefraction-Methoden bei der zooökologischen Zustandsanalyse und Bewertung. Z. Ökol. Natursch. 2: 89-105.

Adema, D.M.M. & Henzen, L. (1989): A comparison of plant toxicities of some industrial chemicals in soil culture and soilless culture. Ecotox. Environ. Safety 18: 219-229.

Adey, M., Walker, P. & Walker, P.T. (1986): Pest Control Safe for Bees. A manual and directory for the tropics and subtropics. London: Intern. Bee Res. Assoc., 224 pp.

Adis, J. (1989): Thirty million arthropod species — too many or too few ? J. Trop. Ecol. 6: 115-118.

AFNOR (Association Francaise de Normalisation) (1984): Détermination de la toxicité d'une substance vis-à-vis des lombriciens (espéce *Eisenia fetida*). Méthode "Artisol". No. X 31-250. Paris.

Ahlers, J. (1991): Einstufung von Stoffen als "Umweltgefährlich" — EG Richtlinie. UWSF.-Z. Umweltchem. Ökotox. 3: 350.

Ahlf, W., Calmano, W., Erhard, J. & Förstner, U. (1989): Comparison of five bioassays techniques for assessing sediment-bound contaminants. Hydrobiologia 188/189: 285-289.

Alawi, M.A., Gharaibeh, S. & Al-Shureiki, Y. (1990): Rückstandsuntersuchungen auf Fenitrothion und Pyrethroide in Wasser, Boden und Pflanzen nach der Heuschreckenbekämpfung in Jordanien 1989. Chemosphere 20: 443-447.

Aldenberg, T. & Slob, W. (1991): Confidence limits for hazardous concentrations based on logistically distributed NOEC toxicity data. In: National Institute of Public Health and Environmental Protection (RIVM). Report No. 719102002.

Allen, J.C. (1985): Soil response to forest clearing in the United States and the tropics: Geological and biological factors. Biotropica 17: 15-27.

Ammon, H.U. (1985): Worm toxicity-test using *Tubifex tubifex*. In: FAO Workshop, Versailles. Les Colloques de l'INRA 31: 303-317.

Anderson, J.P.E. & Domsch, K.H. (1978): A physiological method for the quantitative measurement of microbial biomass in soil. Soil Biol. Chem. 10: 215-221.

Ankley, G.T., Schubauer-Berigan, M.K. & Dierkes, J.R. (1991): Predicting the toxicity of bulk sediments to aquatic organisms with aqueous test fractions: Pore water vs. eluate. Environ. Toxicol. Chem. 10: 1359-1365.

APHA (American Public Health Association) (1989): Standard Methods for the Examination of Water and Wastewater. 17th edition. (eds. Clesceri, L.S. et al.). APHA, Washington, D.C., 8-81-8-85 pp.

Arndt, U., Nobel, W. & Schweizer, B. (1987): Bioindikatoren: Möglichkeiten, Grenzen und neue Erkenntnisse. Stuttgart: Ulmer Verlag, 388 S.

Arndt, U., Franz-Gerstein, C., Hinger, D., Schachner, J., Scherrieble, T., Flores, F., & Strehl, T. (1991): Untersuchung zur Eignung subtropischer Pflanzen als Bioindikatoren in Entwicklungsländern. VDI-Berichte 901: 559-575.

Arunachalam, K.D. & Lakshmanan, M. (1990): Decomposition of ^{14}C-labelled Carbofuran in a Black Tropical Soil under Laboratory Conditions. Soil Biol. Biochem. 22: 407-412.

ASTM (American Society for Testing and Materials) (1987): Standard guide for conducting a terrestrial soil-core microcosm test. E1197-87. Annual Book of ASTM 1104, 743-755.

ASTM (American Society for Testing and Materials) (1988a): Standard practices for conducting acute toxicity tests with fish, macroinvertebrates, and amphibians. E729-88. ASTM, Philadelphia.

ASTM (American Society for Testing and Materials) (1988b): Standard guide for conducting early life-stage toxicity tests with fish. E1241-88. ASTM, Philadelphia.

ASTM (American Society for Testing and Materials) (1990a): Standard guide for conducting static 96 h toxicity tests with microalgae. Standard E1218-90. ASTM, Philadelphia.

ASTM (American Society for Testing and Materials) (1990b): Standard guide for conducting static toxicity tests with *Lemna gibba*. Standard E1415-91. ASTM, Philadelphia.

ASTM (American Society for Testing and Materials) (1990c): Standard guide for conducting static toxicity tests starting with embryos of four species of saltwater bivalve molluscs. Standard E724. ASTM, Philadelphia.

ASTM (American Society for Testing and Materials) (1991): Standard guide for conducting 10-day static sediment toxicity tests with marine and estuarine amphipods. Standard E1367-90. ASTM, Philadelphia.

Ausmus, B.S., Dodson, G.J. & Jackson, D.R. (1978): Behaviour of heavy metals in forest microcosms. III. Effects on litter-soil carbon metabolism. Water, Air and Soil Poll. 10: 19-26.

Auspurg, B. (1985): Kombinationseffekte von Pflanzenschutzmitteln im Boden. Ber. Landwirtschaft 198: 92-105.

Bacci, E. (1992): Value and limitations of evaluative models in field situations. In: EERO Course on Environmental Chemistry of Organic Pollutants, Siena 1992.

Bahadir, M. & Freitag, D. (1989): Zur ökotoxikologischen Bewertung kontaminierter Standorte. Forum Städte-Hygiene 40: 245-248.

Baier, C., Hurle, K. & Kirchhoff, J. (1985): Datensammlung zur Abschätzung des Gefährdungspotentials von Pflanzenschutzmittel-Wirkstoffen für Gewässer. Hamburg: Parey, 294 S.

Balk, I.R. & Koeman, J.H. (1984): Future hazards from pesticide use. The Environmentalist 4 (Suppl. 6): 1-100.

Barnthouse, L.W. (1992): The Role of Models in Ecological Risk Assessment: A 1990's Perspective. Environ. Toxicol. Chem. 11: 1751-1760.

Barnthouse, L.W., Suter, G.W., Rosen, A.E. & Beauchamp, J.J. (1987): Estimating Responses of Fish Populations to Toxic Contaminants. Environ. Toxicol. Chem. 6: 811-824.

Barrett, K. (1992): The BART Group approach to regulatory testing. Asp. Appl. Biol. 31: 165-170.

Bascietto, J., Hinckley, D. & Plafkin, J. (1990): Ecotoxicity and ecological risk assessment. Regulatory applications at EPA. Environ. Sci. Technol. 24: 10-15.

Baudo, R. (1987): Ecotoxicological testing with *Daphnia*. In: *Daphnia*. (Eds. Peters, R.H. & De Bernardi, R.). Mem. 1st. Ital. Idrobiol. 45: 461-482.

Baudo, R., Giesy, J. & Muntau, H. (1990): Sediments: Chemistry and Toxicity of In-Place Pollutants. Lewis Publ., Inc. Ann Arbor. 405 pp.

Bauer, C. (1991): Untersuchungen zur Wirkung zweier Agrochemikalien auf Regenwürmer (Oligochaeta; Lumbricidae). Diploma Thesis University of Frankfurt, 124 S.

Bauer, K. (1964): Studien über Nebenwirkungen von Pflanzenschutzmitteln auf die Bodenfauna. Mittl. Biol. Bundeanstalt für Land- und Forstwirtschaft Berlin-Dahlem 112: 1-42.

Baur, W.H. (1987): Gewässergüte bestimmen und beurteilen. Paul Parey Verlag, Hamburg. 141 S.

Bayne, B.L. (1989): Measuring the biological effects of pollution: the mussel watch approach. Water Sci. Technol. 21: 1089-1100.

BBA (Biologische Bundesanstalt für Land- und Forstwirtschaft) (1980): Richtlinien für die amtliche Prüfung von Pflanzenschutzmitteln. Teil 23-1: Richtlinien für die Prüfung von Pflanzenbehandlungsmitteln auf Bienengefährlichkeit. Erlassen im Dezember 1980.

BBA (Biologische Bundesanstalt für Land- und Forstwirtschaft) (1986a): Richtlinien für die amtliche Prüfung von Pflanzenschutzmitteln. Teil IV, 4-1: Bestimmung des Verbleibs von Pflanzenschutzmitteln im Boden. Erlassen im Dezember 1986.

BBA (Biologische Bundesanstalt für Land- und Forstwirtschaft) (1986b): Richtlinien für die amtliche Prüfung von Pflanzenschutzmitteln. Teil IV, 4-2: Versickerungsverhalten von Pflanzenschutzmitteln. Erlassen im Dezember 1986.

BBA (Biologische Bundesanstalt für Land- und Forstwirtschaft) (1988): Richtlinien für die amtliche Prüfung von Pflanzenschutzmitteln (Nr. IV, 3-10): Prüfung des Rückstandsverhaltens von Pflanzenschutzmitteln in Nachbaukulturen (Nachbaurichtlinie). Erlassen im Mai 1988.

BBA (Biologische Bundesanstalt für Land- und Forstwirtschaft) (1990a): Richtlinien für die Prüfung von Pflanzenschutzmitteln. Teil IV, 4-3: Lysimeteruntersuchungen zur Verlagerung von Pflanzenschutzmitteln in den Untergrund. Erlassen im Februar 1990.

BBA (Biologische Bundesanstalt für Land- und Forstwirtschaft) (1990b): Richtlinien für die Prüfung von Pflanzenschutzmitteln. Teil IV, 5-1: Abbaubarkeit und Verbleib von Pflanzenschutzmitteln im Wasser/Sediment-Sytem. Erlassen im Dezember 1990.

BBA (Biologische Bundesanstalt für Land- und Forstwirtschaft) (1990c): Richtlinien für die Prüfung von Pflanzenschutzmitteln Teil VI, 1-1 (2. Auflage). Auswirkungen auf die Bodenmikroflora. Erlassen im März 1990.

BBA (Biologische Bundesanstalt für Land- und Forstwirtschaft) (1990d): Richtlinien für die Prüfung von Pflanzenschutzmitteln Teil I, 1-2 (2. Auflage). Antrag auf erstmalige/erneute Zulassung eines Pflanzenschutzmittels — Anleitung zum Ausfüllen. Erlassen im November 1990.

BBA (Biologische Bundesanstalt für Land- und Forstwirtschaft) (1991a): Richtlinien für die Prüfung von Pflanzenschutzmitteln. Teil VI, 23 — 2.1.8.: Auswirkungen von Pflanzenschutzmitteln auf Imagines von *Poecilus cupreus* als Vertreter der Familie Carabidae (= Laufkäfer) im Laboratorium. Erlassen im Juni 1991.

BBA (Biologische Bundesanstalt für Land- und Forstwirtschaft) (1991b): Richtlinien für die Prüfung von Pflanzenschutzmitteln. Teil VI, 23-1: Auswirkungen von Pflanzenschutzmitteln auf die Honigbiene. Erlassen im Juni 1991.

BBA (Biologische Bundesanstalt für Land- und Forstwirtschaft) (1993): Richtlinien für die Prüfung von Pflanzenschutzmitteln. Teil VI, 25-1: Prüfung von Ködermitteln, Granulaten und behandeltem Saatgut auf Vogelgefährdung — Annahmeversuche.

BBA (Biologische Bundesanstalt für Land- und Forstwirtschaft) (1994a): Richtlinien für die amtliche Prüfung von Pflanzenschutzmitteln (Nr. VI, 2-2): Auswirkungen von Pflanzenschutzmitteln auf die Reproduktion und das Wachstum von *Eisenia fetida/Eisenia andrei*. Erlassen im Januar 1994.

BBA (Biologische Bundesanstalt für Land- und Forstwirtschaft) (1994b): Richtlinien für die amtliche Prüfung von Pflanzenschutzmitteln (Nr. VI, 2-3): Auswirkungen von Pflanzenschutzmitteln auf Regenwürmer im Freiland. Erlassen im Januar 1994.

Beck, L. (1993): Zur Bedeutung der Bodentiere für den Stoffkreislauf in Wäldern. Biologie in unserer Zeit 23: 286-294.

Beck, L., Dumpert, K., Franke, U., Römbke, J., Mittmann, H.W. & Schönborn, W. (1988): Vergleichende ökologische Untersuchungen in einem Buchenwald nach Einwirkung von Umweltchemikalien. Jül-Spez. 439: 548-701.

Becker, H. (1991): Bodenorganismen — Prüfungskategorien der Forschung. UWSF.-Z. Umweltchem. Ökotox. 3: 19-24.

Beers, J.R., Reeve, M.R. & Grice, R.D. (1977): Controlled ecosystem pollution experiment: effect of mercury on enclosed water columns. Mar. Sci. Comm. 3: 355-394.

Beyer, L., Wachendorf, C., Elsner, D.C. & Knabe, R. (1993): Suitability of dehydrogenase activity assay as an index of soil biological variability. Biol. Fert. Soils 16: 52-56.

Bellan, G. (1981): Manual of methods in aquatic environment research. Part 7. Selected bioassays for the Mediterranean. FAO Fish. Tech. Paper No. 208, 37 pp.

Bellan, G., Desrosiers, G. & Willsie, A. (1988): Use of an Annelid Pollution Index for Monitoring a Moderately Polluted Littoral Zone. Marine Poll. Bull. 19: 662-665.

Belotti, E. (1993): Ein generalisiertes Konzept der Lebensformtypen wirbelloser Bodentiere als Hilfsmittel für den Bodenschutz. Mittl. Deut. Bodenkundl. Ges. 72: 491-494.

Belotti, E. & Babel, U. (1993): Variability in space and time and redundancy as stabilizing principles of forest humus profiles. Europ. J. Soil Biol. 29: 17-27.

Bengtsson, B-E. & Bergström, L. (1987): A Flowthrough Fecundity Test with *Nitocra spinipes* (Harpacticoidea Crustacea) for Aquatic Toxicity. Ecotox. Environ. Safety 14: 260-268.

Bengtsson, G., Berden, M. & Rundgren, S. (1988): Influence of Soil Animals and Metals on Decomposition Processes: A Microcosm Experiment. J. Environ. Qual. 17: 113-119.

Bennett, R.S. & Schafer, D.W. (1988): Procedure for evaluating the potential ability of birds to avoid chemically contaminated food. Environ. Toxicol. Chem. 7: 359-362.

Bergen, B.J., Nelson, W.G. & Pruell, R.J. (1993): Bioaccumulation of PCB Congeners by Blue Mussels (*Mytilus edulis*) Deployed in New Bedford Harbor, Massachusetts. Environ. Toxicol. Chem. 12: 1671-1681.

Bergstrom, L. (1990): Use of lysimeters to estimate leaching of pesticides in agricultural soils. Environ. Poll. 67: 325-348.

Bharathi, C. & Subba Rao, B. (1984): Toxicity of Phosphamidon to the common south Indian earthworm Lampeto mauritii. Bull. Environ. Contam. Toxicol. 32: 295-300

Bias, R. (1993): Biologische Testverfahren — Risikoanalyse und -bewertung. UWSF.-Z. Umweltchem. Ökotox. 5: 91-96.

Bick, H., Hansmeyer, K.H., Olschowy, G. & Schmoock, P. (Hrsg.) (1984): Angewandte Ökologie — Mensch und Umwelt. Band I: Einführung, Räumliche Strukturen, Wasser, Lärm, Luft, Abfall. Stuttgart [etc.]: G. Fischer, xiii+531 S..

Bigler, F. (1989): Gedanken zum nützlingsschonenden Einsatz von Pestiziden im integrierten Pflanzenschutz. Schweiz. Landw. For. 28: 49-55.

Bilyard, G.R. (1987): The Value of Benthic Infauna in Marine Pollution Monitoring Studies. Marine Poll. Bull. 18: 581-581.

Birge, W.J., Black, J.A. & Westerman, A.G. (1985): Short-term fish and amphibian embryo-larval tests for determining the effects of toxicant stress on early life stages and estimating chronic values for single compounds and complex effluents. Environ. Toxicol. Chem. 4: 807-821.

Bitton, G. & Koopman, B. (1992): Bacterial and enzymatic bioassays for toxicity testing in the environment. Rev. Environ. Contam. Toxicol. 125: 1-19.

Blaise, C., Sergy, G., Wells, P., Bermingham, N. & Van Coillie, R. (1988): Biological testing — Development, application, and trends in Canadian Environmental Protection Laboratories. Toxicity Assessment 3: 385-406.

Blanck, H. & Bjornsater, B. (1988): The Algal Microtest Battery. Swedish National Chemicals Inspectorate, Solna, Sweden. 27 pp.

Blume, H-P. (1990): Handbuch des Bodenschutzes. ecomed Verlag, Landsberg. 686 S.

Boese, B.L., Lee, H., Specht, D.T, Randall, R.L. & Winsor, M.H. (1990): Comparison of aqueous and solid-phase uptake for the hexachlorobenzene in the tellinid clam *Macoma nasuta*: A mass balance approach. Environ. Toxicol. Chem. 9: 221-231.

Böhnke, A., Siebers, J. & Nolting, H-G. (1989): Verbleib von Pflanzenschutzmitteln in der Umwelt — Exposition, Bioakkumulation, Abbau — Teil B. BMU Forschungsber. Nr. 05 008/02, 179 S.

Boller, E. (1989): Überlegungen zur Anwendung ökologischer Gesichtspunkte bei der Auswahl von Pestiziden in den Spezialkulturen. Schweiz. Landw. For. 28: 41-47.

Boller, E., Bigler, F., Bieri, M., Häni, F. & Stäubli, A. (1989): Nebenwirkungen von Pestiziden auf die Nützlingsfauna landwirtschaftlicher Kulturen. Schweiz. Landw. For. 28: 3-40.

Borgmann, U., Ralph, K.M. & Norwood, W.P. (1989): Toxicity test procedures for *Hyalella azteca*, and chronic toxicity of cadmium and pentachlorophenol to *H. azteca, Gammarus fasciatus* and *Daphnia magna*. Arch. Environ. Contam. Toxicol. 19: 558-564.

Bostrom, U. & Lofs-Holmin, A. (1982): Testing side-effects of pesticides on soil fauna: A critical literature review. Sveriges Landbrukuniv. Rapp. 12: 1-70.

Bouche, M.B. (1977): Strategies lombriciennes. Ecol. Bull. 25: 122-132.

Bourdeau, P., Haines, J.A., Klein, W. & Krishna Murti, C.R. (eds.) (1989): Ecotoxicology and Climate with Special Reference to Hot and Cold Climates. SCOPE Report 38. J. Wiley & Sons, New York, 392 pp.

Brader, L. (1987): Aspects of pesticide use in developing countries. In: Möglichkeiten, Grenzen und Alternativen des Pflanzenschutzmitteleinsatzes in Entwicklungsländern (hrsg. DSE/ATSAF). Feldafing: DSE/ATSAF, S. 36-44.

Brasse, D. (1990): Einführung der obligatorischen Prüfung der Auswirkung von Pflanzenschutzmitteln auf Nutzorganismen in das Zulassungsverfahren. Nachrichtenbl. Deut. Pflanzenschutzd. 42: 81-86.

Brasse, D. & Rothert, H. (1992): Auswirkungen auf Nutzorganismen. Mitt. Biol. Bundesanstaltf. Land- und Fortswirtschaft, Berlin 284: 113-118.

Brazner, J.C., Heinis, L.J. & Jensen, D.A. (1989): A littoral enclosure for replicated field experiments. Environ. Contam. Toxicol. 8: 1209-1216.

Bresch, H., Beck, H., Ehlermann, D., Schlaszus, H. & Urbanek, M. (1990): A long-term toxicity test comprising reproduction and growth of zebrafish with 4-Chloroaniline. Arch. Environ. Contam. Toxicol. 19: 419-427.

Brezonik, P.L., Baker, L.A., Eaton, J.R., Frost, T.M., Garrison, P., Kratz, T.K., Magnuson, J.J., Perry, J.A., Rose, W.J., Shephard, B.K., Swenson, W.A., Watras, C.J. & Webster, K.W. (1987): Artificial acidification of Little Rock Lake, Wisconsin. Water, Air, Soil Pollut. 31: 115-122.

Brian, P. & Cousens, R. (1989): An equation to describe dose response at low concentration, where there is stimulation of growth at low doses. Water Res. 29: 93-96.

Briggs, G.C. (1981): Theoretical and experimental relationship between soil adsorption, octanol-water partition coefficients, water solubilities, bioconcentration factors and the Parachor. J. Agric. Food Chem. 29: 1050-1059.

Britt, J.K., Dwinell, S.E. & McDowell, T.C. (1992): Matrix decision procedure to assess new pesticides based on relative ground water leaching potential and chronic toxicity. Environ. Toxicol. Chem. 11: 721-728.

Brock, T.D. & Madigan, M.T. (1991): Biology of Microorganisms. 6. Edition, Prentice Hall, Englewood Cliffs, New Jersey.

Brodo, I.M. (1961): Transplant experiments with corticolous lichens using a new technique. Ecology 42: 838-841.

Brooke, D. & Mathiessen, P. (1991): Development and validation of a modified fugacity model of pesticide leaching from farmland. Pest. Sci. 31: 349-361.

Brouwer, A., Murk, A.J. & Koeman, J.B. (1990): Biochemical and physiological approaches in ecotoxicology. Funct. Ecol. 4: 275-281.

Brown, C. (1992): Changing Times in the Notification Business. Chimica Oggi 5/92: 29-31.

Brown, R.A., McMullin, L.C., Jackson, D., Ryan, J. & Coulson, J.M. (1990): Beneficial arthropod toxicity assessments with three insecticides in laboratory, semi-field and field studies. Brighton Crop Protection Conference, 527-534.

Buikema, A.L. & Cairns, J. (1980): Aquatic Invertebrate Bioassays. ASTM Special Technical Publ. 715. ASTM, Philadelphia. 209 pp.

Buikema, A.L., Geiger, J.G. & Lee, D.R. (1980): *Daphnia* Toxicity Tests. In: Aquatic Invertebrate Bioassays. ASTM Special Technical Publ. 715. ASTM, Philadelphia. 209 pp.

Bulich, A.A. (1979): Use of luminescent bacteria for determining toxicity in aquatic environments. In: Aquatic Toxicology, Second Conference. ASTM STP 667 (Eds. Marking, L.L. & Kimerle, R.A.), American Society for Testing and Materials, Philadelphia. 98-106.

Bunyan, P.J., Van den Heuvel, M.J., Stanley, P.I. & Wright, E.N. (1981): An intensive field trial and a multisite surveillance exercise on the use of Aldicarb to investigate methods for the assessment of possible environmental hazards presented by new pesticides. Agroecosystems 7: 239-262.

Burger, J. & Gochfeld, M. (1992): Temporal scales in ecological risk assessment. Arch. Environ. Contam. Toxicol. 23: 484-488.

Burhenne, W.E. (1989): Umweltrecht. Systematische Sammlung der Rechtsvorschriften des Bundes und der Länder. Berlin: E. Schmidt, 8432 S. (Stand 1989), ergänzbare Ausgabe, hrsg. im Auftrag der Interparlamentarischen Arbeitsgemeinschaft.

Burton, G.A. (1991): Assessing the toxicity of freshwater sediments. Environ. Toxicol. Chem. 10: 1585-1627.

Burton, G.A. (ed.) (1992): Sediment Toxicity Assessment. Lewis Publ., Boca Raton, 457 pp.

Burton, M.A.S. (1990): Terrestrial and aquatic bryophytes as monitors of environmental contaminants in urban and industrial habitats. Biol. J. Linn. Soc. 104: 267-280.

Busch, K.-F., Uhlmann, D. & Weise, G. (Hrsg.) (1989): Ingenieurökologie. Jena: VEB G. Fischer, 2. erw. Aufl., 488 S.

Butler, G.C. (ed.) (1978): Principles of Ecotoxicology. Chichester: Wiley, SCOPE Report 12. 350 pp.

Butler, G.C. (1984): Developments in ecotoxicology. Ecol. Bull. 36: 9-12.

Cabridenc, R., Chambon, M., Ducros, M. & Lepailleur, H. (1984): Influence du substrat sur la toxicité des substances chimiques vis-à-vis des vers de terres. In: Actes. Symp. Int. "Ecotoxicologie Terrestre": 285-308.

Cairns, J. (1984): Are single-species toxicity tests alone adequate for estimating environmental hazard ? — Environ. Monitoring Assess. 4: 259-273.

Cairns, J. (ed.) (1985): Multispecies Toxicity Testing. New York: Pergamon.

Cairns, J. (1986a): The myth of the most sensitive species. BioScience 36: 670-672.

Cairns, J. (1986b): What is meant by validation of predictions based on laboratory toxicity tests ? — Hydrobiologia 137: 271-278.

Cairns, J. (1988): Putting the eco in ecotoxicology. Reg. Toxicol. Pharm. 8: 226-238.

Cairns, J. (1989): Where is the ecology in toxicology ? — In: Current Practices in Environmental Science and Engineering (ed. A. Singh & U.S. Sharma). Jodhpur: Divyajyoti Prakashan, pp. 21-40.

Cairns, J. & Cherry, D.S. (1993): Freshwater Multi-species Test Systems. — In: Handbook of Ecotoxicology. (ed. Calow, P.). Blackwell Scientific Publ, Oxford. 101-117.

Cairns, J. & Mount, D.L. (1990): Aquatic toxicology. Environ. Sci. Technol. 24: 154-160.

Cairns, J. & Pratt, J.R. (1986): On the relation between structural and functional analysis of ecosystems. Environ. Toxicol. Chem. 5: 785-786.

Cairns, J. & Pratt, J.R. (1987): Ecotoxicological effect indices: A rapidly evolving system. Wat. Sci. Technol. 11: 1-12.

Cairns, J. & Pratt, J.R. (1989): The scientific basis of bioassays. Hydrobiologia 188/89: 5-20.

Cairns, J., Pratt, J.R. & Niederlehner, B.R. (1985): A provisional multispecies toxicity test using indigenous organisms. J. Testing Eval. 13: 316-319.

Calamari, D. & Bacci, E. (1987): Environmental distribution and fate of pesticides. A predictive approach. In: Toxicology of pesticides: Experimental, Clinical and Regulatory Perspectives (ed. L.G. Costa et al.). Berlin [etc.]: Springer, pp. 171-184, NATO ASI Ser. H, No. 13.

Calamari, D. & Pacchetti, G. (1987): Population parameters as indices of effects of mixtures of chemicals on aquatic populations: the need for methods. In: Methods for Assessing the Effects of Mixtures of Chemicals (ed. V.V. Vouk, A.C. Upton, D.V. Parke & S.C. Asher). Chichester: Wiley, SCOPE 30, SGOMSEC 3, pp. 301-318.

Calamari, D. & Vighi, M. (1990): Quantitative structure activity relationships in ecotoxicology: value and limitations. Reviews Environ. Toxicol. 4: 1-112.

Calamari, D., Bacci, E., Focardi, S., Gaggi, C., Morosini, M. & Vighi, M. (1991): Role of plant biomass in the global environmental partitioning of chlorinated hydrocarbons. Environ. Sci. Technol. 25: 1489-1495.

Calderbank, A. (1989): The occurrence and significance of bound pesticide residues in soil. Rev. Environ. Contam. Toxicol. 108: 71-103.

Callahan, C.A., Menzie, C.A., Burmaster, D.E., Wilborn, D.C. & Ernst, T. (1991): On-site methods for assessing chemical impact on the soil environment using earthworms: A case study at the Baird and McGuire Superfund Site, Holbrook, Massachusetts. Environ. Toxicol. Chem. 10: 817-826.

Callegari Lopes, J.L., Casanova, I.C., Garcia de Figueireido, M.C., Nather, F.C. & Paiva Arlar, W.E. (1992): *Anodontites trapesialis*: A Biological Monitor of Organochlorine Pesticides. Arch. Environ. Contam. Toxicol. 23: 351-354.

Calow, P. (1993): Handbook of Ecotoxicology. Blackwell Scientific Publ, Oxford. 478 pp.

Carl, K.P. (1985): Erfolge der biologischen Bekämpfung in den Tropen. Giessener Beitr. Entwicklungsforsch. I,12: 19-35.

Carpenter, K.E. (1927): The lethal action of soluble metallic salts on fishes. J. Exp. Biol. 4: 378-390.

Carson, D.B., Saeger, V.W. & Gledhill, W.E. (1990): Use of microcosms versus conventional biodegradation testing for estimating chemical persistence. ASTM Spec. Publ. 1096 (Aquat. Toxicol. Risk Assess.: 13th Vol.): 48-59, ASTM Philadelphia.

Casillas, E., Weber, D., Haley, C. & Sol, S. (1992): Comparison of growth and mortality in juvenile sand dollars (*Dendraster excentricus*) as indicators of contaminated marine sediments. Environ. Toxicol. Chem. 11: 559-570.

Chapman, P.M. (1989): Current approaches to developing sediment quality criteria. Environ. Toxicol. Chem. 8: 589-599.

Chapman, P.M. (1990a): Current approaches to developing sediment quality criteria. Environ. Toxicol. Chem. 8: 589-599.

Chapman, P.M. (1990b): The sediment quality triad approach to determining pollution-induced degradation. Sci. Tot. Environ. 97/98: 815-825.

Chapman, P.M. & Morgan, J.D. (1983): Sediment Bioassays with Oyster Larvae. Bull. Environ. Contam. Toxicol. 31: 438-444.

Chapman, P.M., Power, E.A., Dexter, R.N. & Andersen, H.B. (1991): Evaluation of effects associated with an oil platform, using the sediment quality triad. Environ. Contam. Toxicol. 10: 407-424.

ChemG (Chemikaliengesetz) (1990): Gesetz zum Schutz vor gefährlichen Stoffen (Chemikaliengesetz — ChemG). Bundesgesetzblatt 13, I: 521-547.

Chen, S.C., Fitzpatrick, L.C., Goven, A.J., Venables, B.J. & Cooper, A.J. (1991): Nitroblue Tetrazolium Dye Reduction by Earthworm (*Lumbricus terrestris*) Coelomocytes: An Enzyme Assay for Nonspecific Immunotoxicity of Xenobiotics. Environ. Toxicol. Chem. 10: 1037-1043.

Chiang, H.L. (1982): Factors to be considered in refining a general model of economic threshold. Entomophaga 27 (special issue): 99-103.

Clark, J.R. (1989): Field studies in estuarine ecosystems: A review of approaches for assessing contaminant effects. In: Aquatic Toxicology and Hazard Assessment (eds. Cowgill, U.M. & Williams, L.R.) STP 1027, ASTM, Philadelphia. 120-133.

Clark, J.R. & Cripe, C.R. (1993): Marine and Estuarine Multi-species Test Systems. In: Handbook of Ecotoxicology. (ed. Calow, P.). Blackwell Scientific Publ, Oxford. 227-247.

Clark, J.R., Goodman, L.R., Borthwick, P.W., Patrick, J.M., Cripe, G.M., Moody, P.M., Moore, J.C. Lores, J.M. (1989): Toxicity of pyrethroids to marine invertebrates and fish: A literature review and test results with sediment-sorbed chemicals. Environ. Toxicol. Chem. 8: 393-401.

Clark, J.R. & Macauley, J.M. (1990): Comparison of the seagrass *Thalassia testudinum* and its epiphytes in the field and in laboratory test systems. In: Plants for Toxicity Assessment. (eds.: Wang, W., Gorsuch, J.W. & Lower, W.R.). STP 1091, ASTM, Philadelphia.

Clements, W.H., Farris, J.L., Cherry, D.S. & Cairns, J. (1989): The influence of water quality on macroinvertebrate community responses to copper in outdoor experimental streams. Aquat. Toxicol. 14: 249-262.

Coler, R.A., Coler, M.S. & Kosteck, P.T. (1988): Tubificid behavior as a stress indicator. Wat. Res. 22: 263-267.

Connell, D.W. (1986): Ecotoxicology — A new approach to understanding hazardous chemicals in the environment. Search 17: 27-31.

Connell, D.W. & Markwell, R.D. (1990): Bioaccumulation in the Soil to Earthworm System. Chemosphere 20: 91-100.

Cook, K.A. & Greaves, M.P. (1987): Natural Variability in Microbial Activities. In: Pesticide Effects on Soil Microflora. (Somerville, L. & Greaves, M.P. eds.). Taylor & Francis, London. 15-44.

Costa, L.G. (1987): Toxicology of pesticides: A brief history. In: Toxicology of Pesticides: Experimental, Clinical and Regulatory Perspectives (ed. L.G. Costa et al.). Berlin [etc.]: Springer, pp. 3-10, NATO ASI Ser H, No. 13.

Coughtrey, P.J., Jones, C.H., Martin, M.H. & Shales, S.W. (1979): Litter Accumulation in Woodlands Contaminated by Pb, Zn, Cd and Cu. Oecologia (Berlin) 39: 51-60.

Council of Europe (1984): Pesticides. Strassbourg (6th edn.). 100 p.

Cox, P.R. & Smith, R.H. (1990): Rodenticide ecotoxicology: assessing non-target population effects. Funct. Ecol. 4: 315-320.

Cramer, H.H. (1967): Pflanzenschutz und Welternte. Pflanzenschutznachrichten Bayer 20: 524 S.

Cranston, P.S. (1990): Biomonitoring and Invertebrate Taxonomy. Environ. Monit. Assessment 14: 265-274.

Croft, B.A. (1990): Arthropod Biological Control Agents and Pesticides. Wiley Interscience, New York.

Crosby, D.C. & Minyard, J.P. (1991): The Persistant Seventies. In: Regulation of Agrochemicals — A Driving Force in Their Evolution. (eds.: Marco, G.J., Hollingworth, R.M. & Plimmer, J.R.). American Chemical Society, Washington, DC. 9-18.

Crossland, N.O. & Wolff, C.J.M. (1985): Fate and biological effects of pentachlorophenol in outdoor ponds. Environ. Contam. Toxicol. 4: 73-86.

Cullen, P. (1990): Biomonitoring and Environmental Management. Environ. Monit. Assessment 14: 107-114.

Davis, H.C. & Hidu, H. (1969): Effects of turbidity producing substances in seawater on eggs and larvae on three genera of bivalve molluscs. Veliger 11: 316-323.

Davis, J.W. & Madsen, S.S. (1991): The biodegradation of Methylene Chloride in soils. Environ. Toxicol. Chem. 10: 463-474.

Dawson, J. (1990): Crop protection in eastern Europe. Residue Review 86: 65-105.

De Bruijn, J. & Hermens, J.L.M. (1991): Qualitative and quantitative modelling of toxic effects of organophosphorous compounds to fish. The Science of the Total Environment 109/110: 441-455.

De Henau, H. (1993): Biodegradation. In: Handbook of Ecotoxicology. (ed. Calow, P.). Blackwell Scientific Publ, Oxford. 355-377.

Depledge, M. (1990): New approaches in ecotoxicology: Can inter-individual physiological variability be used as a tool to investigate pollution effects ? — Ambio 19: 251-252.

Depledge, M.H. & Andersen, B.B. (1990): A computer-aided physiological monitoring system for continuous, long-term recording of cardiac activity in selected invertebrates. Comp. Biochem. Physiol. 96A: 473-477.

Devillers, J., Chambon, P., Zakarya, D., Chastrette, M. & Chambon, R. (1987): A predictive structure-toxicity model with *D. magna*. Chemosphere 16: 1149-1163.

Devonshire, A.L. & Field, L.M. (1991): Gene amplification and insecticide resistance. Ann. Rev. Entomol. 36: 1-23.

DIN (Deutsche Industrie Norm) (1991): Nr. 38412 Teil 34. Bestimmung der Hemmwirkung von Abwasser auf die Lichtemission von *Photobacterium phosphoreum*. Deutsche Einheitsverfahren zur Wasser-, Abwasser- und Schlamm-Untersuchung: Testverfahren mit Wasserorganismen; Verlag Chemie, Weinheim, 25. Lieferung.

Di Toro, D.M., Zarba, C.S., Berry, W.J., Swartz, R.C., Cowan, C.E., Pavlou, S.P., Allen, H.E., Thomas, N.A. & Paquin, P.R. (1991): Technical basis for establishing sediment quality criteria for nonionic organic chemicals using equilibrum partitioning. Environ. Toxicol. Chem. 10: 1541-1583.

Dive, D., Blaise, C. & Le Du, A. (1991): Standard protocol proposal for undertaking the *Colpidium campylum* ciliate protozoan growth inhibition test. Z. Angew. Zool. 78: 323–336.

Domsch, K.H. (1992): Pestizide im Boden. Mikrobieller Abbau und Nebenwirkungen auf Mikroorganismen. Verlag Chemie, Weinheim. 575 S.

Domsch, K.H., Jagnow, G. & Anderson, T.M. (1983): An ecological concept for the assessment of side-effects of agro-chemicals on soil micro-organisms. Residue Review 86: 65-105.

Donaldson, W.T. (1992): The role of property-reactivity relationships in meeting the EPA's needs for environmental fate constants. Environ. Toxicol. Chem. 11: 887-891.

Dönhardt, A. (1987): Gefährdung des Menschen durch den Einsatz von Pflanzenschutzmitteln in den Entwicklungsländern. In: Möglichkeiten, Grenzen und Alternativen des Pflanzenschutzmitteleinsatzes in Entwicklungsländern (hrsg. DSE/ATSAF). Feldafing: DSE/ATSAF, S. 128-134.

Donker, M. (1992): Physiology of metal adaptation in the isopod *Porcellio scaber*. Dissertation Vrije Universiteit Amsterdam, 117 S.

Donker, M., Eijsackers, H. & Heimbach, F. (1994): Ecotoxicology of Soil Organisms. Lewis Publ., Boca Raton, FL. 470 pp.

Draggan, S. (1976): The microcosms as a tool for estimation of environmental transport of tcxic materials. Intern. J. Environ. Studies 10: 65-70.

DSE/ATSAF (Hrsg.) (1987): Möglichkeiten, Grenzen und Alternativen des Pflanzenschutzmitteleinsatzes in Entwicklungsländern. Bericht der Tagung vom 9.-11.7.1987 in der ZEL in Feldafing. Feldafing: DSE/ATSAF, 241 S. (Leitung: J. Kranz; Berichterstattung: E. Rangkuty; Redaktion: K. Klennert).

Dumpert, K. & Zietz, E. (1984): Platanna (*Xenopus laevis*) as a test organism for determining the embryotoxic effects of environmental chemicals. Ecotox. Environ. Safety 8: 55-74.

Dunger, W. & Fiedler, H.J. (Hrsg.) (1989): Methoden der Bodenbiologie. Stuttgart: G. Fischer, 432 S.

Dürrschmidt, W. (1987): Das Doppelgesicht der Grenzwerte und Höchstmengen. Ökologische Konzepte 26: 19-33.

Dutka, B.J. & Kwan, K.K. (1988): Battery of screening tests approach applied to sediment extracts. Toxicity Assessment 3: 303-314.

Dutka, B.J., Seidl, P. & Munro, D. (1990): Using microbial and toxicant screening-test data to prioritize water-bodies. Surface and ground water flow phenomena. Gulf Publishing, Houston.

DVWK (Deutscher Verband für Wasserwirtschaft und Kulturbau) (1989): Merkblätter zur Wasserwirtschaft: Filtereigenschaften des Bodens gegenüber Schadstoffen. II. Abschätzen des Verhaltens organischer Chemikalien im Boden. Bonn, 31 S.

Eagle, D.J. (1990): Agrochemical damage to UK crops. Pesticide Outlook 1: 14-16.

Eaton, J., Arthur, J., Hermanutz, R., Kiefer, R., Mueller, L., Anderson, R., Erickson, R., Nordling, B., Rogers, J. & Pritchard, H. (1985): Biological effects of continuous and intermittent dosing of outdoor experimental streams with chlorpyrifos. In: Aquatic Toxicology and Hazard Assessment: Eighth Symposium. (eds. Bahner, R.C. Hansen, D.J.). STP 891, ASTM, Philadelphia. 85-118.

Ebing, W., Pflugmacher, J. & Haque, A. (1984): Der Regenwurm als Schlüsselorganismus zur Messung der Bodenbelastung mit organischen Fremdchemikalien. Ber. Landwirtsch. 62: 222-255.

EC (European Community) (1979): Council Directive 79/831: Amending for the sixth time Directive 67/548/EEC on the approximation of the laws, regulations and administrative provisions relating to the classification, packaging and labelling of dangerous substances. Official Journal of the European Communities, L 259.

EC (European Community) (1984): Directive 79/831 Annex V, Part C: Methods for determination of ecotoxicity — level 1. Commission of the European Communities, L. 230.

EC (European Community) (1993): Directive 93/67: Principles for assessment of risks to man and the environment of substances notified in accordance with Council Directive 67/548/EEC. Commission of the European Communities, L. 227.

ECETOC (European Centre for Ecotoxicology and Toxicology of Chemicals) (1990): Hazard Assessment of Chemical Contaminants in Soil. ECETOC Technical Rep. No. 40. 124 pp.

ECETOC (European Centre for Ecotoxicology and Toxicology of Chemicals) (1991): Biodegradation Kinetics. ECETOC Technical Rep. No. 44. 75 pp.

ECETOC (European Centre for Ecotoxicology and Toxicology of Chemicals) (1992): Estimating Environmental Concentrations of Chemicals using Fate and Exposure Models. ECETOC Technical Rep. No. 50. 80 pp.

ECETOC (European Centre for Ecotoxicology and Toxicology of Chemicals) (1993): Environmental Hazard Assessment of Substances. ECETOC Technical Rep. No. 51. 81 pp.

Edwards, C.A. (1977): Environmental aspects of the usage of pesticides in developing countries. Mededelingen van de Faculteit Landbouwweitenschappen, Reijksuniversiteit Gent 42: 853-868.

Edwards, C.A. (1984): Report of the Second Stage in Development of a Standardized Laboratory Method for Assessing the Toxicity of Chemical Substances to Earthworms. Commission of the European Communities, Report No. EUR 9360 EN.

Edwards, C.A. (1986): Agrochemicals as environmental pollutants. In: Control of Pesticide Applications and Residues in Food (ed. B. v. Hofsten & G. Ekstroem). Upsala: Swedish Sci. Pr., pp. 1-19.

Edwards, C.A. & Bohlen, P.J. (1992): The effects of toxic chemicals on earthworms. Rev. Environ. Contam. Toxicol. 125: 23-99.

Edwards, P.J., Wilkinson, W. & Coulson, M. (1984): A laboratory toxicity test for carabid beetles. British Crop Protection Conference, Brighton. 359-362.

Egeler, P. (1993): Aufnahme und Elimination von Lindan, 4-Chloranilin und 4-Nitroanilin bei den Tubificiden *Limnodrilus hoffmeisteri* und *Tubifex tubifex*. Diploma thesis, University of Mainz, 86 S.

Eide, I., Jensen, A. & Melsom, S. (1979): Application of in situ cage cultures of phytoplankton for monitoring heavy metal pollution in two Norwegian fjords. J. Exp. Mar. Biol. 37: 271-286.

Eijsackers, H. (1978): Side-effects of the herbicide 2,4,5-T affecting mobility and mortality of the springtail *Onychiurus quadriocellatus* (Collembola). Z. Angew. Entomol. 86: 349-372.

Ellenberg, H. (1984): Standardmodelle und Bezugsflächen von Landökosystemen zum Testen chemischer Belastungen an konkurrierenden Pflanzen und Nahrungsketten. Jül.-Spez. 274.

Ellenberg, H., Mayer, R. & Schauermann, J. (Hrsg.) (1986): Ökosystemforschung. Ergebnisse des Sollingprojektes. Stuttgart: Ulmer, 507 S.

Elmgren, R. & Frithsen, J.B. (1982): The use of experimental ecosystems for evaluating the environmental impact of pollutants: A comparison of an oil spill in the Baltic Sea and two long-term, low-level oil addition experiments in mesocosms. In: Marine Mesocosms. Biological and Chemical Research in Experimental Ecosystems. (eds. Grice, G.D. & Reeve, M.R.). Springer Verlag, New York. 153-165.

El-Sebae, A.H. (1989): Fate and undesirable effects of pesticides in Egypt. In: Ecotoxicology and Climate (ed. P. Bourdeau et al.). New York: Wiley, SCOPE, pp. 359-371.

Enserink, E.L., Maas-Diepeveen, J.L. & Van Leeuwen, L.J. (1991): Combined effects of metals: An ecotoxicological evaluation. Water Res. 25: 679-687.

EPA (Environmental Protection Agency) (1974): Marine Algal Assay Procedure Bottle Test. National Environmental Research Center, U.S. EPA, Corvallis. 43 pp.

EPA (Environmental Protection Agency) (1976): Bioassay procedures for the ocean disposal permit program. Environmental Research Laboratory, Office of Research and Development, U.S. EPA, Gulf Breeze. 96 pp.

EPA (Environmental Protection Agency) (1982a): Pesticide Assessment Guidelines Standards, Subdivision N, Section 162: Metabolism Studies, Chemistry: Environmental Fate. Office of Pesticide and Toxic Substances, Washington, D.C.

EPA (Environmental Protection Agency) (1982b): Pesticide Assessment Guidelines Standards, Subdivision N, Section 162: Mobility Studies, Chemistry: Environmental Fate. Office of Pesticide and Toxic Substances, Washington, D.C.

EPA (Environmental Protection Agency) (1982c): Pesticide Assessment Guidelines Standards, Subdivision E, Section 71: Avian and Mammalian Testing. Office of Pesticide Programs, Washington, D.C.

EPA (Environmental Protection Agency) (1984): Estimating "concern levels" for concentrations of chemical substances in the environment. In: Environmental Effect Branch Health and Environmental Review Division. 43 pp.

EPA (Environmental Protection Agency) (1985): Toxic substances control act test guidelines: environmental effects testing guidelines, 40 CFR Part 797. Fed. Reg. 50 (188), 39389.

EPA (Environmental Protection Agency) (1987): Toxic substances control act test guidelines. Subpart D — Microcosm Guidelines. Federal Register (USA) 52 (187), 36363 36371.

EPA (Environmental Protection Agency) (1988): Short-Term Methods for Estimating the Chronic Toxicity of Effluents and Receiving Waters to Marine and Estuarine Organisms. Report No. 600/4-87/028, U.S. EPA, Cincinnati.

EPA (Environmental Protection Agency) (1989): Oyster Bioconcentration Test. Code of Federal Regulations, 40CFR Ch. 1, Section 797.1830, dated 7 January 1989.

EPA (Environmental Protection Agency) (1992): Framework for Ecological Risk Assessment. Risk Assessment Forum Report No. 630/R-92/001. U.S. EPA, Washington.

EPA (Environmental Protection Agency) (1993): A review of ecological assessment case studies from a risk assessment perspective. Risk Assessment Forum Report No. 630/R-92/005. U.S. EPA, Washington.

EPPO (European and Mediterranean Plant Protection Organization/Council of Europe) (1993): Decision-making scheme for the environmental risk assessment of plant protection products. Bull. OEPP/EPPO Bull. 23: 1-157.

EPS (Environmental Protection Series) (1990): Biological Test Method: Acute Lethality Test using *Daphnia spp.*. Report EPS 1/RM/11, Environment Canada, 57 pp.

Ernst, W. (1984): Marine Ecotoxicological Testing in the Federal Republic of Germany. In: Ecotoxicological Testing in the Marine Environment. (eds.: Persoone, G., Jaspers, E. & Claus, C.). State Univ. Ghent and Inst. Mar. Scient. Res., Bredene, Belgium. Vol. 1: 85-109.

Esser, H.O., Hemingway, R.J., Klein, W., Sharp, D.B., Vonk, J.W. & Holland, P.T. (1988): Recommended approach to the evaluation of the environmental behaviour of pesticides. Pure & Appl. Chem. 60: 901-932.

Everts, J.W., Kortenhoff, B.A., Hoogland, H., Vlug, H.J., Jocque, R. & Koeman, J.H. (1985): Effects on Non-target Terrestrial Arthropods of Synthetic Pyrethroids Used for the Control of the Tsetse Fly (*Glossina spp.*) in Settlement Areas of the Southern Ivory Coast, Africa. Arch. Environ. Contam. Toxicol. 14: 641-650.

Everts, J.W., Aukema, B., Hengeveld, R. & Koeman, J.H. (1989): Side-Effects of Pesticides on Ground-Dwelling Predatory Arthropods in Arable Ecosystems. Environ. Pollut. 59: 203-225.

Ewell, W.S., Gorsuch, J.W., Kringle, R.O., Robillard, K.A. & Spiegel, R.C. (1986): Simultaneous evaluation of the acute effects of chemicals on seven aquatic species. Environ. Toxicol. Chem. 5: 831-840.

Faber, J. (1991): Functional classification of soil fauna: A new approach. Oikos 62: 110-117.

Faber, J.H. & Verhoef, H.A. (1991): Functional differences between closely-related soil arthropods with respect to decomposition processes in the presence or absence of tree roots. Soil Biol. Biochem. 23: 15-23.

FAO (Food and Agriculture Organization) (1988): Guidelines on the registration of biological pest control agents. Food and Agriculture Organization of the United Nations, Rome, Italy, 8 pp.

FAO (Food and Agriculture Organization) (1990a): International Code of Conduct on the Distribution and Use of Pesticides. Resolution 10/85 of the 23rd FAO Conference of November 1985 (Amended Version). Rome, Italy.

FAO (Food and Agriculture Organization) (1990b): Environmental Effects of Chemical Locust and Grasshopper Control. A Pilot Study of Project ECLO/SEN/003/NET. Migratory Pests Group, Plant Protection and Production Division (AGP). Rome, Italy.

FAO (Food and Agriculture Organization) (1992): Draft Code for the Import and Release of Biological Control Agents. Meeting Report AGP/1992/M/3 of the Expert Consultation on Guidelines for the Introduction of Biological Control Agents, 1991. Rome, Italy, 21 pp.

Farke, H., Schulz-Baldes, M. Ohm, K. & Gerlach, S.A. (1984): Bremerhaven caisson for intertidal field studies. Mar. Ecol. Prog. Ser. 16: 193-197.

Fava, J.A., Adams, W.J., Larson, R.J., Dickson, G.W., Dickson, K.L. & Bishop, W.E. (1987): Research Priorities in Environmental Risk Assessment. SETAC Workshop Report, Breckenridge, Colorado. 103 pp.

FDA (Food and Drug Administration) (1988): Environmental assessment technical assistance document 4.12. Earthworm Subacute Toxicity. Center for Food Safety and Applied Nutrition Environmental Impact Section and Center for Veterinary Medicine Environmental Staff. Wahington, D.C. 12 pp.

Felton, J.C., Oomen, P.A. & Stevenson, J.H. (1986): Toxicity and hazard of pesticides to honeybees: harmonization of test methods. Bee World 67: 114-124.

Ferriere, G., Fayolle, L. & Bouche, M.B. (1981): Un nouvel outil, essentiel pour l'ecophysiologie et l'ecotoxicologie: l'élevage des lombriciens en sol artificiel. Pedobiologia 22: 196-201.

Figge, K. (1992): Facilities for the examination of the degradation and distribution of chemical compounds in sections of terrestrial ecosystems. Brighton Crop Protection Conf. Mono. 53: 83-99.

Figge, K., Klahn, J. & Koch, J. (1985): Chemische Stoffe in Ökosystemen. Schriftenreihe Ver. Wasser-, Boden-, Lufthygiene 61: 1-234.

Figge, K., Klahn, J. & Koch, J. (1986): Kinetic distribution model for chemicals based on results from a standard environmental system. Ecotox. Environ. Safety 11: 320-338.

Finney, D.J. (1971): Probit Analysis. Cambridge University Press. 333 pp.

Fischer, R. & Heimbach, F. (1991): Pflanzenschutzmittel und Naturhaushalt — Prüfung, Gefährdungsabschätzung und Risikobeurteilung. IV. Möglichkeiten und Grenzen der Verwendung von Labordaten. Gesunde Pflanzen 43: 19-23.

Fletcher, J.S., Johnson, F.L. & McFarlane, J.C. (1990): Influence of greenhouse versus field testing and taxonomic differences on plant sensitivity to chemical treatment. Environ. Toxicol. Chem. 9: 769-776.

FONDS (1992): Textheft zur Folienserie Pflanzenschutz. Fonds der Chemischen Industrie zur Förderung der Chemie und Biologischen Chemie im Verband der Chemischen Industrie, Frankfurt/Main (Hrsg.), 96 S.

Forbes, T.L. & Forbes, V.E. (1993): A critique of the use of distribution-based models in ecotoxicology. Functional Ecology 7: 249-254.

Forget, G. (1991): Pesticides and the Third World. J. Toxicol. Environ. Health 32: 11-32.

Förstner, U. (1990): Umweltschutztechnik. Spiegel Verlag, Berlin.

Förstner, U., Calmano, W., Ahlf, W. & Kersten, M. (1987): Ansätze zur Beurteilung der "Sedimentqualität" in Gewässern. Vom Wasser 73: 25-42.

Francis, B.M. & Zaber, D.J. (1990): Developmental and reproductive toxicology of xenobiotics. Reviews Environ. Toxicol. 4: 167-240.

Fransson, M.A. (ed.) (1985): Standard methods for the examination of water and wastewater. Washington D.C.: American Publ. Health Ass., 16th edn., 1034 p.

Franz, J.M. & Krieg, A. (1982): Biologische Schädlingsbekämpfung. Pareys Studientexte 12. Verlag Paul Parey, Berlin + Hamburg, 252 S.

Fränzle, O. (1983): Ökosystemforschung: allgemeine Grundlagen und Definition, trophische Strukturen, biozönotische Gesetze und Thermodynamik. In: Ökosystemforschung als Beitrag zur Beurteilung der Umweltwirksamkeit von Chemikalien. (ed. Deutsche Forschungsgemeinschaft). Verlag Chemie, Weinheim, 21-29.

Frederickson, J.K., Bolton, H., Bentjen, S.A., McFadden, K.M., Li, S.W. & Van Voris, P. (1990): Evaluation of intact soil-core microcosms for determining potential impacts on nutrient dynamics by genetically engineered microorganims. Environ. Toxicol. Chem. 9: 551-558.

Freitag, D., Ballhorn, L., Geyer, H. & Korte, F. (1985): Environmental Hazard Profile of Organic Chemicals. Chemosphere 14: 1589-1616.

Führ, F., Kloskowski, R. & Burauel, P-W. (1985): Bedeutung der gebundenen Rückstände. Ber. Landwirt. 198: 106-115.

Führ, F., Steffens, W., Mittelstaedt, W. & Brumhard, B. (1991): Lysimeter experiments with ^{14}C-labelled pesticides — an agroecosystem approach. In: Pesticide Chemistry. Frehse, E. (ed.), VCH Verlag, Weinheim. 37-48.

Ganzelmeier, H. (1986): Abtrift beim Einsatz von Sprüh- und Staubgeräten in Obst- und Ackerbau. Grundl. Landtechnik 36: 174-179.

Garcia, J-E., Pestemer, W. & Günther, P. (1992): Degradation of Isoxaben in a wheat culture soil under field conditions: Comparative study of an instrumental (HPLC) vs. a bioassay method. Nachrichtenbl. Deut. Pflanzenschutzd. 44: 105-108.

GDCh (Gesellschaft Deutscher Chemiker) (1993): Grundsätze der ökotoxikologischen Bewertung von Chemikalien und Sachstand der heutigen Praxis. Positionspapier der Gesellschaft Deutscher Chemiker, 116 S.

Gearing, J.N. (1989): The role of aquatic microcosms in ecotoxicologic research as illustrated by karge marine systems. In: Ecotoxicology: Problems and Approaches (eds. Levine, S.A., Harwell, M.A., Kelly, J.R. & Kimball, K.D.). Springer-Verlag, New York. 411-470.

Gersich, F.M., Blanchard, F.A., Applegath, S.L. & Park, C.N. (1986): The precision of daphnid (*Daphnia magna*) static acute toxicity tests. Arch. Environ. Contam. Toxicol. 15: 741-749.

Ghabbour, S. & Imam, M. (1967): The effect of 5 herbicides on three oligochaete species. Rev. Ecol. Biol. Soil 4: 119-122.

Giesy, J.P. (1980): Microcosms in Ecological Research. DOE CONF-781101, National Technical Information Service, Springfield. 1110 pp.

Giesy, J.P. & Graney, R.L. (1989): Recent developments in and intercomparisons of acute and chronic bioassays and bioindicators. Hydrobiologia 188/189: 21-60.

Giesy, J.P. & Hoke, R.A. (1990): Freshwater sediment toxicity bioassessments: rationale for species selection and test design. J. Great Lakes Res. 15: 539-569.

GIFAP (1989): Guidelines for Personal Protection when Using Pesticides in Hot Climates. Brussels.

Gillett, J.W. (1989): The role of terrestrial microcosms and mesocosms in ecotoxicologic research. In: Ecotoxicology: Problems and Approaches, (ed. S.A. Levin et al.). Berlin: Springer Verlag, pp. 367-410.

Gillett, J.W. & Witt, J.M. (1980): Chemical evaluation: Projected application of microcosms technology. Microcosms in Ecological Research. (ed. Giesy, J.R.). DOE CONF-781101, National Technical Information Service, Springfield. 1008-1033.

Gledhill, W.E. (1987): Microbial Toxicity and Degradation Test Methodology: An Industrial Perspective. Toxicity Assessm. 2: 89-96.

Goh, K.S., Hernandez, J., Powell, S.J., Garretson, C., Troiano, J., Ray, M. & Greene, C.D. (1991): Enzyme Immunoassay for the Determination of Atrazine Residues in Soil. -Bull. Environ. Contam. Toxicol. 46: 30-36.

Gonzalez, R.H. (1976): Plant protection in Latin America. PANS Pest Articles News Sum. 22: 26-34.

Gonzalez, H., Lodenius, M. & Otero, M. (1989): Water Hyacinth as indicator of heavy metal pollution in the tropics. Bull. Environ. Contam. Toxicol. 43: 910-914.

Goodman, D. (1975): The theory of diversity-stability relationships in ecology. Q. Rev. Biol. 50: 237-266.

Graefe, U. (1993): Die Gliederung von Zersetzergesellschaften für die standortsökologische Ansprache. Mittl. Deut. Bodenkundl. Ges. 69: 95-98.

Grant, I.F. (1988): Appropriate technology for measuring environmental effects of insecticides in the tropics. In: Field methods for the study of environmental effects of pesticides (ed. M.P. Greaves, B.D. Smith & P.W. Greig-Smith). Thornton Heath: BCPC, pp. 147-154.

Grant, I.F. (1989a): Monitoring insecticide side-effects in large-scale treatment programmes: Tsetse spraying in Africa. In: Pesticides and Non-Target Invertebrates. (ed. P.C. Jepson). Wimborne, UK: Intercept, pp. 43-69.

Grant, I.F. (1989b): Environmental effects of desert locust control. FAO Plant Protect. Bull. 37: 27-35.

Graham, R.L., Hunsaker, C.T., O'Neill, R.V. & Jackson, B.L. (1991): Ecological risk assessment at the regional scale. Ecol. Applications 1: 196-206.

Grau, R., Pflüger, W. & Schmuck, R. (1992): Acceptance studies to assess the hazard of pesticides formulated as dressings, baits and granules, to birds. Brighton Crop Protection Conf. 787-792.

Gray, E.C. (1991): A Short History of Pesticide Re-registration. In: Regulation of Agrochemicals — A Driving Force in Their Evolution. (Marco, G.J., Hollingworth, R.M. & Plimmer, J.R. eds.). American Chem. Soc., Washington, DC. 45-54.

Gray, J.S., Aschan, M., Carr, M.R., Clarke, K.R., Green, R.H., Pearson, T.H., Rosenberg, R. & Warwick, R.M. (1988): Analysis of community attributes of the benthic macrofauna of Frierfjord/Langesundfjord and in a mesocosm experiment. Mar. Ecol. Prog. Ser. 46: 151-165.

Greaves, M.P., Greig-Smith, P.W. & Smith, B.D. (eds.) (1988): Field methods for the study of environmental effects of pesticides. Proc. Symp. Brit. Crop Protec. Council, Churchill Coll., Cambridge, UK March 28-30, 1988, BCPC Monograph No. 40, Thornton Heath: BCPC, 370 pp.

Greig-Smith, P.W. (1990a): The Boxworth-Project. Pesticide Outlook 1: 16-19.

Greig-Smith, P.W. (1990b): Intensive studies versus extensive monitoring in pesticide field trials. In: Pesticide Effects on Terrestrial Wild Life (eds. Somerville, L. & Walker, C.H.). Taylor & Francis, London.

Greig-Smith, P.W. (1992): A European Perspective on Ecological Risk Assessment, Illustrated by Pesticide Registration Procedures in the United Kingdom. Environ. Toxicol. Chem. 11: 1673-1689.

Groombridge, B. (ed.) (1992): Global Biodiversity. Status of the Earth's Living Resources. World Conservation Monitoring Centre. Chapman & Hall, London etc.

Gruttke, H., Kratz, W., Weigmann, G. & Haque, A. (1988): Terrestrial model food chain and environmental chemicals. I. Transfer of Sodium [^{14}C] Pentachlorophenate between springtails and carabids. Ecotox. Environ. Safety 15: 253-259.

Guarra, R.T., Bueno, C.R. & Schubart, H.O. (1982): Avaliacao preliminar sobre os efeitos da aplicacao do herbicida Paraquat e aracao convencional na mesofauna do solo regiao de Manaus-Am. Acta Amazonica 12: 7-13.

Guhl, W. (1991): Beitrag zur biologischen Bewertung von Umweltchemikalien mit Laborökomodellen. II. Vergleichbarkeit der Flußmodelle mit realen Gewässern. Z. Angew. Zool. 78: 323-336.

Günther, P. & Pestemer, W. (1990): Risk assesssment for selected xenobiotics by bioassay methods with higher plants. Environ. Management 14: 381-388.

Gupta, P.K. & Durve, V.S. (1987): A study on the temperature dependence of the acute toxicity of Pentachlorphenol to a freshwater pond snail, Viviparus bengalensis L. Rev. Biol. (Lisbon) 13: 103-112.

Gustafsson, D.I. (1989): Ground water ubiquity score: A simple method for assessing pesticide leachability. Environ. Toxicol. Chem. 8: 339-357.

Gydesen, H. (1984): Mathematical models of the transport of pollutants in ecosystems. Ecol. Bull. 36: 17-25.

Haeckel, E. (1866): Generelle Morphologie der Organismen. Reimer, Berlin, Bd. 2.

Hagens, M. & Westheide, W. (1987): Subletale Schädigungen bei *Enchytraeus minutus* (Oligochaeta, Annelida) durch das Insektizid Parathion: Veränderungen in der Ultrastruktur von Chloragog- und Darmzellen in Abhängigkeit von der Belastungsdauer. Verh. Ges. Ökol. XVI: 423-426.

Haider, K. (1985): Der Einfluß von Bodenparametern auf den Abbau von Pflanzenschutzmitteln. Ber. Landwirtschaft 198: 81-91.

Hallberg, G.R. (1989): Pesticide pollution of ground water in the humid United States. Agric. Ecosyst. Environ. 26: 299-367.

Haltrich, W.G. (1991): Ökologische Untersuchungen und Grenzwerte — Anforderungen und Aussagefähigkeit. UWSF.-Umweltchem. Ökotox. 3: 8-11.

Hamburger, B. (1983): Stand der Normung von Biotesten. Vom Wasser 61: 227-235.

Hamer, M.J., Maund, S.J. & Hill, I.R. (1992): Laboratory methods for evaluating the impact of pesticides on water/sediment organisms. Proc. Brighton Crop Protection Conference 487-496.

Hamilton, W.E. & Dindal, D.L. (1989): Influence of earthworms and leaf litter on edaphic variables in sewage-sludge-treated soil microcosms. Biol. Fert. Soils 7: 129-133.

Handbook on the use of pesticides in the Asia-Pacific Region (1987): Asian Development Bank, Manila: 294 S.

Hanlon, R.D.G. & Anderson, J.M. (1979): The effects of Collembola grazing on microbial activity in decomposition of litter. Oecologia 38: 93-99.

Hansen, D.J. (1989): Status of the development of water quality criteria and advisories. In: Water Quality Standards for the 21st Century. (ed.: U.S. EPA Office of Water, Criteria and Standards Division). EPA, Washington. 163-169.

Hansen, D.J. & Tagatz, M.E. (1980): A laboratory test for assessing impacts of substances on developing communities of benthic estuarine organisms. In: Aquatic Toxicology. (eds.: Eaton, J.G., Parrish, P.R. & Hendricks, A.C.). STP 707, ASTM, Philadelphia. 40-57.

Hansen, S.R. (1981): Screening for toxic effects on interspecies interactions: a mechanistic or an empirical approach ? — Arch. Environ. Contam. Toxicol. 10: 597-603.

Haque, A. & Pflugmacher, J. (1985): Einflüsse von Pflanzenschutzmitteln auf Regenwürmer. Ber. Landwirtschaft 198: 176-188.

Hardy, A.R. (1986): The Boxworth project — a progress report. Brighton Crop Protection Conference, 1215-1224.

Hardy, A.R. (1987): Ecotoxicology of pesticides: The laboratory and field evaluation of the environmental hazard presented by new pesticides. In: Toxicology of Pesticides: Experimental, Clinical and Regulatory Perspectives (ed. L.G. Costa et al.). Berlin [etc.]: Springer, pp. 185-196, NATO ASI Ser. H, No. 13.

Harland, B.J., Mathiessen, P. & Murray-Smith, R. (1990): The prediction of fate and exposure of organic chemicals in aquatic sediments. Proc. European Sci. Foundation Strasbourg, 125-147.

Harms, H. & Sauerbeck, D. (1984): Organische Schadstoffe in Siedlungsabfällen: Herkunft, Gehalt und Umsetzung in Böden und Pflanzen. Angew. Botanik 58: 97-108.

Harris, H.J., Sager, P.E., Regier, H.A. & Francis, G.R. (1990): Ecotoxicology and ecosystem integrity: The Great Lakes examined. Environ. Sci. Technol. 24: 598-603.

Hart, T.B. (1987): Pesticides — Risk assessment and safe use. In: Toxicology of Pesticides: Experimental, Clinical and Regulatory Perspectives (ed. L.G. Costa et. al.). Berlin [etc.]: Springer, pp. 215-239, NATO ASI Ser. H, No. 13.

Hart, W.B., Doudoroff, P. & Greenbank, J. (1945): The Evaluation of Industrial Wastes, Chemicals and Other Substances to Freshwater Fish. Philadelphia: Waste Control Laboratory, Atlantic Refining Co.

Harwell, C.C. (1989): Regulatory framework for ecotoxicology. In: Ecotoxicology. Problems and Approaches (ed. S.A. Levin et al.). New York [etc.]: Springer, pp. 497-516.

Harwell, M.A. & Harwell, C.C. (1989): Environmental decision-making in the presence of uncertainty. In: Ecotoxicology: Problems and Approaches (ed. S.A. Levin et al.). New York [etc.]: Springer, pp. 517-540.

Harwell, M.A., Harwell, C.C. & Kelly, J.R. (1986): Regulatory endpoints, ecological uncertainties, and environmental decision-making. Oceans 86 Proceedings 3: 993-998, Washington, DC: Marine Technology Society.

Hassan, S.A. (1985): Standard methods to test the side-effects of pesticides on natural enemies of insects and mites developed by the IOBC/WPRS Working Group "Pesticides and Beneficial Organisms". Bull. OEPP/EPPO Bull. 15: 214-255.

Hassan, S.A. (1988): Guidelines for Testing the Effects of Pesticides on Beneficials: Short Description of Test Methods. Bull. IOBC/WPRS Bull. 11: 1-143.

Hassan, S.A. (1992): Guidelines for testing the effects of pesticides on beneficial organisms: Description of test methods. Bull. IOBC/WPRS Bull. 15: 1-186.

Hawker, D.W. & Connell, D.W. (1985): Relationships between partition coefficient, uptake rate constant, clearance rate constant and time to equilibrium for bioaccumulation. Chemosphere 14: 1205-1219.

Hawkins, C.P. & MacMahon, J.A. (1989): Guilds: the multiple meanings of a concept. Ann. Rev. Ent. 34: 423-451.

Hawkins, W.E., Overstreet, R.M. & Walker, W.W. (1988): Carcinogenetic tests with small fish species. Aquat. Toxicol. 11: 113-128.

Heck, M., Kratz, W., Nüss, D., Rink, U. & Weigmann, G. (1992): Untersuchungen zur Bioindikatoreignung von Bodentieren und bodenökologischen Prozessen für die Bewertung des Zustandes urbaner Waldökosysteme in Berlin. Report F + E-Vorhaben Ballungsraumnahe Waldökosysteme. Senat Berlin 172 S.

HCN (Health Council of the Netherlands) (1992): Risk of dangerous substances exported to developing countries. Publ. No. 92/2E. 135 pp.

Heidler, G. (1987): Wirkung von Pflanzenschutzmitteln auf aquatische Ökosysteme — Prüfung und Bewertung im Zulassungsverfahren. Nachrichtenbl. Deut. Pflanzenschutzd. 39: 161-165.

Heimbach, F. (1984): Correlations between three methods for determining the toxicity of chemicals to earthworms. Pestic. Sci. 15: 605-611.

Heimbach, F. (1992): Correlation between data from laboratory and field tests for investigating the toxicity of pesticides to earthworms. Soil Biol. Biochem. 24: 1749-1754.

Heimbach, U. (1992): Standard laboratory method to test effects of pesticides on *Poecilus cupreus* (Coleoptera, Carabidae). In: Proc. Intl. Symp. Ecotoxicology. GSF-Ber. 1/92: 105-109.

Heinisch, E. & Klein, S. (1989): Einsatz chemischer Pflanzenschutzmittel — ein Spannungsfeld von Ökonomie und Ökotoxikologie. Nachrichten Mensch + Umwelt, 17: 53-66.

Heinrich, D. & Hergt, M. (1990): dtv-Atlas zur Ökologie. Deutscher Taschenbuch Verlag, München, 283 S.

Heise, M., Schuphan, I. & Ebing, W. (1988): Einsetzbarkeit künstlicher Modellökosysteme zum Testen von Umweltchemikalien. Report für das Umweltbundesamt (Berlin, FRG) F + E Nr. 106 05 033, 100 S.

Heitefuss, R. (1987): Pflanzenschutz. Grundlagen der praktischen Phytomedizin. Stuttgart: Thieme, 2. Aufl., 342 S.

Henschke, R.B., Nücken, E. & Schmidt, F.R.J. (1989): Fate and dispersal of recombinant bacteria in a soil microcosm containing the earthworm Lumbricus terrestris. Biol. Fert. Soils 7: 374-376.

Herman, P.M. & Heip, C. (1988): On the use of meiofauna in ecological monitoring: Who needs taxonomy ? — Mar. Poll. Bull. 19: 665-658.

Hermens, J.L.M. (1989): Quantitative Structure-Activity Relationships of Environmental Pollutants. In: Handbook of Environmental Chemistry: Reactions and Processes. Hutzinger, O. (ed.). Springer Verlag, Berlin. 111-162.

Hermens, J., Busser, F., Leeuwangh, P. & Musch, A. (1985): Quantitative structure-activity relationships and mixture toxicity of organic chemicals in Photobacterium phosphoreum. The Microtox test. Ecotox. Environ. Safety 9: 17-25.

Hermosin, M.C. & Cornejo, J. (1989): Assessing factors related to pesticide adsorption by soils. Toxicol. Environ. Chem. 25: 45-55.

Herzel, F. (1987): Einstufung von Pflanzenschutzmitteln aus der Sicht des Trinkwasserschutzes. Nachrichtenbl. Deut. Pflanzenschutzd. 39: 97-104.

Herzel, F. & Schmidt, G. (1979): Prüfung des Versickerungsverhaltens von Pflanzenschutzmitteln in Lysimetern und Kleinsäulen. WaBoLu-Berichte 3/1979: 1-16.

Herzel, F. & Schmidt, G. (1991): Zur Prüfung des Versickerungsverhaltens von Pflanzenschutzmitteln in Modellversuchen. IV. Entstehung von Metaboliten während der Alterung. Nachrichtenbl. Deut. Pflanzenschutzd. 43: 13-16.

Hilbig, V., Joermann, G. & Solecki, R. (1994): Gefährdungspotential chemischer Pflanzenschutzmittel für Vögel. Risikoabschätzung im Zulassungsverfahren nach dem Pflanzenschutzgesetz, Teil I. Bundesgesundheitsblatt 94: 151- 157.

Hildebrandt, A., Schön, H., Hammer, W. & Hille, M. (1991): Veränderungen im chemischen Pflanzenschutz des Ackerbaus in den 80er Jahren. Gesunde Pflanzen 43: 71-78.

Hill, I.R. (1989): Aquatic Organisms and Pyrethroids. Pestic. Sci. 27: 429-465.

Hill, I.R., Heimbach, F., Leeuwangh, P. & Matthiessen, P. (1993): Freshwater Field Tests for Hazard Assessment of Chemicals. Lewis Publ., Boca Raton, FL. 561 pp.

Hock, B. & Elstner, E.F. (Hrsg.) (1988): Schadwirkungen auf Pflanzen. Lehrbuch der Pflanzentoxikologie. Mannheim etc.: BI-Wissenschaftsverlag, 2. Aufl., 348 S.

Hoekstra, J.A. & van Ewijk, P.H. (1993): Alternatives for the no-observed-effect level. Environ. Tox. Chem. 12: 187-194.

Hoffman, D.J., Rattner, B.A. & Hall, R.J. (1990): Wildlife Toxicology. Environ. Sci. Technol. 24: 276-283.

Högger, C.H.; Ammon, H-U.; Bieri, M. (1992): The Daniel Funnel Test in a Sequence of Earthworm Tests of Agricultural Pesticides. In: Ecotoxicology of Earthworms. (Greig-Smith, P., Becker, H., Edwards, P.J., Heimbach, F. eds.): 220-224. Intercept, Andover.

Holcombe, G.W., Phipps, G.L., Sulaiman, A.H. & Hoffman, A.D. (1987): Simultaneous multiple species testing: Acute toxicity of 13 chemicals to 12 diverse freshwater amphibian, fish, and invertebrate families. Arch. Environ. Contam. Toxicol. 16: 697-710.

Hollihn, K-U., Fischer, R., Fritsch, H., Guth, J.A., Johnen, B.G., Knauf, W., Pflüger, W., Säle, M. & Voget, M. (1991): Pflanzenschutzmittel und Naturhaushalt — Prüfung, Gefährdungsabschätzung und Risikobeurteilung. Teil III. Stufenkonzept für Prüfschemata als Grundlage der Gefährdungsabschätzung und der Risikobeurteilung. Gesunde Pflanzen 43: 221-231.

Hollis, J.M. & Carter, A.D. (1990): A comparison of laboratory, lysimeter and field study techniques to determine environmental fate of pesticides in the U.K. Brighton Crop Protection Conference, 1005-1010.

Holst, R.W. & Ellwanger, T.C. (1982): Pesticide Assessment Guidelines. Subdivision J. Hazard Evaluation: Non-target Plants, EPA 540/9-82-020. Office of pesticides and toxic substances. U.S.-EPA Washington, DC.

Hooftman, R.N. & Vink, G.J. (1980): The Determination of Toxic Effects of Pollutants with the Marine Polychaete Worm *Ophryotrocha diadema*. Ecotox. Environ. Safety 4: 252-262.

Hope, B.K. (1993): Ecological considerations in the practice of ecotoxicology. Environ. Toxicol. Chem. 12: 205-206.

Hopkin, S.P. (1989): Ecophysiology of Metals in Terrestrial Invertebrates. Elsevier Applied Science, London.

Hopkin, S.P. (1990): Species-specific differences in the net assimilation of zinc, cadmium, lead, copper and iron by the terrestrial isopods *Oniscus asellus* and *Porcellio scaber*. J. Appl. Ecol. 27: 460-474.

Hopkin, S.P. (1993a): In Situ Biological Monitoring of Pollution in Terrestrial and Aquatic Ecosystems. In: Handbook of Ecotoxicology. (ed. Calow, P.). Blackwell Scientific Publ, Oxford. 397-427.

Hopkin, S.P. (1993b): Ecological implications of "95% protection levels" for metals in soil. Oikos 66: 137-141.

Huismans, J.W. (1980): The international register of potentially toxic chemicals (IRPTC). Ecotox. Environ. Safety 4: 393-403.

Houx, N.W.H. & Dekker, A. (1987): A test system for the determination of the fate of pesticides in surface water. Int. J. Environ. Anal. Chem. 29: 37-59.

Howard, P.H. (1993): Progress Report: Determining "Real World" Biodegradation Rates. Environ. Toxicol. Chem. 12: 1135-1137.

Huang, P. (1992): Entwurf einer vorläufigen Richtlinie zur Prüfung der Wirkung von Pflanzenbehandlungsmitteln auf *Isotoma tigrina* Nicolet (Collembola). In: Hassan, S.A.: Guidelines for testing the effects of pesticides on beneficial organims: Description of test methods. IOBC/WPRS Bulletin XV/3: 122-130.

Hulzebos, E.M., Adema, D.M.M., Dirven-van Breemen, E.M., Henzen, L., van Dis, W.A., Herbold, H.A., Hoekstra, J.A., Baerselman, R. & van Gestel, C.A.M. (1993): Phytotoxicity studies with *Lactuca sativa* in soil and nutrient solution. Environ. Tox. Chem. 12: 1079-1094.

Hutzinger, O. (1991): Was ist ein Schadstoff ? — UWSF — Z. Umweltchem. Ökotox. 3: 259.

Igbedioh, S.O. (1991): Minimizing environmental and health effects of agricultural pesticides in developing countries. Ambio 20: 219-221.

Iglisch, I. (1985): Bodenorganismen für die Bewertung von Chemikalien. Z. Angew. Zool. 72: 395-431.

Iglisch, I. (1986): Hinweise zur Entwicklung von Testverfahren zum Nachweis subakuter Wirkungen von Chemikalien. Z. Angew. Zool. 73: 199-218.

Ingersoll, C.G. & Nelson, M.K. (1990): Testing sediment toxicity with *Hyalella azteca* (Amphipoda) and *Chironomus riparius* (Diptera). In: Aquatic Toxicology and Risk Assessment: 13th Vol. ASTM STP 1096 (eds. Landis, W.G. & Van der Schalie, W.H.), ASTM, Philadelphia. 93-109.

Inglesfield, C. (1989): Pyrethroids and terrestrial non-target organisms. Pestic. Sci. 27: 387-428.

ISO (International Standards Organization) (1988): Water Quality — Marine Algal Growth Inhibition Test with *Skeletonema costatum* and *Phaeodactylum tricornutum*. Draft Proposal. ISO, Delft. 10 pp.

ISO (International Standards Organization) (1989): Terms in the field of soil quality/soil pollution (Draft). ISO Document ISO/TC 190 SC 1 "Terminology"/ WG 2 Terminology — Soil Pollution" Doc. N1, 24 p.

ISO (International Standards Organization) (1992a): Guidance on the collection, handling and storage of soil for the assessment of aerobic microbial processes in the laboratory. (Draft). ISO Document ISO/TC 190 SC 2 "Soil Quality — Sampling — Part 6, 6 p.

ISO (International Standards Organization) (1992b): Guidance on laboratory tests for biodegradation of organic chemicals in soil. Part 1: Aerobic conditions. (Draft). ISO Document ISO/TC 190 SC 4 "Soil Quality", 13 p.

ISO (International Standards Organization) (1992c): Determination of the effects of root growth on soil flora. Part 1: Method for the measurement of inhibition of root growth. (Draft). ISO Document ISO/TC 190 SC 4 "Soil Quality", 12 p.

IVA (Industrieverband Agrar) (Hrsg.) (1993): Jahresbericht 1992/93. Frankfurt/ Main, 64 S.

Iyatoma, K., Tamura, T. et al. (1958): Toxicity of endrin to fish. Prog. Fish Cult. 20: 155-162.

Jagnow, G. (1984): Nebenwirkungen von Pflanzenschutzmitteln auf Mikroorganismen und mikrobielle Umsetzungen im Boden. Angew. Botanik 58: 61-66.

Jahns, H.M. (1989): Problems in Lichen Taxonomy. Abh. Naturw. Ver. Hamburg (NF) 28: 247-252.

Jancke, G. (1989): Modellversuche zur subakuten und subletalen Wirkung von Herbiziden auf Collembolen in Hinblick auf ein Testsystem für Umweltchemikalien. Zool. Beitr. N.F. 32: 261-299.

Janczyk, H.J. (1983): Lysimeter-Anlage der Bayer AG, Pflanzenschutz Anwendungstechnik in Leverkusen. Pflanzenschutz-Nachrichten Bayer 36: 1-19.

Janssen, C.R. & Persoone, G. (1992): Routine Aquatic Toxicity Testing: Some Problems and New Approaches. In: Biological Indicators for Environmental Monitoring. Bonotto, S., Nobili, R. & Revoltella, R.P. (eds.). Sereno Symposia Review 27: 195-205.

Jeffrey, D.W. (1990): Biomonitoring of Catastrophes. Environ. Monit. Assessment 14: 131-138.

Jellinek, S.D. & Gray, E.C. (1992): Some lessons from the U.S. Re-registration Program. In: Proc. Brighton Crop Protection Conf. 1992: 1261-1266.

Jenkinson, D.S. & Oates, J.M. (1979): A method for measuring adenosine triphosphate in soil. Soil Biol. Biochem. 11: 193-199.

Jensen, K., Albrechtsen, H.J., Nielsen, J. & Kruse, B. (1988): The use of ecocores to evaluate biodegradation in marine sediments. Water, Air, Soil Pol. 39: 89-99.

Jepson, P.C. (1993): Insects, Spiders and Mites. In: Handbook of Ecotoxicology. (ed. Calow, P.). Blackwell Scientific Publ, Oxford. 299-325.

Jepson, P.C., Duffield, S.J., Thacker, J.R.M., Thomas, C.F.G. & Wiles, J.A. (1990): Predicting the side-effects of pesticides on beneficial invertebrates. Brighton Crop Protection Conference, 957-962.

Jepson, P.C. & Mead-Briggs, M. (1992a): A discussion of methods used in semi-field studies to evaluate pesticide toxicity to beneficial invertebrates. In: Guidelines for testing the effects of pesticides on beneficial organisms: Description of test methods. (ed. Hassan, S.A.). IOBC/WPRS Bull. XV/3: 4-18.

Jepson, P.C. & Mead-Briggs, M. (1992b): The field approach to pesticide ecotoxicity testing against beneficial invertebrates in cereals. In: Guidelines for testing the effects of pesticides on beneficial organisms: Description of test methods. (ed. Hassan, S.A.). IOBC/WPRS Bull. XV/3: 183-186.

Johnen, B.G. (1990): Assessment and regulation of pesticide residues in drinking water. Pesticide Outlook 1: 9-12.

Jury, W.A., Focht, D.A. & Farner, W.J. (1987): Evaluation of pesticide ground water pollution potential from standard indices of soil-chemical adsorption and biodegradation. J. Environ. Qual. 16: 422-428.

Kanazawa, J. (1989): Relationship between the soil sorption constants for pesticides and their physico-chemical properties. Environ. Toxicol. Chem. 8: 477-484.

Kanne, R. (1991): Aquatische Ökotoxikologie. UWSF.-Z. Umweltchem. Ökotox. 3: 16-18.

Kappeler, T.U. (1979): The world ecotoxicology watch. Environ. Sci. Technol. 13: 412-415.

Kapustka, L.A. & Reporter, M. (1993): Terrestrial Primary Producers. In: Handbook of Ecotoxicology. (ed. Calow, P.). Blackwell Scientific Publ, Oxford. 278-298.

Karg, W. (1993): Raubmilbenarten in Boden- sowie Vegetationsschichten und ihre Eignung als Indikatoren für Agrochemikalien. Anz. Schädlingskde. Pflanzenschutz, Umweltschutz 66: 126-131.

Karickhoff, S.W. & Morris, K.R. (1985): Impact of Tubificid Oligochaetes on Pollutant Transport in Bottom Sediments. Environ. Sci. Technol. 19: 51-56.

Karr, J.R. (1991): Biological integrity: A long-neglected aspect of water resource management. Ecol. Appl. 1: 66-84.

Karr, J.R. (1993): Defining and Assessing Ecological Integrity: Beyond Water Quality. Environ. Toxicol. Chem. 12: 1521-1531.

Karr, J.R., Heidinger, R.C. & Helmer, E.H. (1985): Sensitivity of the index of biotic integrity to changes in chlorine and ammonia levels from wastewater treatment facilities. J. Water Poll. Control Fed. 57. 912-915.

Kaus, V. (1993): Die Zweitanmelderproblematik im Pflanzenschutzrecht. Europäische Hochschulschriften, Reihe II, Rechtswissenschaft, Frankfurt. Bd. 1471, 284 S.

Kaushik, C.P. (1991): Persistence and Metabolism of HCH and DDT in Soil under Subtropical Conditions. Soil Biol. Biochem. 23: 131-134.

Kearney, P.C. (1991): Impact of Regulations on the ACS Division of Pesticide Chemistry. In: Regulation of Agrochemicals. A Driving Force in Their Evolution. (eds.: Marco, G.J., Hollingworth, R.M. & Plimmer, J.R.). American Chemical Society, Washington D.C.. 19-26.

Kegel, B. (1989): Laboratory experiments on the side-effects of selected herbicides and insecticides on the larvae of three sympatric *Poecilus*-species (Col., Carabidae). J. Appl. Ent. 108: 144-155.

Keith, J.R. (1990): Possible solutions for the treatment of existing chemicals in the U.S. and Canada. Paper presented at the Congress Eurochemie '92 in Frankfurt/ Main 1990.

Kemp, W.M., Lewis, M.R., Cunningham, J.J., Stevenson, J.C. & Boynton, W.R. (1980): Microcosms, macrophytes, and hierachies: environmental research in the Chesapeake Bay. In: Microcosms in Ecological Research. (ed.: Giesy, J.P.). U.S. Dept. of Energy Symposium Series 52, NTIS, Springfield.

Kemp, W.M., Means, J.C., Jones, T.W. & Stevenson, J.C. (1982): Herbicides in Chesapeake Bay and their effects on submerged aquatic vegetation: A synthesis of research supported by U.S. EPA Chesapeake Bay Program. In: Chesapeake Bay Program Technical Studies: A Synthesis. (eds.: Macalaster, E.G., Barker, D.A. & Kasper, N.E.). U.S. EPA, Washington. 503-567.

Kenaga, E.E. (1981): Comparative toxicity of 131,596 chemicals to plant seeds. Ecotox. Environ. Savety 5: 469-475.

Kenaga, E.E. & Goring, C.A. (1980): Relationship between Water Solubility, Soil Sorption, Octanol-Water Partitioning and Bioconcentration of chemicals in biota. In: Eaton, J.G. et al. (eds.). Aquatic Toxicology 707, ASTM, Philadelphia.

Kenaga, E.E. & Moolenaar, R.J. (1979): Fish and *Daphnia* toxicity as surrogates for aquatic vascular plants and algae. Environ. Sci. Technol. 13: 1479-1480.

Kersting, K. (1984): Properties of an aquatic microecosystem. IV. Development and use of an aquatic microecosystem as a test system for toxic substances. Int. Rev. Ges. Hydrobiol. 69: 567-607.

Kersting, K. & Van Wijngaarden, R. (1992): Effects of Chlorpyrifos on a microecosystem. Environ. Toxicol. Chem. 11: 365-372.

Kettrup, A., Steinberg, C. & Freitag, D. (1991): Ökotoxikologie — Wirkungserfassung und Bewertung von Schadstoffen in der Umwelt. UWSF.-Z.Umweltchem.Ökotox. 3: 370-377.

Kidd, H., Hartley, D. & Ekström, G. (1989): World Directory of Pesticide Control Organisations. The Royal Society of Chemistry, Cambridge, UK. 311 pp.

Kimball, K.D. & Levin, S.A. (1985): Limitations of laboratory bioassays: The need for ecosystem-level testing. BioScience 35: 165-171.

Kitano, M. (1990): Current state and future trends in chemicals legislation in Japan. Vortrag Kongress Eurochemie '92, Frankfurt, 1990.

Klein, A.W., Klein, W., Kördel, W. & Weiss, M. (1988): Structure-activity relationships for selecting and priority setting of existing chemicals. A computer-assisted approach. Environ. Toxicol. Chem. 7: 455-462.

Klein, A.W., Apel, P. & Goedicke, J. (1992): UBA-Principles on Criteria and Procedures for Environmental Assessment of Pesticides. Chemosphere 24: 793-816.

Klein, M. (1991): Application and validation of pesticide leaching models. Pestic. Sci. 31: 389-398.

Klein, W. (1991): Bewertung und Beurteilung von Chemikalien im Boden — Informationsbedarf und Datenlage. UWSF.-Z. Umweltchem. Ökotox. 3: 25-27.

Klein, W., Geyer, H., Freitag, D. & Rohleder, H. (1984): Sensitivity of schemes for ecotoxicological hazard ranking of chemicals. Chemosphere 13: 203-211.

Klingauf, J. (1987): Möglichkeiten und Grenzen des biologischen Pflanzenschutzes in Entwicklungsländern. In: Möglichkeiten, Grenzen und Alternativen des Pflanzenschutzmitteleinsatzes in Entwicklungsländern (hrsg. DSE/ATSAF). Feldafing: DSE/ATSAF, S. 104-114.

Klöpffer, W. (1989): Persistenz und Abbaubarkeit in der Beurteilung des Umweltverhaltens anthropogener Chemikalien. UWSF — Z. Umweltchem. Ökotox. 2: 43-51.

Kloskowski, R. & Führ, F. (1988): Charakterisierung und Bioverfügbarkeit von gebundenen Pflanzenschutzmitteln im Boden. Wissenschaft und Umwelt 2: 112-121.

Knacker, T., Marcinkowski, A. & Schallnass, H. (1989): Ecotoxicological effects of artificial smokes on a terrestrial microcosm. Arch. Toxicol. Suppl. 13: 398-401.

Knacker, T. & Morgan, E. (1994): Test Guideline — Terrestrial Model Ecosystem. In: UBA-Workshop on Terrestrial Model Ecosystems. Appendix UBA-Texte 54/94: 1–35.

Knacker, T., Schallnaß, H., Marcinkowski, A., Förster, B. & Vincena, R. (1990): Einsetzbarkeit seminatürlicher (terrestrischer) Systeme für die Bewertung der Umweltgefährlichkeit nach dem ChemG. Battelle Report für das Umweltbundesamt F+E-Vorhaben Nr. 106 03 069, Frankfurt.

Knacker, T., Römbke, J., Förster, B., Marcinkowski, A., & Schallnaß, H. (1991): Die Wirkung von Pestiziden auf die Mesofauna und den CO_2-Gas-Austausch eines terrestrischen Ökosystems. Battelle Report für das Umweltbundesamt F+E-Vorhaben Nr. 106 03 069/01, Frankfurt.

Kneitz, G. (1983): Aussagefähigkeit und Problematik eines Indikatorkonzepts. Verh. Deutsch. Zool. Ges. 1983: 117-119.

Knirsch, J. (1993): Pestizid-Lebenszyklus-Analyse dreier Pestizide im Niger. Schriftenreihe der GTZ Nr. 240, Eschborn, 330 S.

Koch, R. (1989): Umweltchemie und Ökotoxikologie — Ziele und Aufgaben. Z. Umweltchemie Ökotox. 1: 41-43.

Koch, R. & Nagel, M. (1988): Quantitative structure-activity relationships in soil ecotoxocology. The Science of the Total Environment 77: 269-276.

Koch, R. & Wagner, B.O. (1989): Umweltchemikalien: physikalisch-chemische Daten, Toxizitäten, Grenz- u. Richtwerte, Umweltverhalten. Weinheim etc.: VCH, 423 S.

Koch, W., Sauerborn, J., Kunisch, M. & Pülschen, L. (Hrsg.) (1990): Agrarökologie und Pflanzenschutz in den Tropen und Subtropen. Weikersheim: Margraf, PLITS Jhg. 8, Heft 2, 339 S.

Köhler, H-R., Triebskorn, R., Stöcker, W., Kloetzel, P-M. & Alberti, G. (1992): The 70 kD Heat Shock Protein (hsp 70) in Soil Invertebrates: A Possible Tool for Monitoring Environmental Toxicants. Arch. Environ. Contam. Toxicol. 22: 334-338.

Koeman, J.H. (1977): Effects of Tsetse fly control measures on non-target organisms. Meded. Fac. Landbouw. Rijskuniv. Gent 42: 889-896.

Koeman, J.H., Den Boer, W.M.J., Feith, A.F., De Jongh, F.H., Spliethoff, P.C., Naisa, B.K. & Spielberger, U. (1978): Three years observation on side-effects of helicopter applications of insecticides used to exterminate *Glossina*-species in Nigeria. Environ. Poll. 15: 31-59.

Koetering-Walker, C. & Buck, J.D. (1989): The effect of bacteria and bioturbation by *Clymenella torquata* on oil removal from sediment. Water Air Soil Poll. 43: 413-424.

Kohn, G.K. (1991): Agrochemicals and the Regulatory Process Before 1970. In: Regulation of Agrochemicals — A Driving Force in Their Evolution. (eds.: Marco, G.J., Hollingworth, R.M. & Plimmer, J.R.). American Chemical Society, Washington, DC. 1-8.

Kokta, C. (1992): Measuring Effects of Chemicals in the Laboratory: Effect Criteria and Endpoints. In: Ecotoxicology of Earthworms. (ed. Greig-Smith, P., Becker, H., Edwards, P.J. & Heimbach, F.). Intercept, Andover, pp. 213-216.

Kolkwitz, R. & Marsson, M. (1908): Ökologie der pflanzlichen Saprobien. Ber. Dtsch. Bot. Ges. 26a: 505-519.

Kolkwitz, R. & Marsson, M. (1909): Ökologie der tierischen Saprobien. IR 2: 126-162.

Kollig, H.P. & Kitchens, B.E. (1990): Problems associated with published environmental fate data. Toxicol. Environ. Chem. 28: 95-103.

Kooijman, S.A.L.M. (1987): A safety factor for LC_{50} values allowing for differences among species. Water Res. 21: 269-276.

Kooijman, S.A.L.M., Hansveit, A.O. & Oldersma, H. (1983): Parametric analysis of population growth in bioassay. Water Res. 17: 527-538.

Könemann, H. (1986): Quantitative structure-activity relationships in aquatic toxicology. In: Organic Micropollutants in the Aquatic Environment (ed. A. Bjorseth & G. Angeletti). Dordrecht: Reidel, pp. 465-474.

König, K. (1985): Nebenwirkungen von Pflanzenschutzmitteln auf die Fauna des Bodens. Nachrichtenbl. Deut. Pflanzenschutzd. 37: 8-12.

Kononen, D.W., Hochstein, J.R. & Ringer, R.K. (1986): A quantitative method for evaluating avian food avoidance behavior. Environ. Toxicol. Chem. 5: 823-830.

Kördel, W. (1992): Prediction of the Fate of Pesticides in Soil. Proc. Intern. Symp. Ecotox. GSF-Bericht 1/92: 71-81.

Kördel, W., Herrchen, M. & Klein, W. (1991): Experimental assessment of pesticide leaching using undisturbed lysimeters. Pestic. Sci. 31: 337-348.

Korte, F. (1980): Chemikalien im Ökotest. Umschau 80: 643-648.

Korte, F. (1987): Lehrbuch der ökologischen Chemie. Stuttgart: Thieme, 353 S.

Korte, F., Freitag, D., Geyer, H., Klein, W., Kraus, A.G. & Lahaniatis, E. (1978): Ecotoxicological profile analysis. A concept for establishing ecotoxicologic priority lists for chemicals. Chemosphere 1: 79-102.

Krall, S. & Nasseh, O. (1990): Erforschung biologisch-integrierter Methoden zur Bekämpfung schädlicher Heuschrecken. In: Agrarökologie und Pflanzenschutz in den Tropen und Subtropen (hrsg. W. Koch et al.). Weikersheim: Margraf, S. 215-224, PLITS 8(2).

Krasel, G., Maas, G. & Pestemer, W. (1992): Versuchsapparatur zur Messung der Verflüchtigung von Pflanzenschutzmitteln von Oberflächen. Nachrichtenbl. Deut. Pflanzenschutzd. 44: 1-5.

Krebs, C.J. (1985): Ecology. The Experimental Analysis of Distribution and Abundance. New York: Harper & Row, 3rd edn., 800 p.

Kruger, E.L., Somasundaram, L., Kanwar, R.S. & Coats, J.R. (1993): Movement and Degradation of [^{14}C]Atrazine in Undisturbed Soil Columns. Environ. Toxicol. Chem. 12: 1969-1975.

Kubiak, R., Führ, F. & Mittelstaedt, W. (1990): Comparative studies on the formation of bound residues in soil in outdoor and laboratory experiments. Intern. J. Environ. Anal. Chem. 39: 47-57.

Küchler, A.W. & Zonneveld, I.S. (1988): Vegetation mapping. Kluwer Academic Publ., Dordrecht, Boston, London. 635 pp.

Kuhnt, G. & Muntau, H. (1992): EURO-SOILS: Identification, Selection, Treatment, Characterization. Commission of the European Communities, Joint Research Centre, Ispra, 150 pp.

Kusk, K.O. & Nyholm, N. (1991): Evolution of a phytoplankton toxicity test for water pollution assessment and control. Arch. Environ. Contam. Toxicol. 20: 375-379.

Lahr, J., De Jongh, H.H. (1990): The impact of permethrin, bioresmethrin, carbosulfan and pyraclofos on the non-target aquatic fauna — a literature survey. AID Environment, Amsterdam.

Lamont, B. & Grant, K. (1979): A comparison of 21 measures of site dissimilarity. In: Multivariate Methods of Ecology (ed. L. Orloci et al.). Fairland, Maryl.: Int. Co-op. Publ. House, pp. 101-126.

Lange, A.W. (1991): Umweltgefährlichkeit von Stoffen — Beurteilung durch die Bewertungsstelle Umweltbundesamt. UWSF.-Z. Umweltchem. Ökotox. 3: 5-7.

Laskowski, D.A., Swann, R.L., McCall, P.J. & Bidlack, H.D. (1983): Soil Degradation Studies. Residual Reviews 85: 139-147.

Lavelle, P. (1984): The soil system in the humid tropics. Biol. Intern. 9: 2-7.

Leake, C. (1991): Lysimeter Studies. Pestic. Sci. 31: 363-373.

Leber, G. (1987): Was bedeutet der "Code of Conduct" der FAO für die Pflanzenschutzmittel herstellende und exportierende Industrie ? — In: Möglichkeiten, Grenzen und Alternativen des Pflanzenschutzmitteleinsatzes in Entwicklungsländern (ed. DSE/ATSAF). Feldafing: DSE/ATSAF, S. 45-52.

Le Blanc, F. & De Sloover, J. (1970): Relation between industrialization and the distribution and growth of epiphytic lichens and mosses in Montreal. Can. J. Bot. 48: 1485-1486.

Lee, K.E. (1985): Earthworms: Their ecology and relationships with soils and land use. Academic Press, Sydney, Australia. 411 pp.

Leffler, J.W. (1981): Aquatic Microcosms and Stress Criteria for Assessing Environmental Impact of Organic Chemicals. Office of Pesticides and Toxic Substances, U.S. Environmental Protection Agency, Washington, 50 pp.

Leisinger, K.M. (1989): Ländliche Entwicklung und Pflanzenschutz in der Dritten Welt. Frankfurt: Industrieverband Agrar, 64 S.

Leng, M.L. (1991): Consequences of Re-registration on Existing Pesticides.- In: Regulation of Agrochemicals — A Driving Force in Their Evolution. (Marco, G.J., Hollingworth, R.M. & Plimmer, J.R. eds.). American Chem. Soc., Washington, DC. 27-44.

Levin, S.A., Harwell, M.A., Kelly, J.R. & Kimball, K.D. (eds.) (1989a): Ecotoxicology: Problems and Approaches. New York [etc.]: Springer, 547 pp.

Levin, S.A., Harwell, M.A., Kelly, J.R. & Kimball, K.D. (1989b): Ecotoxicology: Problems and Approaches. In: Ecotoxicology: Problems and Approaches (ed. S.A. Levin et al.). New York [etc.]: Springer, pp. 3-7.

Levine, S.N., Rudnick, D.T., Kelly, J.R., Morton, R.D. & Buttel, L.A. (1990): Pollution dynamics as influenced by seagrass beds: Experiments with tributyltin in *Thalassia* microcosms.- Mar. Environ. Res. 30: 297-322.

Lewis, M.A. (1986): Comparisons of the effects of surfactants on freshwater phytoplankton communities in experimental enclosures and on algal population growth in the laboratory. Environ Toxicol. Chem. 5: 319-332.

Lewis, M.A. (1993): Freshwater Primary Producers. In: Handbook of Ecotoxicology. (ed. Calow, P.). Blackwell Scientific Publ, Oxford. 28-50.

Lewis, R.A. (1985): Richtlinien für den Einsatz einer Umweltprobenbank in der Bundesrepublik Deutschland auf ökologischer Grundlage. Saarbrücken: Univ. d. Saarlands.

Lewis, R.A. & Klein, B. (1990): A brief history of specimen banking. Storage, institutions and applications. Toxicol. Environ. Chem. 27: 251-266.

Livingston, R.J. & Meeter, D.A. (1985): Correspondence of laboratory and field results: What are the criteria for verification? — In: Multispecies Toxicity Testing (ed. J. Cairns). New York: Pergamon, pp. 76-88.

Lobel, P.B., Belkhode, S.P., Jackson, S.E. & Longerich, H.P. (1990): Recent taxonomic discoveries concerning the mussel *Mytilus*: implications for biomonitoring. Arch. Environ. Contam. Toxicol. 19: 508-512.

Lökke, H. (1994): Ecotoxicological Extrapolation: Tool or Toy ? — In: Ecotoxicology of Soil Pollution. Eijsackers et al. (eds.). Lewis Publ., Chelsea, Michigan. 411-425.

Lökke, H. & Van Gestel, C.A.M. (1993): Manual of: SECOFASE. Development, improvement and standardization of test systems for assessing sublethal effects of chemicals on fauna in the soil ecosystem. Natl. Environmental Research Inst., Silkeborg.

Luff, M.L., Clements, R.O. & Bale, J.S. (1990): An integrated approach to assessing the effects of some pesticides in grassland. Brighton Crop Protection Conference, 143-152.

Luoma, S.N. & Ho, K.T. (1993): Appropriate Uses of Marine and Estuarine Sediment Bioassays. In: Handbook of Ecotoxicology. (ed. Calow, P.). Blackwell Scientific Publ, Oxford. 193-226.

Lupi, C., Bucchi, A.R., Piccioni, A. & Zapponi, G.A. (1988): The environmental behavior of chemicals in soil: Atrazine as an example. Ecotox. Environ. Safety 16: 133-142.

Lussier, S.M., Gentile, J.H. & Walker, J. (1985): Acute and chronic effects of heavy metals and cyanide on *Mysidopsis bahia* (Crustacea: Mysidacea). Aquat. Toxicol. 7: 25-35.

Lyman, W.J. (1982): Adsorption Coefficient for Soils and Sediments. In: Handbook of chemical property estimation methods. (ed. Lyman, W.J.). McGraw-Hill, New York.

Maas, G. (1987): Wirkungen von Pflanzenschutzmitteln auf Bodenmikroorganismen sowie Bewertung der Ergebnisse. Mittl. Biol. Bundesanstalt f. Land- und Forstwirtschaft Berlin 234: 42-46.

MacArthur, R.H. (1969): Patterns of communities in the tropics. Biol. J. Linn. Soc. 1: 19-30.

MacKay, D. (1988): On low, very low and negligible concentrations. Environ. Toxicol. Chem. 7: 1-3.

MacKay, D. (1991): Multimedia Environmental Models. The Fugacity Approach. Lewis Publ., Chelsea, Michigan. 257 pp.

Madsen, L. (1984): Problems connected with the use of algal tests in ecotoxicology. Ecol. Bull. 36: 165-170.

MAFF (Ministry of Agriculture, Fisheries and Food) (1979): Pesticides Safety Precautions Scheme. Laboratory testing for toxicity to honeybees and testing for toxicity to foraging bees under field conditions. MAFF, London.

Majewski, M.S., Glotfelty, D.E., Paw U, K.T. & Seiber, J.N. (1990): A field comparison of two methods for measuring DCPA soil evaporation rates. Environ. Toxicol. Chem. 10: 301-312.

Maki, A.W., Dickson, K.L. & Brungs, W.A. (1987): Introduction. In: Effects of Sediment-bound Chemicals in Aquatic Sediments. Dickson, K.L., Maki, A.W., & Brungs, W.A. (eds.). SETAC Special Publ. Series, Pergamon Press, New York. XV-XXI.

Malkomes, H.-P. (1985): Einflüsse von Pflanzenschutzmitteln auf Bodenmikroorganismen und ihre Leistungen. Ber. Landwirtschaft 198: 134-146.

Malkomes, H.P. (1987): Respiration and dehydrogenase as side-effect indicators. In: Pesticide Effects on Soil Microflora. Somerville, L.; Greaves, M.P. (eds.). Taylor & Francis, London. 81-96.

Malkomes, H.-P. & Wöhler, B. (1983): Vergleich von Testverfahren zur Erfassung einiger Nebenwirkungen von Pflanzenschutzmitteln auf Bodenmikroorganismen am Beispiel eines Herbizids. Nachrichtenbl. Deutsch. Pflanzenschutzd. 35: 86-92.

Maltby, L. & Calow, P. (1989): The application of bioassays in the resolution of environmental problems; past, present and future. Hydrobiologia 188/189: 65-76.

Maltby, L., Naylor, C. & Calow, P. (1990): Effect of stress on a freshwater benthic detrivore: scope for growth in *Gammarus pulex*. Ecotox. Environ. Safety 19: 285-291.

Mapes, J.P., McKenzie, K.D., McClelland, L.R., Movassaghi, S., Anantha Reddy, A., Allen, R.L. & Friedman, S.B. (1992): Penta RISc™Soil — A rapid, on-site screening test for Pentachlorophenol. Bull. Environ. Contam. Toxicol. 49: 334-341.

Marco, G.J., Hollingworth, R.M. & Plimmer, J.R. (1991): Regulation of Agrochemicals. A Driving Force in Their Evolution. American Chemical Society, Washington, D.C. 188 pp.

Marschner, A. (1992): Phytotoxizitätsuntersuchungen an Wildkräutern und einer Kulturpflanze. Schriftenreihe des Vereins für Wasser-, Boden- und Lufthygiene. G. Fischer Verlag, Stuttgart.

Mathes, K., Weidemann, G. & Beck, L. (1991): Indikatoren für Ökosystembelastung. Berichte aus der Ökologischen Forschung 2, Forschungszentrum Jülich GmbH.

Mathies, M. (1991): Expositionsmodelle. UWSF. Schadstoff.Ökotox. 3: 37-41.

Matthiessen, P. (1985): Contamination of wildlife with DDT insecticide residues in relation to tsetse fly control operations in Zimbabwe. Environ. Poll. Ser. B 10: 189-211.

Mayer, F.L. & Ellersiek, M.R. (1986): Manual of acute toxicity: Interpretation and data base for 410 chemicals and 66 species of freshwater animals. Resource Publ. 160 U.S. Fish and Wildlife Service, Dept. of the Interior, Washington, D.C.

Mayer, F.L. & Ellersiek, M.R. (1988): Experiences with single-species tests for acute toxic effects on freshwater animals. Ambio 17: 367-375.

Mayfield, C.I. (1993): Microbial Systems. In: Handbook of Ecotoxicology. (ed. Calow, P.). Blackwell Scientific Publ, Oxford. 9-27.

McArthur, M.L.B., Fox, G.A., Peakall, D.B. & Philogène, B.J.R. (1983): Ecological significance of behavioral and hormonal abnormalities in breeding ring doves fed an organochlorine chemical mixture. Archives Environ. Contamin. Toxicol. 12: 343-353.

Mearns, A.J., Swartz, R.C., Cummins, J.M., Dinnel, P.A., Plesha, P. & Chapman, P.M. (1986): Inter-Laboratory Comparison of a Sediment Toxicity Test Using the Marine Amphipod, *Rhepoxynius abronius*. Mar. Environ. Research 19: 13-37.

Metcalfe, J.L. (1989): Biological water quality assessment of running waters based on macroinvertebrate communities: history and present status in Europe. Environ. Poll. 60: 101-139.

Metcalf, R.L. (1977): Model ecosystem approach to insecticide degradation: A critique. Ann. Rev. Entomol. 22: 241-261.

Micklitz, H-W. (1991): Entwicklungslinien der Geschichte des deutschen Pflanzenschutzrechts. In: Kolloquium über Pflanzenschutz. Rehbinder, E. (ed.). Werner Verl., Düsseldorf. 44-68.

Middaugh, D.P., Goodman, L.R. & Hemmer, M.J. (1993): Methods for Spawning, Culturing and Conducting Toxicity Tests with Early Life Stages of Estuarine and Marine Fishes. In: Handbook of Ecotoxicology. (ed. Calow, P.). Blackwell Scientific Publ, Oxford. 167-192.

Miguel, A.H. (1991): Environmental Pollution Research in South America. Environ. Sci. Technol. 25: 590-594.

Miller, W.E., Peterson, S.A., Greene, J.C. & Callahan, C.A. (1985): Comparative toxicology of laboratory organisms for assessing hazardous waste sites. J. Environ. Qual. 14: 569-574.

Milles, D. (1991): Von Schädlingen und Schädigungen — Zur Geschichte der Pestizidzulassung. In: Kolloquium über Pflanzenschutz. Rehbinder, E. (ed.). Werner Verl., Düsseldorf. 17-43.

Mineau, P. (1991): Cholinesterase Inhibiting Insecticides. Their impact on wild life and the environment. Elsevier, Amsterdam.

Mineau, P. & Peakall, D.B. (1987): An evaluation of avian impact assessment techniques following broad-scale forest insecticide sprays. Environ. Toxicol. Chem. 6: 781-791.

Moewus, F. (1949): Der Kressewurzeltest, ein neuer quantitativer Wuchsstofftest. Biol. Zentralblatt 68: 118-139.

Moltmann, J.F. (1989): Der Einfluß des Oryctes-Baculovirus auf die Populationsdynamik des Indischen Nashornkäfers, Oryctes rhinoceros L. Maraun Verlag, Frankfurt. 214 S.

Moreth, L. & Naton, E. (1992): Richtlinie zur Prüfung der Nebenwirkung von Pflanzenschutzmitteln auf *Aleochara bilineata* (Col., Staphylinidae) (erweiterter Laborversuch). Bull. Sect. Reg. Ouest Palearct. 15: 152-158.

Morgan, E. & Knacker, T. (1994): The Role of Laboratory Terrestrial Model Ecosystems in the Testing of Potentially Harmful Substances. Ecotoxicology. 213–233.

Moriarty, F. (1988): Ecotoxicology. The Study of Pollutants in Ecosystems. London [etc.]: Academic Pr., 2nd edn., 289 p.

Moriarty, F. & Walker, C.H. (1987): Bioaccumulation in food chains — a rational approach. Ecotox. Environ. Safety 13: 208-215.

Morton, R.D., Duke, T.D., Macauley, J.M., Clark, J.R., Price, W.A., Hendricks, S.J., Owsley-Montgomery, S.L. & Plaia, G.R. (1986): Impact of drilling fluids on seagrasses: An experimental community approach. In: Community Toxicity Testing (ed.: Cairns, J.). STP 920, ASTM, Philadelphia. 199-212.

Müller, P. (1987): Umweltverträglichkeiten des Einsatzes von Pflanzenschutzmitteln in Entwicklungsländern. In: Möglichkeiten, Grenzen und Alternativen des Pflanzenschutzmitteleinsatzes in Entwicklungsländern (hrsg. DSE/ATSAF). Feldafing: DSE/ATSAF, S. 150-169.

Müller, P. (1988): Ökotoxikologische Wirkungen von chlorierten Kohlenwasserstoffen, Phosphorsäureestern, Carbamaten und Pyrethroiden im nordöstlichen Sudan. Eschborn: GTZ, Projekt 86.3504.7-100, FB 152, 71 S.

Müller, P., Nagel, P. & Flacke, W. (1981): Ecological side-effects of Dieldrin application against Tse-Tse flies in Adamaoua, Cameroon. Oecologia 50: 187-194.

Mueller, T., Jones, R.E., Bush, P.B. & Banks, P.A. (1992): Comparison of PRZM and GLEAMS computer model predictions with field data for Alachlor, Metribuzin and Norflurazon leaching. Environ. Toxicol. Chem. 11: 427-436.

Munawar, M. & Munawar, I.F. (1987): Phytoplankton bioassays for evaluating toxicity of in situ sediment contaminants. Hydrobiologia 149: 87-105.

Murphy, L.S. & Belastock, R.A. (1980): The effect of environmental origin on the response of marine diatoms for chemical stress. Limnol. Oceanogr. 25: 160-165.

Nacci, F., Jackim, E. & Walsh, R. (1986): Comparative evaluation of three rapid marine toxicity tests: Sea urchin early embryo growth test, sea urchin cell toxicity test and Microtox. Environ. Toxicol. Chem. 5: 521-525.

Nagel, R. (1994): Fish and environmental chemicals — a critical evaluation of tests. In: Fish in Ecotoxicology and Ecophysiology. (eds. Braunbeck, Segner & Hanke). VCH Verlagsges., Weinheim (in press).

Nagel, R. & Loskill, R. (1991): Bioaccumulation in Aquatic Systems. VCH Verlagsges., Weinheim. 239 pp.

Nährig, D. (1991): Systematische Untersuchungen der höheren Straten von Hecken mit verschiedenen Methoden dargestellt am Beispiel der Spinnenfauna. Beih. Verh. Ges. Ökol. 2: 107-116.

NAS (National Academy of Sciences) (1981): Testing for the Effects on Ecosystems. Washington, DC: National Academy Press, 103 p.

Nash, R.G. (1983): Determining environmental fate of pesticides with micro-agroecosystems. Residue Rev. 85: 199-215.

Naton, E. (1989): Die Prüfung der Nebenwirkung von Pflanzenschutzmitteln auf *Aleochara bilineata* Gyll. (Col., Staphylinidae). Anz. Schädlingskde., Pflanzenschutz, Umweltschutz 62: 1-6.

Nebeker, A.V., Cairns, M.A., Gakstatter, J.H., Malueg, K.W., Schuytema, G.S. & Krawczyk, D.F. (1984): Biological methods for determining toxicity of contaminated freshwater sediments to invertebrates. Environ. Toxicol. Chem. 3: 617-630.

Nebelung, K. & Bode, E. (1988): Dekomposition und Pflanzenschutz — methodische Ansätze zur Prüfung von Mittelauswirkungen im Freiland. Mittl. Biol. Bundesanst. 245. 404-405.

Nellessen, J.E. & Fletcher, J.S. (1993): Assessment of Published Literature Pertaining to the Uptake/Accumulation, Translocation, Adhesion and Biotransformation of Organic Chemicals by Vascular Plants. Environ. Toxicol. Chem. 12: 2045-2051.

NEN (Nederlandse Norm) (1988a): Soil — Determination of the acute toxicity of chemical substances to earthworms No. 5794. Nederlands Normalisatie-instituut, Delft.

NEN (Nederlandse Norm) (1988b): Soil — Determination of the influence of chemicals on soil nitrification No. 5795. Nederlands Normalisatie-instituut, Delft.

Nendza, M. & Seidel, J.K. (1988): Quantitative structure-toxicity relationships for ecotoxicologically relevant biotestsystems and chemicals. Chemosphere 17: 1585-1602.

NETAC (National Environmental Technology Applications Corporation) (1991): Protocols for Evaluating Products Proposed for Use in Bioremediation of Oil Spill. University of Pittsburgh, Pittsburgh, PA.

Neumann, D. (1982): Schwachstellen beim Einsatz von Bioindikatoren und weitere Forschungsziele. Decheniana 26: 193-196.

Neuruer, H. (1975): Bioteste in der Herbologie. Zeitschr. PflanzenKrh. PflSchutz 82: 316-328.

Niederlehner, B.R., Pontasch, K.W., Pratt, J.R. & Cairns, J. (1990): Field evaluation of predictions of environmental effects from a multispecies-microcosm toxicity test. Arch. Environ. Contam. Toxicol. 19: 62-71.

Norton, S.B., Rodier, D.J., Gentile, J.H., Van der Schalie, W.H., Wood, W.P. & Slimak, M.W. (1992): A framework for ecological risk assessment at the EPA. Environ. Toxicol. Chem. 11: 1663-1672.

Notenboom, J. & Boessenkoll, J-J. (1993): Acute toxicity testing with the ground water copepod *Parastenocaris germanica* (Crustacea). Proc. 1st Int. Conf. Ground Water Ecol. US EPA & AWRA (1992): 301–309.

Nusch, E.A. (1991): Ökotoxikologische Testverfahren — Anforderungsprofile in Abhängigkeit vom Anwendungszweck. UWSF.-Umweltchem. Ökotox. 3: 12-15.

Nyholm, N. (1991): The European system of standardized legal tests for assessing the biodegradability of chemicals. Environ. Toxicol. Chem. 10: 1237-1246.

Nyholm, N. & Källquist, T. (1989): Methods for growth inhibition toxicity tests with freshwater algae. Environ. Toxicol. Chem. 8: 689-704.

Nylander, M.W. (1866): Les lichens du Jardin du Luxembourg. Bull. Soc. Bot. de France 13: 364-371.

Odum, E.P. (1971): Fundamentals of Ecology. W.B. Saunders Co., Philadelphia.

Odum, E.P. (1984): The mesocosm. BioScience 34: 558-562.

OECD (Organisation for Economic Development) (1981a): OECD Guideline for Testing of Chemicals No. 304 A: Inherent Biodegradability in Soil. Paris.

OECD (Organization for Economic Co-operation and Development) (1981b): Guideline for Testing of Chemicals No. 106: Adsorption/Desorption. OECD, Paris.

OECD (Organization for Economic Co-operation and Development) (1981c): Guideline for Testing of Chemicals No. 305: Fish Bioaccumulation. OECD, Paris.

OECD (Organization for Economic Co-operation and Development) (1984): Guidelines for Testing of Chemicals. OECD, Paris.

OECD (Organization for Economic Co-operation and Development) (1986): Existing Chemicals. Systematic Investigation. Priority Setting and Chemical Reviews. OECD, Paris.

OECD (Organization for Economic Co-operation and Development) (1988): Guidelines for Testing of Chemicals No. 305 E. Bioaccumulation: Flow-through Fish Test. OECD, Paris.

OECD (Organization for Economic Co-operation and Development) (1989a): Report of the OECD Workshop on ecological effects assessment. OECD Environment Monograph No. 26, Paris.

OECD (Organization for Economic Co-operation and Development) (1989b): Compendium for Environmental Exposure Assessment Methods for Chemicals. OECD Environment Monograph No. 27, Paris.

OECD (Organization for Economic Co-operation and Development) (1992): Report of the OECD Workshop on the extrapolation of laboratory aquatic toxicity data to the real environment. OECD Environment Monograph No. 59, Paris.

Oerke, E.C., Dehne, H.W., Schönbeck, F. & Weber, A. (1994): Crop Production and Crop Protection. Estimated losses in major food and cash crops. Elsevier Science, Amsterdam. 808 pp.

Ohba, K. & Ishibashi, N. (1984): A nematode, *Caenorhabditis elegans*, as test organism for nematicide evaluation. J. Pest. Sci. 9: 91-96.

Okkerman, P.C., Van der Plassche, E.J., Emans, H.J.B. & Canton, J.H. (1993): Validation of Some Extrapolation Methods with Toxicity Data Derived from Multiple Species Experiments. Ecotox. Environ. Safety 25: 341-359.

O'Neill, R.V., DeAngelis, D.L., Waide, J.B. & Allen, T.F.H. (1986): A Hierarchical Concept of Ecosystems. Princeton: Princeton Univ. Press, 253 p., [Monographs in Population Biology, No. 23].

Oomen, P.A., Romeijn, G. & Wiegers, G.L. (1991): Side-effects of 100 pesticides on the predatory mite *Phytoseiulus persimilis*, collected and evaluated according to the EPPO guideline. Bull. OEPP/EPP Bull. 21: 701-712.

Ottow, J.C.G. (1985): Einfluß von Pflanzenschutzmitteln auf die Mikroflora von Böden. Naturwiss. Rundschau 38: 181-189.

Owesen, H.A. (1976): Artendiversität in der Ökologie. SFB 95, Rep. 16, Kiel.

Paasivirta, J. (1988): Baskets, Ecorisk and Ecotoxa. Chemosphere 17: 1655-1662.

Paasivirta, J. (1991): Chemical Ecotoxicology. Lewis Publ., Chelsea, Michigan. 210 pp.

Pagga, U. (1984): Die Prüfung der biologischen Abbaubarkeit von Chemikalien. GIT-Suppl. 4: 9-14.

PAN (Pesticide Action Network) (1987): Monitoring and Reporting the Implementation of the International Code of Conduct on the Use and Distribution of Pesticides. Final Report. Nairobi, Kenya: Environ. Liaison Centre, 41 p.

PAN (Pestizid Aktions-Netzwerk) (Hrsg.) (1989): Pestizideinsatz in Entwicklungsländern. Gefahren und Alternativen. Weikersheim: Margraf. Von Schwab, A., Görgen, R., Jäger-Mischke, I., Prexler-Schwab, S., Stoll, G., 275 S.(Tropical Agroecology Nr. 3).

Paoletti, M.G., Favretto, M.R., Stinner, B.R., Purrington, F.F. & Bater, J.E. (1991): Invertebrates as bioindicators of soil use. Agriculture, Ecosystems and Environment 34: 341-362.

Parkerton, T.F., Connolly, J.P., Thomann, R.V. & Uchrin, C.G. (1993): Do aquatic effects or human health end points govern the development of sediment-quality criteria for nonionic organic chemicals ? — Environ. Toxicol. Chem. 12: 507-523.

Parlar, H. & Angerhöfer, D. (1991): Chemische Ökotoxikologie. Springer Verlag, Berlin.

Pascoe, G.A. (1993): Wetland Risk Assessment. Environ. Toxicol. Chem. 12: 2293-2307.

Pastre, P., Smolikowski, S. & Thewys, G. (1989): Locust and grasshopper control: Deltamethrin file. Roussel Uclaf, Div. Agrovet, Paris.

Paul, E. A. & Clark, F.E. (1989): Soil Microbiology and Biochemistry, Academic Press Inc., San Diego, New York, Berkeley, Boston, London, Sydney, Toky, Toronto.

Payne, A.S. & Hall, R.H. (1979): A method for measuring algal toxicity and its application to the safety assessment of chemicals. In: Aquatic Toxicology. (eds. Marking, L.L. & Kimmerle, R.A.). ASTM, Philadelphia. 171-180.

Pedersen, O.C. (1984): An insecticide bioassay for *Myzus persicae* validated by Probit Analysis and esterase measurement. Ecol. Bull. 36: 50-56.

Peichl, L., Reiml, D., Ritzl, I. & Schmidt-Bleek, F. (1987): Übersicht biologischer Wirkungs-Testsysteme zur Beobachtung unerwarteter Umweltveränderungen. Biosonden. GSF-Bericht 28/87: 1-51.

Pelfrene, A.F. (1987): Pesticide use exposure and regulation in developed and developing countries. In: Toxicology of Pesticides: Experimental, Clinical and Regulatory Perspectives (ed. L.G. Costa et al.). Berlin [etc.]: Springer, pp. 253-260, NATO ASI Ser. H, No. 13.

Perkins, E.J. (1979): The need for sublethal studies. Phil. Trans. Roy. Soc. London 286: 425-442.

Perkow, W. (Hrsg.) (1989): Wirksubstanzen der Pflanzenschutz- und Schädlingsbekämpfungsmittel. Hamburg [etc.]: Parey, Loseblatt-Sammlung.

Perry, J.A. & Troelstrup, N.H. (1988): Whole ecosystem manipulation: A productive avenue for test system research ? — Environ. Toxicol. Chem. 7: 941-951.

Persoone, G. & Janssen, C.R. (1993): Freshwater Invertebrate Toxicity Tests. In: Handbook of Ecotoxicology. (ed. Calow, P.). Blackwell Scientific Publ, Oxford. 51-65.

Persoone, G., Van de Vel, A., Van Stertegem, M. & De Nayer, B. (1989): Predictive value of laboratory tests with aquatic invertebrates: Influence of experimental conditions. Aquatic Toxicol. 14: 149-166.

Pestemer, W. (1988): Vergleich des OECD-Entwurfs "Terrestrial Plant Growth Test" mit Ergebnissen aus Freilandversuchen. Jül.-Spez. 441: 93-109.

Pestemer, W. & Auspurg, B. (1986): Eignung eines Testpflanzensortiments zur Risikoabschätzung von Stoffwirkungen auf höhere Pflanzen im Rahmen des Chemikaliengesetzes. Nachrichtenblatt Deut. Pflanzenschutzd. 38: 120-125.

Pestemer, W. & Günther, P. (1989): Mathematisch-statistische Modellierung von Dosis-Wirkungs-Beziehungen und des Einflusses der Expositionszeit zur Bewertung phytotoxischer Effekte auf höhere Pflanzen. BMFT Forschungsbericht (FKZ: 0339084 B).

Pestemer, W. & Nordmeyer, H. (1990): Modelluntersuchungen mit ausgewählten Pflanzenschutzmitteln im Bodenprofil im Hinblick auf die Beurteilung einer Grundwasserbelastung. Mittl. Biol. Bundesanstalt 259: 1-80.

Petersen, P. & Luxton, M. (1982): A comparative analysis of soil fauna populations and their role in decomposition processes. Oikos 39: 287-338.

Petzold, R. (1990): Legislation of Plant Protection in the Federal Republic of Germany. Gesunde Pflanzen 42: 225-257.

Petzold, R. (1991): EG-Harmonisierung der Zulassung von Pflanzenschutzmitteln. Gesunde Pflanzen 43: 350-354 + 380-382.

Pfarl, C., Ditzelmüller, G., Loidl, M. & Streichsbier, F. (1990): Microbial degradation of xenobiotic compounds in in soil columns. FEMS Microbiol. Ecol. 73: 255-262.

Pfeiffer, W.C., De Lacerda, L.D., Malm, O., Souza, C.M.M., Da Silveira, E.G. & Bastos, W.R. (1989): Mercury concentrations in inland waters of gold-mining areas in Rondonia, Brazil. Sci. Total Environ. 87/88: 233-240.

Phillips, D.J.H. (1993): Bioaccumulation. In: Handbook of Ecotoxicology. (ed. Calow, P.). Blackwell Scientific Publ, Oxford. 378-396.

Pieters, A.J. (1977): Pesticide regulations in international perspective. Meded. Fac. Landbouw. Rijksuniv. Gent 42: 831-838.

Pilli, A., Carle, D.O., Kline, E., Pickering, Q. & Lazorchak, J. (1988): Effects of pollution on freshwater organisms. J. WPCF 60: 994-1065.

Plachter, H. (1991): Naturschutz. G. Fischer Verl., Stuttgart. 463 S.

Pontasch, K.W., Niederlehner, B.R. & Cairns, J. (1989): Comparisons of Single-species, Microcosm and Field Responses to a Complex Effluent. Environ. Toxicol. Chem. 8: 521-532.

Portier, R.J. (1985): Comparison of environmental effect and biotransformation of toxicants on laboratory microcosm and field microbial communities. In: Validation and Predictability of Laboratory Methods for Assessing the Fate and Effects of Contaminants in Aquatic Ecosystems. (eds.: Boyle, T.P.). STP 865, ASTM, Philadelphia. 14-30.

Pratt, J.R., Mitchell, J., Ayers, R. & Cairns, J. (1989): Comparison of estimates of effects of a complex effluent at different levels of biological organization. In: Aquatic Toxicology and Environmental Fate: Eleventh Volume. (eds. Suter, G.W. & Lewis, M.A.) STP 1007, ASTM, Philadelphia. 174-188.

Price, C. & Hikino, H. (1984): The Effect of Soil Contaminants on Different Plant Species. In: Actes du Symp. Intl. sur l'Ecotoxicologie terrestre (eds. N.N.), Les Arcs. 175-181.

Pritchard, P.H. (1982): Model ecosystems. In: Environmental Risk Analysis for Chemicals (ed. R.A. Conway). pp. 257-353.

Pritchard, P.H. (1987): Assessing the biodegradation of sediment associated chemicals. In: Fate and Effects of Sediment-Bound Chemicals in Aquatic Systems: Proceedings of the Sixth Pellston Workshop, Florissant (August 12-17, 1984). (eds.: Dickson, K.L., Maki, A.W. & Brungs, W.A.). Pergamon Press, New York. 109-135.

Puzika, H. & Knie, J. (1994): Der Dynamische Daphnientest. Erfahrungen und praktische Hinweise. Landesumweltamt Nordrhein-Westfalen, Essen, 44 S.

Ramade, F. (1987a): Ecotoxicology. Chichester: Wiley, 2nd edn., 262 p.

Ramade, F. (1987b): Proposal of ecotoxicological criteria for the assessment of the impact of pollution on environmental quality. Toxicol. Environ. Chem. 13: 189-203.

Ramamoorthy, S. & Baddaloo, E. (1991): Evaluation of Environmental Data for Regulatory and Impact Assessment. Studies in Environ. Sci. 41: 238.

Ramesh, A., Tanabe, S., Murase, H., Subramanian, A.N. & Tatsukawa, R. (1991): Distribution and Behaviour of Persistent Organochlorine Insecticides in Paddy Soil and Sediments in the Tropical Environment: A Case Study in South India. Environ. Poll. 74: 293-307.

Rand, G.M. & Petrocelli, S.R. (1985): Introduction. In: Fundamentals of Aquatic Toxicology. Rand, G.M. & Petrocelli, S.R. (eds.). Hemisphere, New York. 1-28.

Rapport, D.J., Regier, H.A. & Hutchinson, T.C. (1985): Ecosystem behavior under stress. Amer. Nat. 125: 617-640.

Ratcliffe, D.A. (1980): The Peregrine Falcon. Calton: Poyser.

Rawson, D.M. (1993): Bioprobes and Biosensors. In: Handbook of Ecotoxicology. (ed. Calow, P.). Blackwell Scientific Publ, Oxford. 428-437.

Reish, D.J. (1985): The use of the polychaetous annelid *Neanthes arenaceodentata* as a laboratory experimental animal. Tethys 11: 335-341.

Reish, D.J. (1988): The Use of Toxicity Testing in Marine Environmental Research. In: Marine Organisms as Indicators. Soule, D.F. & Kleppel, G.S. (eds.). Springer Verlag, New York. 231-246.

Reynoldson, T.B. & Day, K.E. (1993): Freshwater Sediments. In: Handbook of Ecotoxicology. (ed. Calow, P.). Blackwell Scientific Publ, Oxford. 83-100.

Reynoldson, T.B., Thompson, S.P. & Bamsey, J.L. (1991): A sediment bioassay using the tubificid oligochaete worm *Tubifex tubifex*. Environ. Toxicol. Chem. 10: 1061-1071.

Richter, G. (1953): Die Auswirkung von Insektiziden auf die terricole Makrofauna. Nachrichtenbl. Deut. Pflanzenschutzd. 7: 61-72.

Riepert, F. (1984): Ist Risikoabschätzung von Stoffwirkungen auf Ökosysteme mit Hilfe einfacher standardisierter Prüfungen an einzelnen Arten möglich ? — Angew. Botanik 58: 217-226.

Riepert, F. (1991): Effects of soil pollutants on Collembola: Determination of the inhibition of reproduction. Biol. Bundesanst. f. Land- und Forstwirtschaft, Berlin, Germany. Revised Version 91/10.

Rinkus, S.J. & Legator, M.S. (1979): Chemical characterization of 465 known or suspected carcinogens and their correlation with mutagenic activity in the Salmonella typhimurium system. Cancer Res. 39: 3282-3318.

Rippen, G. (Hrsg.) (1987 ff.): Handbuch der Umwelt-Chemikalien. Stoffdaten — Prüfverfahren — Vorschriften. Landsberg: ECOMED, 2. Aufl., Lose-Blatt-Sammlung.

Roberts, T.R. (1984): Non-extractable pesticide residues in soils and plants. Pure Appl. Chem. 56: 945-956.

Robertson, J.L. & Worner, S.P. (1990): Population toxicology: Suggestions for laboratory bioassays to predict pesticide efficacity. J. Econ. Entomol. 83: 8-12.

Rodgers, J.H., Clark, J.R., Dickson, K.L. & Cairns, J. (1980): Nontaxonomic analyses of structure and function of aufwuchs communities in lotic microcosms. In: Microcosms in Ecological Research. (ed. Giesy, J.R.). DOE CONF-781101, National Technical Information Service, Springfield. 625-644.

Römbke, J. (1989): *Enchytraeus albidus* (Enchytraeidae, Oligochaeta) as a test organism in terrestrial laboratory systems. Arch. Toxicol. Suppl. 13: 402-405.

Römbke, J. (1991): Entwicklung eines Reproduktionstests an Bodenorganismen — Enchytraeen. Bericht im Auftrag des Umweltbundesamts, F+E-Vorhaben Nr. 106 03 051/01, Battelle Institut Frankfurt.

Römbke, J., Vickus, P. & Bauer, C. (1992): Experiences and Problems with the OECD-Earthworm Acute Test in Routine Testing. In: Ecotoxicology of Earthworms. Greig-Smith, P., Becker, H., Edwards, P.J., Heimbach, F. (eds.): 209-212. Intercept, Andover.

Römbke, J. & Heimbach, U. (1994): Experiences derived from the Carabid Beetle Laboratory Test. Pestic. Sci. (in press).

Römbke, J., Bauer, C. et al. (1993): Grundlagen für die Beurteilung des ökotoxikologischen Gefährdungspotentials von Altstoffen im Medium Boden — Entwicklung einer Teststrategie. Bericht für das Umweltbundesamt Nr. 106 04 103, Battelle-Institut e.V., Frankfurt. 496 S.

Römbke, J., Knacker, T., Förster, B. & Marcinkowski, A. (1994): Comparison of Effects of Two Pesticides on Soil Organisms in Laboratory Tests, Microcosms and in the Field. In: Ecotoxicology of Soil Pollution. Eijsackers et al. (eds.). Lewis Publ., Chelsea, Michigan. 229-240.

Rossel, D & Tarradellas, J. (1991): Dehydrogenase Activity of Soil Microflora: Significance in Ecotoxicological Tests. Environ. Toxicol. Water Qual. 6: 17-33.

Rothert, H., Brasse, D. & Bode, E. (1990): Abwägungs- und Entscheidungsprozesse bei der Zulassung von Pflanzenschutzmitteln im Hinblick auf die terrestrische Fauna. Gesunde Pflanzen 42: 29-33.

Rudolph, P. (1987): Grundlagen einer ökotoxikologischen Bewertung der Prüfungen an Vögeln nach dem Chemikaliengesetz. Ökol. Vögel 9: 131-141.

Rudolph, P. & Boje, R. (1986): Ökotoxikologie: Beurteilung von Umweltgefährdungen nach dem Chemikaliengesetz. Landsberg: ECOMED, 106 S.

Runge, M., Schaefer, M., Werner, W., Conrady, D., Neemann, G. & Stickan, W. (1989): Erarbeitung von Indikatoren für die Auswirkungen von chemischen Belastungen auf Grünlandökosysteme. Jül-Spez. 503: 1-218.

Sabljic, A. & Piver, W.T. (1992): Quantitative modelling of environmental fate and impact of commercial chemicals. Environ. Toxicol. Chem. 11: 961-972.

Samiullah, Y. (1990): Prediction of the Environmental Fate of Chemicals. Essex: Elsevier Applied Science, 284 p.

Samsoe-Petersen, L. (1987): Laboratory method for testing side-effects of pesticides on the rove beetle *Aleochara bilineata* — adults (Col., Staphylinidae). Entomophaga 32: 73-81.

Samsoe-Petersen, L. (1990): Sequences of Standard Methods to Test Effects of Chemicals on Terrestrial Arthropods. Ecotox. Envir. Safety 19: 310-319.

Sarbhoy, R.K., Sharma, A. & Singh, R.M. (1991): The *Pisum* test: An alternative in environmental studies; the relative toxicity of pesticides. J. Environ. Biol. 12: 137-141.

Sauerbeck, D. (1983): Welche Schwermetallgehalte in Pflanzen dürfen nicht überschritten werden, um Wachstumsbeeinträchtigungen zu vermeiden? — Landwirtsch. Forsch. 39: 108-129.

Sauphanor, B., Blaisinger, P., & Sureau, F. (1992): Methode de Laboratoire pour evaluer l'Effet des Pesticides sur *Forficula auricularia* (Dermaptera: Forficulidae). In: Guidelines for testing the effects of pesticides on beneficial organims: Description of test methods. (ed.: Hassan, S.A.). IOBC/WPRS Bulletin XV/3: 117-121.

Schäfer, M. & Tischler, W. (1983): Wörterbuch der Biologie: Ökologie. Jena: VEB G. Fischer, 354 S.

Schaeffer, D.J., Cox, D.K. & Deem, R.A. (1987): Variability of test systems used to assess ecological effects of chemicals. Water Sci. Technol. 19: 39-46.

Schaeffer, D.J., Seastedt, T.R. et al. (1990): Field bioassessments for selecting test systems to evaluate military training lands in Tallgrass Prairie. Environ. Managem. 14: 81-93.

Schafer, E.W., Bowles, W.A. & Hurlbut, J. (1983): The Acute Oral Toxicity, Repellency and Hazard Potential of 998 Chemicals to One or More Species of Wild and Domestic Birds. Arch. Environ. Contam. Toxicol. 12: 355-382.

Scheunert, I. (1991): Langzeitverhalten von Chemikalien im Boden. UWSF.-Umweltchem. Ökotox. 3: 28-32.

Scheunert, I. & Klein, W. (1985): Predicting the movement of chemicals between environmental compartments (air-water-soil-biota). In: Appraisal of Tests to Predict the Environmental Behaviour of Chemicals (ed. P. Sheehan, F. Korte, W. Klein & P. Bourdeau). SCOPE 25, Chichester: Wiley, pp. 285-332.

Schimmel. S.C., Hansen, D.J. & Forester, J. (1974): Effects of Aroclor 1254 on laboratory-reared embryos and fry of sheepshead minnows (*Cyprinodon variegatus*). Trans. Am. Fish. Soc. 103: 582-586.

Schindler, D.W., Mills, K.H., Malley, D.F., Findlay, D.L., Shearer, J.A., Davies, I.J., Turner, M.A., Linsey, G.A. & Cruikshank, D.R. (1985): Long-term ecosystem stress: The effects of years of experimental acidification on a small lake. Science 228: 1395-1401.

Schinkel, K. (1985): Prüfung der Beständigkeit von Pflanzenschutzmitteln im Rahmen des Zulassungsverfahrens. Ber. Landwirtschaft 198: 9-20.

Schlosser, H.J. (1988): Auswertung ökotoxikologischer Forschungen zur Belastung von Ökosystemen durch Chemikalien. Jülich: KFA-Jülich, 247 S., zugl. Diss. 1988.

Schlosser, H. & Riepert, F. (1992): Entwicklung eines Prüfverfahrens für Chemikalien an Bodenraubmilben (Gamasina). Teil 2: Erste Ergebnisse mit Lindan und Kaliumdichromat in subletaler Dosierung. Zool. Beitr. N.F. 34: 413-434.

Schmidt, G. (1986): Use of grasshoppers as test animals for the ecotoxicological evaluation of chemicals in the soil. Agriculture, Ecosystems and Environment 16: 175-188.

Schmitz, M. (1991): Pestizidbelastung und Gewässerschutz. In: Kolloquium über Pflanzenschutz. Rehbinder, E. (ed.). Werner Verl., Düsseldorf. 93-103.

Schöller, H. (1993): Zur Problematik von Bioindikator-Modellen am Beispiel der Flechten. Natur und Museum 123: 292-314.

Schreiber, B. & Brink, N. (1989): Pesticide toxicity using protozoans as test organisms. Biol. Fert. Soils 7: 289-296.

Schubert, R. (1991a): Bioindikation in terrestrischen Ökosystemen. Stuttgart: G. Fischer, 2. Aufl., 327 S.

Schubert, R. (Hrsg.) (1991b): Lehrbuch der Ökologie. Jena: VEB G. Fischer, 3. Aufl., 595. S.

Schüepp, H., Dehn, B., Rüegg, J., Niggli, U. & Bodmer, M. (1990): Nebenwirkungen von Pflanzenschutzmitteln auf das Bodenökosystem. Bericht Eidgen. Forschungsanst. für Obst-, Wein- und Gartenbau 42: 161 S.

Schuphan, I. (1985): Ermittlung des quantitativen ökochemischen und ökotoxikologischen Verhaltens von Pflanzenschutzmitteln mit Hilfe kontrolliert belüfteter Vegetationskammern. Ber. Landwirtschaft 198: 21-34.

Schuster, E. & Schröder, D. (1990): Side-effects of sequentially- and simultaneously-applied pesticides on non-target soil microorganisms: Laboratory experiments. Soil Biol. Biochem. 22: 375-383.

Scott, M.G. & Hutchinson, T.C. (1990): The use of lichen growth abnormalities as an early warning indicator of forest dieback. Environ. Monit. Assess. 15: 213-218.

Seager, J. & Maltby, L. (1989): Assessing the impact of episodic pollution. Hydrobiologia 188/189: 633-640.

SETAC (Society of Environmental Toxicology and Chemistry) (1991): Workshop on Aquatic Microcosms for Ecological Assessment of Pesticides (Wintergreen 1991). SETAC, Washington. 56 pp.

SETAC (Society of Environmental Toxicology and Chemistry) (1992): Summary and Recommendations of the European Workshop on Freshwater Field Tests (Potsdam, 25-26 June, 1992). SETAC Europe, Brussels. 37 pp.

Shannon, L., Harrass, M., Yount, J.D. & Walbridge, C.T. (1986): A comparison of mixed flask culture and standardized laboratory model ecosystems for toxicity testing. In: Community Toxicity Testing. (ed. Cairns, J.). STP 920, ASTM, Philadelphia. 135-157.

Sheehan, P.J. (1984a): Effects on Community and Ecosystem Structure and Dynamics. In: Effects of Pollutants at the Ecosystem Level (ed. P.J. Sheehan, D.R. Miller, G.C. Butler & P. Bourdeau). SCOPE 22, New York: Wiley, pp. 51-99.

Sheehan, P.J. (1984b): Functional Changes in the Ecosystem. In: Effects of Pollutants at the Ecosystem Level (ed. P.J. Sheehan, D.R. Miller, G.C. Butler & P. Bourdeau). SCOPE 22, New York: Wiley, pp. 101-145.

Sheehan, P.J., Korte, F., Klein, W. & Bourdeau, P. (eds.) (1985): Appraisal of Tests to Predict the Environmental Behaviour of Chemicals. Chichester [etc.]: Wiley, SCOPE Report 25.

Sheehan, P.J., Miller, D.R., Butler, G.C. & Bourdeau, P. (eds.) (1984): Effects of Pollutants at the Ecosystem Level. New York [etc.]: Wiley, SCOPE Report 22, 443 p.

Sheppard, S.C. & Evenden, W.G. (1992): Bioavailability Indices for Uranium: Effect of Concentration in Eleven Soils. Arch. Environ. Contam. Toxicol. 23: 117-124.

Siedentop, S. (1993): Entwicklung eines Streuabbautests zur Prüfung von Pflanzenschutzmittelauswirkungen auf die Bodenmesofauna. Dissertation Universität Braunschweig. 114 S.

Siegfried, B.D. (1993): Comparative toxicity of pyrethroid insecticides to terrestrial and aquatic insects. Environ. Toxicol. Chem. 12: 1683-1689.

Slooff, W. & Canton, J.H. (1983): Comparison of the susceptibility of 22 freshwater species to 8 chemical compounds. II. (Semi)chronic tests. Aquatic Toxicology 4: 113-128.

Slooff, W. & Van de Meent, D. (1990): Ecotoxicological Risk Levels, Natural Background Levels and Environmental Quality Objectives for Surface Water and Soil. In: GSF-Bericht 1/92: Proc. Int. Symp. Ecotox. Munich, 341-348.

Slooff, W., Van Oers, J.A.M. & De Zwart, D. (1986): Margins of uncertainty in ecotoxicological hazard assessment. Environ. Toxicol. Chem. 5: 841-852.

Smart, L.E., Stevenson, J.H. & Walters, J.H.H. (1989): Development of field trial methodology to assess short-term effects of pesticides on beneficial arthropods in arable crops. Crop Protection 8: 169-180.

Smith, C.N., Parrish, R.S. & Brown, D.S. (1990): Conducting field studies for testing pesticide leaching models. Internat. J. Environ. Anal. Chem. 39: 3-21.

Smith, E.P. & Cairns, J. (1993): Extrapolation methods for setting ecological standards for water quality: statistical and ecological concerns. Ecotoxicology 2: 203-219.

Smith, I.M. (1990): EPPO — Europe's regional plant protection organization. Pesticide Outlook 1: 33-35.

Smith, R.H. (1993): Terrestrial Mammals. In: Handbook of Ecotoxicology. (ed. Calow, P.). Blackwell Scientific Publ, Oxford. 339-352.

Snell, T.W. & Persoone, G. (1989): Acute Toxicity Bioassays using Rotifers. I. A Test for Brackish and Marine Environments with *Brachionus plicatilis*. Aquat. Toxicol. 14: 65-81.

Solbe, J.F. de L.G. (1993): Freshwater Fish. In: Handbook of Ecotoxicology. (ed. Calow, P.). Blackwell Scientific Publ, Oxford. 66-82.

Solomon, K.R. & Lieber, K. (1988): Fate of pesticides in aquatic mesocosm studies — an overview of methodology. Brighton Crop Protection Conf. 139-148.

Solomon, K.R., Yoo, J.Y., Lean, D., Kauschik, N.K., Day, K.E. & Stephenson, G.L. (1986): Methoxychlor distribution, dissipation, and effects in freshwater limnocorals. Environ. Toxicol. Chem. 5: 577-586.

Somerville, L. & Greaves, M.P. (1987): Pesticide Effects on Soil Microflora. Taylor & Francis, London. 240 pp.

Somerville, L. & Walker, C.H. (1990): Pesticide Effects on Terrestrial Wildlife. Taylor & Francis, London.

Sortkjaer, O. (1984): Macrophytes and macrophyte communities as test systems in ecotoxicological studies of aquatic systems. Ecol. Bull. 36: 75-80.

Spencer, D.F. & Green, R.W. (1981): Effects of nickel on seven species of freshwater species. Environ. Pollut. Ser. A 25: 241-247.

Spencer, W., Jury, W. & Farmer, W. (1984): Importance of volatilization as a pathway for pesticide loss from forest soil. ACS Symposium Series 238: 193-210.

Steinemann, A., Stamm, E. & Frei, B. (1990): Chemodynamics in research and development of new plant protection agents. Pesticide Outlook 1: 3-7.

SRU (Sachverständigenrat für Umweltfragen) (1987): Umweltgutachten. Kohlhammer GmbH, Stuttgart.

Steinemann, A., Stamm, E. & Frei, B. (1990): Chemodynamics in research and development of new plant protection agents. Pesticide Outlook 1: 3-7.

Stephan, C.E. (1977): Methods for Calculating an LC_{50}. In: Aquatic toxicology and hazard evaluation. Mayer, F.L. & Hamelink, J.L. (eds.). American Society for Testing and Materials STP 634, Philadelphia.

Stephan, C.E., Mount, D.I., Hansen, D.J., Gentile, J.H., Chapman, G.A. & Brungs, W.A. (1985): Guidelines for deriving numerical national water quality criteria for the protection of aquatic organisms and their uses. U.S. Environmental Protection Agency. PB85-227049, Springfield.

Stern, V.M. (1973): Economic thresholds. Ann. Rev. Entomol. 18: 259-280.

Stevenson, J.H. & Walters, J.H.H. (1983): Evaluation of pesticides for use with biological control. Agriculture, Ecosystems and Environment 10: 201-215.

Stöcker, G. (1979): Ökosystem — Begriff und Konzeption. Arch. Naturschutz, Landwirtschaftsforsch. 19: 157-176.

Stoeppler, M. (1984): Bedeutung von Umweltprobenbanken — Anorganisch-analytische Aufgabenstellungen und erste Ergebnisse des Deutschen Umweltprobenbankprogramms. Fresenius Zeit. Anal. Chemie 317: 228-235.

Strand, J.A., Templeton, W.L., Lichatowich, J.A. & Apts, C.W. (1971): Development of toxicity test procedures for marine phytoplankton. In: Proceedings of Joint Conference on Prevention and Control of Oil Spills. American Petroleum Institute, Washington. 279-286.

Stratton, G.W. (1986): Medium composition and its influence on solvent-pesticide interactions in laboratory bioassays. Bull. Environ. Contamin. Toxicol. 36: 807-814.

Streit, B. (1989): Zum Problem der Bioindikatoren aus zoologisch-ökologischer Sicht. Geomethodica 14: 19-45.

Streit, B. (1991): Lexikon Ökotoxikologie. Verlag Chemie, Weinheim. 726 S.

Streit, B. (1992): Bioaccumulation processes in ecosystems. Experientia 48: 955-970.

Sunderland, S.L., Santelmann, P.W., Baughman, T.A. (1991): A rapid, sensitive soil bioassay for Sulfonylurea herbicides. Weed Sci. 39: 296-298.

Suter, G.W. (1989): Ecological endpoints. In: Ecological assessment of hazardous waste sites: A field and laboratory reference. Warren-Hicks, W., Parkhurst, B.R. & Baker, S.S. (eds.). EPA Report 600-3-89-013. Environmental Protection Agency, Corvallis.

Suter, G.W. (1990): Endpoints for regional ecological risk assessments. Environ. Management 14: 9-23.

Suter, G.W. (1993): A Critique of Ecosystem Health Concepts and Indexes. Environ. Toxicol. Chem. 12: 1533-1539.

Swartz, R.C. (1987): Toxicological methods for determining the effects of contaminated sediment on marine organisms. In: Fate and Effects of Sediment-bound Chemicals in Aquatic Systems. (eds.: Dickson, K.L., Maki, A.W. & Brung, W.A.). Ch. 14. Pergamon Press, New York.

Swartz, R.C., Schults, D.W., Dewitt, T.H., Ditsworth, G.R. & Lamberson, J.O. (1990): Toxicity of fluoranthene in sediment to marine amphipods: A test of the equilibrium partitioning approach to sediment quality criteria. Environ. Toxicol. Chem. 9: 1071-1080.

Swift, M.J., Cook, A.G., Critchley, B.R., Critchley, U., Moore, R.L., Perfect, T.J., Russel-Smith, A. & Yeadon, R. (1977): Persistent pesticides and tropical soil fertility. Meded. Fac. Landbouw. Rijksuniv. Gent 42: 845-852.

Tagatz, M.E. & Deans, C.A. (1983): Comparison of field- and laboratory-developed estuarine benthic communities for toxicant-exposure studies. Water Air Soil Poll. 20: 199-209.

Tagatz, M.E., Stanley, R.S., Plaia, G.R. & Deans, C.H. (1987): Responses of estuarine macrofauna colonizing sediments contaminated with fenvalerate. Environ. Toxicol. Chem. 6: 21-25.

Tanabe, S. (1991): Fate of Toxic Chemicals in the Tropics. Mar. Poll. Bull. 22: 259-260.

Tardiff, R.G. (1992): Methods to Assess Adverse Effects of Pesticides on Non-target Organisms. SCOPE 49: ICPS Joint Symposia 16, J. Wiley & Sons, Chichester, 270 pp.

Taub, F.B. (1976): Demonstration of pollution effects in aquatic microcosms. Int. J. Environ. Studies. 10: 23-33.

Taub, F.B., Kindig, A.C. & Conquest, L.L. (1986): Preliminary results of interlaboratory testing of a standardized aquatic microcosm. In: Community Toxicity Testing. (ed. Cairns, J.). STP 920, ASTM, Philadelphia. 93-120.

Thain, J.E. (1990): Protocol for conducting the oyster embryo test. In: Utility of experimental measures of biological effects for monitoring marine sewage-sludge disposal sites. Aquatic Environment Monitoring Report, No. 24: 39-43. Ministry of Agriculture, Fisheries and Food, Lowestoft.

The Agrochemicals Handbook (1992): Cambridge: The Royal Society of Chemistry.

Theiling, K.M. & Croft, B.A. (1988): Pesticide Side-Effects on Arthropod Natural Enemies: a Database Summary. Agriculture, Ecosystems and Environment 21: 191-218.

Thompson, A.R.; Gore, F.L. (1972): Toxicity of twenty-nine insecticides to *Folsomia candida*: Laboratory studies. J. Econ. Entomol. 65: 1255-1260.

Thompson, D.G., Holmes, S.B., Wainio-Keizer, K., MacDonald, L. & Solomon, K.R. (1993): Impact of Hexazinon and Metsulfuron Methyl on the Zooplankton Community of a Boreal Forest Lake. Environ. Toxicol. Chem. 12: 1709-1717.

Thursby, G.B., Anderson, B.S., Walsh, G.E. & Steele, R.L. (1993): A review of the current status of marine algal toxicity testing in the United States. In: Environmental Toxicity and Risk Assessment: Aquatic, Plant and Terrestrial. (eds. Gorsuch, J.W. & Lewis, M.A.). STP 1179, ASTM, Philadelphia. 362-377.

Torstensson, L. (1988): Testing pesticide fate and side-effects in the terrestrial environment. Toxicity Assessment: An Internat. J. 3: 407-414.

Tosato, M.L., Vigano, L., Skagerberg, B. & Clementi, S. (1991): A New Strategy for Ranking Chemical Hazards. Framework and Application. Environ. Sci. Technol. 25: 695-702.

Touart, L.W. (1988): Aquatic Mesocosm Tests to Support Pesticide Registration. EPA 540/09-88-035. National Technical Information Service, Springfield. 35 pp.

Treidler, S. & Winter, G. (1991): Materielle Voraussetzungen der Zulassung von Pflanzenschutzmitteln. In: Kolloquium über Pflanzenschutz. Rehbinder, E. (ed.). Werner Verl., Düsseldorf. 158-164.

Truhaut, R. (1977): Ecotoxicology: Objectives, principles and perspectives. Ecotox. Environ. Safety 1: 151-173.

Turbak, S.C., Olson, S.B. & McFeters, G.A. (1986): Comparison of algal systems for detecting water-borne herbicides and metals. Water Res. 20: 91-96.

Türk, R., Wirth, V. & Lange, O.L. (1974): CO_2-Gaswechseluntersuchungen zur SO_2-Resistenz von Flechten. Oecologia 15: 33-65.

Turtle, E.E. (1977): Agricultural pesticides and the environment in the developing world. Meded. Fac. Landbouw. Rijskuniv. Gent 42: 839-844.

UBA (Umweltbundesamt) (Hrsg.) (1984): Chemikaliengesetz, Prüfung und Bewertung der Umweltgefährlichkeit von Stoffen. UBA, Berlin. 157 S.

UBA (Umweltbundesamt) (Hrsg.) (1986a): Chemikaliengesetz — Prüfung und Bewertung von Stoffen auf ihre Umweltgefährlichkeit: Grundzüge der Bewertung. UBA-Texte 26/86, Berlin. 30 S.

UBA (Umweltbundesamt) (Hrsg.) (1986b): Chemikaliengesetz — Prüfung und Bewertung von Stoffen auf ihre Umweltgefährlichkeit: Entwicklung von Testverfahren an marinen Arten für ökotoxikologische Untersuchungen nach dem ChemG. UBA-Texte 27/86, Berlin. 207 S.

UBA (Umweltbundesamt) (Hrsg.) (1990): Grundzüge der Bewertung von Neuen Stoffen nach der Chemikaliengesetz — 1. Fortschreibung 1990.

UBA (Umweltbundesamt) (Hrsg.) (1992): Bewertung der Umweltgefährlichkeit von Alten Stoffen nach dem Chemikaliengesetz (ChemG). Chemikaliengesetz Heft 10, Texte 12/92. Umweltbundesamt, Berlin.

Van Assche, F. & Clusters, H. (1990): A Biological Test System for the Evaluation of the Phytotoxicity of Metal-Contaminated Soils. Environ. Poll. 66: 157-172.

Van der Kooij, L.A., Van de Meent, D., Van Leeuwen, C.J. & Bruggeman, W.A. (1991): Deriving quality criteria for water and sediment from the results of aquatic toxicity tests and product standards: Application of the equilibrum partitioning method. Water Res. 25: 697-705.

Van Gestel, C.A.M. (1991): Earthworms in Ecotoxicology. Ph.D. Thesis, University of Utrecht.

Van Gestel, C.A.M. & Ma, W-C. (1988): Toxicity and bioaccumulation of chlorophenols in earthworms, in relation to bioavailability in soil. Ecotox. Environ. Safety 15: 289-297.

Van Gestel, C.A.M., Van Dis, W.A., Driven-Van Breemen, E.M., Sparenburg, P.M. & Baerselmann, R. (1991): Influence of Cadmium, Copper and Pentachlorophenol on Growth and Sexual Development of *Eisenia andrei* (Oligochaeta; Annelida). Biol. Fert. Soils 12: 117-121.

Van Gestel, C.A.M., Dirven-van Breemen, E.M., Baerselmann, R., Emans, H.J.B., Janssen, J.A.M., Postuma, R. & van Vliet, P.J.M. (1992): Comparison of sublethal and lethal criteria for nine different chemicals in standardized toxicity tests using the earthworm *Eisena fetida*. Ecotox. Environ. Safety 23: 206-230

Vanhaecke, P. & Persoone, G. (1984): The ARC-Test: A standardized short-term routine toxicity test with *Artemia* nauplii. Methodology and evaluation. In: Ecotoxicological Testing for the Marine Environment. (eds. Persoone, G., Jaspers, E. & Claus, C.). State Univ. Ghent and Inst. Mar. Scient. Res., Bredene, Belgium. Vol. 2: 143-157.

Van Hattum, B., Timmermanns, K.R. & Govers, H.A. (1991): Abiotic and biotic factors influencing in situ trace metal levels in macroinvertebrates in freshwater ecosystems. Environ. Toxicol. Chem. 10: 275-292.

Van Kessel, W.H.M., Brocades-Zaalberg, R.W. & Seinen, W. (1989): Testing environmental pollutants on soil organisms: A simple assay to investigate the toxicity of environmental pollutants on soil organisms, using cadmium chloride and nematodes. Ecotox. Environ. Safety 18: 181-190.

Van Leeuwen, C.J. (1990): Ecotoxicological effects assessment in The Netherlands. Environ. Management 14: 779-792.

Van Leeuwen, C.J., Van der Zandt, P.T.J., Aldenberg, T., Verhaar, H.J.M. & Hermens, J.L.M. (1992): Application of QSARs, extrapolation and equilibrum partitioning in aquatic effects assessment. I. Narcotic industrial pollutants. Environ. Toxicol. Chem. 11: 267-282.

Van Straalen, N.M. (1993): An ecotoxicologist in politics. Oikos 66: 142-143.

Van Straalen, N.M. & Denneman, C.A.J. (1989): Ecotoxicological evaluation of soil quality criteria. Ecotox. Environ. Safety 18: 241-251.

Van Straalen, N.M., Schobben, J.H. & Traas, T.P. (1992): The use of ecotoxicological risk assessment in deriving maximum acceptable half-lives of pesticides. Pestic. Sci. 34: 227-231.

Van Straalen, N.M. & Van Gestel, C.A.M. (1993a): Soil Invertebrates and Micro-Organisms. In: Handbook of Ecotoxicology. (ed. Calow, P.). Blackwell Scientific Publ, Oxford. 251-277.

Van Straalen, N.M. & Van Gestel, C.A.M. (1993b): Ecotoxicological Test Methods using Terrestrial Arthropods. Discussion Paper for the OECD Test Guidelines Programme. Vrije Universiteit Amsterdam, Dept. of Ecology and Ecotoxicology, Report No. D93002. 63 pp.

Van Straalen, N.M., Leeuwangh, P. & Stortelder, P.B.M. (1994): Progressing Limits for Soil Ecotoxicological Risk Assessment. In: Ecotoxicology of Soil Pollution. Eijsackers et al. (eds.). Lewis Publ., Chelsea, Michigan. 397-409.

Van Voris, P., Tolle, D.A., Arthur, M.F., Chesson, J. & Zwick, T.C. (1984): Development and Validation of a Terrestrial Microcosm Test System for Assessing Ecological Effects of Utility Wastes. Electric Power Research Institute. 140 pp.

Van Wensem, J. (1989): A terrestrial micro-ecosystem for measuring effects of pollutants on isopod-mediated litter decomposition. Hydrobiologia 188/189: 507-516.

Van Wensem, J., Akkerhuis, G.A. & Van Straalen, N.M. (1991): Effects of the fungicide Triphenyltin hydroxide on soil fauna mediated litter decomposition. Pest. Sci. 32: 307-316.

Vasseur, P., Ferrard, J.F. & Babut, M. (1991): The biological aspects of the regulatory control of industrial effluents in France. Chemosphere 22: 625-633.

Venter, J.M. & Reinecke, A.J. (1985): Dieldrin and growth and development of the earthworm, *Eisenia fetida* (Oligochaeta). Bull. Environ. Contam. Toxicol. 35: 652-659.

Vetter, W. & Luckas, B. (1991): Meeressäuger als Bioindikatoren zur Erfassung der globalen Belastung mit chlorierten Kohlenwasserstoffen. UWSF.-Z. Umweltchem. Ökotox. 3: 324-327.

Vollner, L. & Ghods-Esphahani, A. (1988): Controlled release insecticide formulations for tropical application. In: Pesticides: Food and environmental implications. Intl. Atomic Energy Agency, Vienna. 29-43.

Volz, P. (1962): Beiträge zu einer pedozoologischen Standortslehre. Pedobiologia 1: 242-290.

Von Lersner, H. (1982): Das 3. Medium: Der Schutz des Bodens als umweltpolitische Aufgabe. Natur + Recht 4: 201-205.

Von Loeper, H. (1986): FAO-Code of Conduct — Aus der Sicht der Pflanzenschutzmittelindustrie. Mittl. BBA 232: 76-87.

Von Oepen, B. (1990): Sorption organischer Chemikalien an Böden. Verl. W. Maraun, Frankfurt, 123 S.

Von Törne, E. (1990): Asessing feeding activities of soil-living animals. I. Bait lamina tests. Pedobiologia 34: 89–101.

Vouk, V.B., Butler, G.C., Upton, A.C., Parke, D.V. & Asher, S.C. (eds.) (1987): Methods for Assessing the Effects of Mixtures of Chemicals. Chichester: Wiley, SCOPE 30, SGOMSEC 3.

Wagner, C. & Lökke, H. (1991): Estimation of ecotoxicological protection levels from NOEC toxicity data. Water Res. 25: 1237-1242.

Walker, C.H. (1988): Some biochemical responses of vertebrates to sublethal doses of insecticides. Aspects Appl. Biol. 17: 257-263.

Walker, C.H. (1993): Birds. In: Handbook of Ecotoxicology. (ed. Calow, P.). Blackwell Scientific Publ, Oxford. 326-338.

Walker, J.D. (1990): Bioconcentration, chemical fate, and environmental effects testing under section 4 of the Toxic Substances Control Act. Toxicity Assessment 5: 61-75.

Walker, W.W., Cripe, C.R., Pritchard, P.H. & Bourquin, A.W. (1988): Biological and abiotic degradation of xenobiotic compounds in in vitro estuarine water and sediment/water systems. Chemosphere 17: 2255-2270.

Walsh, G.E. (1988): Methods for Toxicity Tests of Single Substances and Liquid Complex Wastes with Marine Unicellular Algae. U.S. EPA, EPA/600/8-87/043. Gulf Breeze. 64 pp.

Walsh, G.E. (1993): Primary Producers. In: Handbook of Ecotoxicology. (ed. Calow, P.). Blackwell Scientific Publ, Oxford. 119-144.

Walsh, G.E., Weber, D.E., Brashers, L.K. & Simon, T.L. (1990): Artificial sediments for use in tests with wetland plants. Environ. Exp. Bot. 30: 391-396.

Walter, H. & Breckle, S.W. (1983): Ökologische Grundlagen in globaler Sicht. Ökologie der Erde, Bd. 1. Stuttgart: Gustav Fischer, 238 S.

Walton, B.T. (1989): Insects as indicators of toxicity, bioaccumulation and bioavailability of environmental contaminants. Environ. Toxicol. Chem. 8: 649-658.

Ware, G. (1986): Fundamentals of Pesticides. Fresno: Thomson, 2nd edn., 274 p.

Warren-Hicks, W., Parkhurst, B.R. & Baker, S.S. (1989): Ecological assessment of hazardous waste sites: A field and laboratory reference. EPA Report 600-3-89-013. Environmental Protection Agency, Corvallis.

Warwick, R.M., Platt, H.M., Clarke, K.R., Agard, J. & Gobin, J. (1990): Analysis of macrobenthic and meiobenthic community structure in relation to pollution and disturbance in Hamilton Harbour, Bermuda. J. Exp. Mar. Biol. Ecol. 138: 119-142.

Weber, A. & Barben, H. (1984): Beurteilung ökologisch relevanter Daten neuer organischer Chemikalien. Chimia 38: 443-455.

Weidemann, G. (1990): Indikation, Beurteilung und Bewertung in der Ökotoxikologie. Mittl. Dtsch. Ges. Allg. Angew. Ent. 7: 577-581.

Weidemann, G., Koehler, H. Mathes, K. & Schulz-Berendt, V. (1987): Konzeption und Anlage eines Ökosystems zur ökotoxikologischen Beurteilung von Chemikalien. Verh. Ges. f. Ökol. 16: 393-398.

Weidemann, G. & Mathes, K. (1991): Beurteilung und Bewertung der Wirkung von Pflanzenschutzmitteln auf den Naturhaushalt. In: Kolloquium über Pflanzenschutz. Rehbinder, E. (ed.). Werner Verl., Düsseldorf. 85-92.

Weidemann, G., Mathes, K. & Koehler, H. (1988): Bezugsökosystem unter Pestizid-Belastung. Jül.-Spez. 439: 7-222.

Weigel, H.-J. & Jäger, H.-J. (1988): Zur Ökotoxikologie von Luftschadstoffen. 1. Experimentelle Verfahren zur Schadstoffexpedition von Pflanzen. Landbauforsch. Völkenrode 38: 103-115.

Welp, G. & Brummer, G. (1985): Der Fe(III)-Reduktionstest — ein einfaches Verfahren zur Abschätzung der Wirkung von Umweltchemikalien auf die mikrobielle Aktivität im Boden. Z. Pflanzenern. Bodenk. 148: 10-23.

Westheide, W., Bethge-Beilfuß, D. & Gebbe, J. (1991): Effects of benomyl on reproduction and population structure of enchytraeid oligochaetes (Annelida) — sublethal tests on agar and soil. Comp. Biochem. Physiol. 100C: 221-224.

Wevers, M.J., Liss, W.J. & Warren, C.E. (1988): Utility of laboratory streams for ecosystem toxicity studies. Environ. Management 12: 19-27.

Weischet, W. (1977): Die ökologische Benachteiligung der Tropen. Stuttgart: Teubnaer Verlag, 127 S.

White, D.S. & Keilty, T.J. (1988): Burrowing Avoidance Assays of Contaminated Detroit River Sediments, Using the Freshwater Oligochaete *Stylodrilus heringianus* (Lumbriculidae). Arch. Environ. Contam. Toxicol. 17: 673-681.

Whitehurst, I.T. (1991): The *Gammarus : Asellus* ratio as an index of organic pollution. Water Res. 25: 333-339.

Widdows, J. (1993): Marine and Estuarine Invertebrate Toxicity Tests. In: Handbook of Ecotoxicology. (ed. Calow, P.). Blackwell Scientific Publ, Oxford. 145-166.

Wiedman, S.J. & Appelby. A.P. (1972): Plant growth stimulation by sublethal concentrations of herbicides. Weed Res. 12: 65-74.

Wiederholm, T. & Dave, G. (1989): Toxicity of metal polluted sediments to *Daphnia magna* and *Tubifex tubifex*. Hydrobiologia 176/177: 411-417.

Wiles, J., Jepson, P.C., Salt, D.W. & Ford, M.G. (1991): Evaluating the hazard of pesticides to target and non-target terrestrial invertebrates. Pestic. Sci. 31: 98-99.

Winner, R.W., Boesez, M.W. & Farrell, M.P. (1980): Insect community structure as an index of heavy-metal pollution in lotic systems. Can. J. Fish. Aquat. Sci. 37: 647-655.

Winteringham, F.P.W. (1988): Chemical and radioactive residues in soil: A global perspective. In: Pesticides: Food and environmental implications. Intl. Atomic Energy Agency, Vienna. 29-43.

Woelke, C.E. (1972): Development of a receiving water quality bioassay criterion based on the 48-hour Pacific oyster (*Crassostrea gigas*). Department of Fisheries Technical Report No. 9, Washington. 93 pp.

Wood, B.J. & Chung, G.F. (1990): Warfarin resistance of *Rattus tiomanicus* in oil palms in Malaysia and the associated increase of *Rattus diardii*. In: Proc. 14th Vertebrate Pest Conf. (eds. Davis, L.R. & Marsh, R.E.), University of California, Davis. 129-134.

Worthing, C.R. & Walker, S.B. (1987): The pesticide manual: A world compendium. Thornton Heath: The British Crop Protection Council, 1082 p.

Yameogo, L., Abban, E.K., Elouard, J-M., Traore, K. & Calamari, D. (1993): Effects of Permethrin as *Simulium* larvicide on non-target aquatic fauna in an African river. Ecotoxicology 2: 157-174.

Zimmermann, R-D. & Rudolph, E. (1986): Klon-Fichten — Ein neuer Bioindikator im aktiven Monitoring. Allg. Forst Zeitschrift 1/2: 13-14.

Zitko, V. (1990): Priority ranking of chemicals for risk assessment. The Science of the Total Environment 92: 29-39.

Zurhake, F. (1989): Entwicklung neuer Pflanzenschutzwirkstoffe. Biol. in unserer Zeit 19: 189-195.

INDEX

A

D

E

F

G

H

I

L

M

Q

R

S

T

U

V

W

X

Y

Z